Fluid Mechanics Aspects of Fire and Smoke Dynamics in Enclosures

This guide provides the essential knowledge for understanding flows in fire and smoke dynamics in enclosures, covering combustion, heat transfer and fire suppression in more detail than other introductory books. It moves from the basic equations for turbulent flows with combustion, through a discussion of the structure of flames, to fire and smoke plumes and their interaction with enclosure boundaries. This is then applied to fire dynamics and smoke and heat control in enclosures.

This new edition provides considerably more on the fluid mechanics of the effect of water, and on fire dynamics modelling using Computational Fluid Dynamics.

- Written by world-leading experts in the field
- Presents worked examples taken from practical, everyday fire-related problems
- Covers a broad range of topics, from the basics to state-of-the-art computer simulations of fire and smoke-related fluid mechanics, including the effect of water
- Provides extensive treatment of the interaction of water sprays with a fire-driven flow
- Contains a chapter on CFD (Computational Fluid Dynamics), the increasingly popular calculation method in the field of fire safety science

The book serves as a comprehensive guide at the undergraduate and starting researcher level on fire and smoke dynamics in enclosures, with an emphasis on fluid mechanics.

Fluid Mechanics Aspects of Fire and Smoke Dynamics in Enclosures

Second Edition

Bart Merci and Tarek Beji

CRC Press
Taylor & Francis Group
Boca Raton London New York Leiden

CRC Press is an imprint of the
Taylor & Francis Group, an **informa** business

A BALKEMA BOOK

Second edition published 2023
by CRC Press/Balkema
Schipholweg 107C, 2316 XC Leiden, The Netherlands
e-mail: enquiries@taylorandfrancis.com
www.routledge.com – www.taylorandfrancis.com

CRC Press/Balkema is an imprint of the Taylor & Francis Group, an informa business

Second edition © 2023 Bart Merci and Tarek Beji

First edition published by CRC Press/Balkema 2016

Library of Congress Cataloging-in-Publication Data
A catalog record has been requested for this book

ISBN: 978-1-032-06887-9 (hbk)
ISBN: 978-1-032-06584-7 (pbk)
ISBN: 978-1-003-20437-4 (ebk)

DOI: 10.1201/9781003204374

Typeset in Times
by codeMantra

Contents

Preface

Fire is a fascinating phenomenon. It is often associated with hazards, damage, losses and destruction, but in a more poetic context, it is also related to passion and love. It is the passion for fire that led us to write the first edition of this book, and now we hope to provide a useful update in the second edition. The major updates, other than corrections here and there, are the further elaboration of Chapters 7 and 8, incorporating recent research activities.

We tried to maintain the balance in mentioning and explaining state-of-the-art science, while not getting lost in providing too much detail, and respecting that some state-of-the-art findings have not yet sufficiently matured to be included in a book. Moreover, it is not possible, nor desirable, to gather all knowledge in a book. Our intention is rather to provide insight into the wealth of knowledge, keeping primarily master students and early-career researchers in mind as the reading audience.

The question can be posed whether there is a need for another textbook on fire. Indeed, there are already excellent textbooks on fire dynamics and fundamentals of fire science on the market. In that sense, the book you are about to read benefits from standing on the shoulders of giants. The intended added value is that fluid mechanics aspects are given a central role. Indeed, in our humble opinion, fluid mechanics has received less attention than heat transfer, combustion, pyrolysis, etc. This book aims at providing a comprehensive discussion of fluid mechanics in the context of fire. We have the advantage of being able to rely upon our years of experience in teaching courses on fire dynamics, smoke and heat control, and modelling of turbulence and combustion, in fire safety engineering educational programmes at Ghent University (Belgium). As such, a didactical structure has evolved, explaining fire-related phenomena and observations, starting from fundamental equations. Starting from the fundamentals, smoke and fire plumes are reviewed first. With this as the starting point, fire and smoke dynamics in enclosures are discussed, so that smoke and heat control in case of fire can be tackled. One chapter is devoted to the effect of water on fire and smoke dynamics, and the final chapter provides guidelines and modelling aspects for CFD (Computational Fluid Dynamics), the most advanced computational tool at hand for fire calculations.

We would like to thank everyone who has ever collaborated in research with us, but also everyone who has ever discussed with us, as almost always something is learnt from a discussion. This includes the large number of talented young students that have challenged us while teaching. Likewise, and a fortiori, we would like to express our sincere gratitude to every teacher that crossed our path and enlightened us with new knowledge and stimulated us to discover new knowledge by ourselves. It is the long and steady process of growing up this way that enabled us to start writing the book. We also thank all the funding agencies that enabled us to do research and thank Ghent University, in particular the Faculty of Engineering and Architecture, for having established educational programmes on fire safety engineering. However, most of all, we thank the persons that are closest to our heart, as we could not have created the slightest part of this book without their everlasting love and support: thank you very much, beloved grand-parents, parents, parents-in-law and, of course, special thanks to our wives (Vicky, resp. Aicha) and children (Camille and Morgane, resp. Yahia and Idris).

We wish you many moments of pleasure, reading this book, and hope you can learn from it.

Sincerely,
Bart and Tarek

Authors

Prof. Bart Merci obtained his PhD, entitled 'Numerical Simulation and Modelling of Turbulent Combustion', from the Faculty of Engineering at Ghent University in the year 2000. As a post-doctoral fellow of the Fund for Scientific Research – Flanders (FWO-Vlaanderen), he specialized in numerical simulations of turbulent non-premixed combustion, with focus on turbulence–chemistry interaction and turbulence–radiation interaction. He reoriented his research towards fire safety science, taking the fluid mechanics aspects as the central research topic. He became lecturer at Ghent University in 2004, Full Professor in 2012 and Senior Full Professor in 2020. He is the head of the research unit 'Fire Safety Science and Engineering' in the Department of Structural Engineering and Building Materials. Since 2009, Bart Merci has coordinated the 'International Master of Science in Fire Safety Engineering', with Lund University and the University of Edinburgh as partners. He was the President of The Belgian Section of The Combustion Institute from 2009 until 2016. Since 2016, he has been Editor-in-Chief of *Fire Safety Journal*. He is a member of the Executive Committee of the International Association for Fire Safety Science. He is co-author of more than 150 journal papers.

Prof. Tarek Beji obtained his Ph.D. degree in 2009 from the University of Ulster (UK) with a thesis entitled 'Theoretical and Experimental Investigation on Soot and Radiation in Fires'. In 2011, he joined Ghent University (UGent) as a post-doctoral researcher and worked on the topic of Fire Forecasting. In 2014, he was awarded an FWO post-doctoral fellowship and performed research on 'Numerical Modelling of Water Sprays in Fire-Driven Flows'. In 2019, Tarek Beji was promoted to Assistant Professor in Fire Dynamics at Ghent University (UGent) in the Department of Structural Engineering and Building Materials where he is consolidating his research activities around the topic of 'Numerical Modelling of Multi-Phase Aspects in Fire-Driven Flows' with a particular interest in liquid sprays, pool fires and soot modelling.

1 Introduction

1.1 THE CANDLE FLAME

When attempting to explain in scientific, yet accessible, terms the definition and main underlying phenomena of a fire to a non-expert or uninitiated audience, one could consider first the example of a **candle flame** (see Figure 1.1).

> When you light a candle (with a lighter for example), the wax melts near the wick. The liquid (i.e., molten wax) is absorbed by the wick and then vaporized by the flame heat. When the wax vapor meets the oxygen in the air, the burning occurs (sustained by the flame heat) generating thus light and (more) heat.

Although this example might seem at first too simplistic, it actually conveys the main principles of fire behaviour. In fact, Michael Faraday, in a series of lectures entitled *A Chemical History of a Candle* [1], paid an appealing tribute to the scientific and pedagogical values of a candle by stating:

> There is not a law under which any part of this universe is governed which does not come into play and is touched upon in these phenomena.

A tribute that extends beyond the topic of fire.

A candle flame clearly illustrates indeed what is often referred to as the **fire triangle** (see Figure 1.2) that is composed of (a) fuel, (b) oxidizer and (c) heat. The fuel is the wax, the oxidizer is the oxygen in the air, and heat is initially brought by the ignition (the lighter in this case) and then sustained by the flame.

FIGURE 1.1   Illustration of a laminar candle flame.

DOI: 10.1201/9781003204374-1

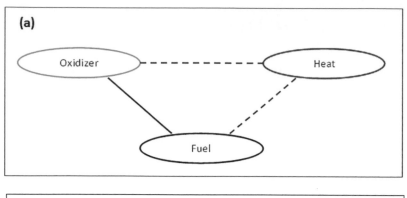

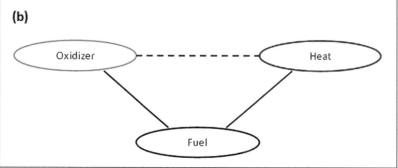

FIGURE 1.2 The fire triangle. Dashed lines indicate weak interaction. (a) Gaseous fuel and (b) liquid fuel or solid fuel.

1.2 THE IMPORTANCE OF CHEMISTRY, HEAT TRANSFER AND FLUID MECHANICS IN FIRES

A fundamental understanding of the interaction between each of the three *fire ingredients* (i.e., fuel, oxidizer and heat) requires at least three scientific disciplines, namely chemistry, heat transfer and fluid mechanics. Chemistry addresses the details of the chemical oxidation process (i.e., combustion) of the vaporized fuel molecules by the oxygen molecules in the air. Such details include, for example, the nature and quantities of the chemical species that are produced. Heat transfer allows to understand, for instance, the heat-up process of the fuel and subsequently the rate at which it is vaporized. Fluid mechanics provides a tool to quantify the amount of air (and thus oxygen) that is supplied to the fuel vapour. In presenting the triangle, often no distinction is made with respect to the relative importance of the interactions. Yet, in the majority of fires involving solid or liquid fuel, the interaction between fuel and oxidizer, and between fuel and heat, is much more important than the interaction between heat and oxidizer.

1.2.1 CHEMISTRY

A fire is primarily an exothermic oxidation reaction that provides the heat required to sustain the fire. This rapid chemical process often involves chemical time scales (nanoseconds to microseconds) that are substantially shorter than (a) turbulent mixing and residence times (milliseconds to seconds), and (b) characteristic times of heat transfer towards and inside the fuel (seconds to minutes). That is why Drysdale [2] emphasizes the fact that: *although fire is [primarily] a manifestation of a chemical reaction, the mode of burning may depend more on the physical state and distribution of the fuel and its environment, than on its chemical nature.* A fire involves, however, slow chemical processes such as NO_x and soot formation. Chemistry is nevertheless not the rate-limiting factor in fire development in enclosures. Needless to say, chemistry is essential when toxicity is considered. Toxicity, however, is not addressed in this book. The reader is referred to specialized literature, such as [3], for more detailed information on this topic.

1.2.2 HEAT TRANSFER

As mentioned above, heat transfer is key in the context of any fire. The heat supplied to the fuel can transform the fuel from solid, resp. liquid, phase into combustible gaseous phase, by supplying the required latent heat of pyrolysis, resp. vaporization, for this endothermic process (Section 2.14). It is therefore not surprising that heat transfer has been given a central role in numerous textbooks, such as [2,4,5]. In the example of the candle flame, heat transfer from the flame to the top surface of the stick of wax is essential in the melting of the wax and the vaporization of the generated liquid. Figure 1.3 illustrates the different several heat transfer processes (i.e., conduction, convection and radiation) for a more complex and 'practical' case, namely a sofa fire. This case is discussed in the following.

1.2.2.1 Flame, Smoke Plume and Surroundings

A fraction of the heat released by the flame (located in the corner of the sofa) is transferred by radiation to the smoke plume as well as the surroundings. The largest fraction is transferred by convection to the smoke plume in an upward motion driven by buoyancy (see Chapter 4).

1.2.2.2 Heat Transfer at the Level of the Fuel Surface

Heat transfer from the flame to the fuel surface occurs primarily by convection and radiation. The predominance of radiation over convection depends mainly on the level of sootiness (i.e.,

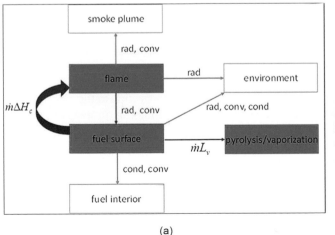 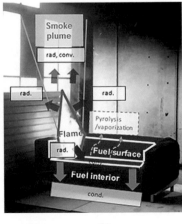

(a)   (b)

FIGURE 1.3 Illustration of heat transfer processes in a free fire plume. (a) A schematic: green arrows indicate heat transfer which is not used in the fire process. 'cond' refers to conduction, 'conv' refers to convection and 'rad' refers to radiation. The curved arrow refers to the mass flow rate of combustible gases. For simplification purposes, in-depth radiation in the fuel is not included in the figure. (b) Example of a fire on a sofa.

propensity to produce soot) of the fuel and the size of the fire. Very often soot promotes thermal radiation. It could nevertheless hinder in some cases heat transfer by radiation to the fuel surface, leading to what is referred to as thermal blockage, in particular for large fires, where much soot can be produced close to the fuel surface when oxygen cannot easily reach the combustible gases. Furthermore, part of the heat, incident onto the fuel surface, is not used in the pyrolysis/evaporation process, because it is 'lost' by in-depth conduction into the solid or liquid fuel and/or by internal convection into the liquid fuel. Also lateral conduction and convection heat losses are possible. In addition to that, the fuel surface radiates towards the environment. For the sake of completeness, it is mentioned that there can also be 'in-depth radiation', i.e., radiation stemming from the flames which is not absorbed or reflected at the fuel surface, but is absorbed directly inside the fuel. This heats up the fuel and thus can contribute to the fire process. For simplification purposes, in-depth radiation in the fuel is not included in Figure 1.3.

The heat balance at the level of the fuel surface involves also an endothermic process that consumes an amount of the energy produced by the fire to pyrolyse (resp. vaporize) the solid (resp. liquid) fuel. This required amount of energy is called heat of pyrolysis (resp. latent heat of vaporization) for a solid (resp. liquid) and is noted as L_v. In ideal conditions (in terms of heat and oxygen concentrations), the vaporized fuel is burned, releasing an energy content per unit time of $\dot{Q}_f = \dot{m}_F \Delta h_c$, where $\dot{m}_F$ is the mass flow rate of vaporized combustible gases and Δh_c is the theoretical heat of combustion (i.e., the amount of energy released per kg of fuel upon complete combustion). This is commonly called the (theoretical) total heat release rate (HRR). In reality, combustion is never complete, and hence, a correction is introduced (see Section 3.5). If this total HRR is larger than the energy flows (per unit time) as sketched in Figure 1.3a, more heat (per unit time) is available for the pyrolysis/evaporation process and a larger mass flow rate $\dot{m}_F$ of combustible gases is released, so the flames and the fire become larger ('grow'). The opposite is the case when the total HRR is less than the total amount of energy flows (per unit time) as sketched in Figure 1.3a, and the flames and fire will become smaller ('decay') or even extinguish. A steady situation prevails in case of a balance of the energy flows. This is discussed in detail in Chapters 3 and 5.

1.2.2.3   Enclosure Effect on Heat Transfer

When a fire occurs in an enclosed space, additional elements need to be considered in the heat transfer analysis. The very first element is the build-up of a smoke layer due to the bounding surfaces (i.e. ceiling and walls). The smoke layer interacts essentially with (a) the smoke plume, (b) ceiling and walls, and (c) the cold lower layer.

Heat contained in the smoke plume is essentially transferred by convection to the smoke layer. However, at later stages in the enclosure fire development, flames might be immersed into the smoke layer, which increases radiative heat transfer and leads to a direct energy transfer by convection. The smoke layer heats up the ceiling and parts of the walls by convection and radiation. Furthermore, a fraction of the smoke layer heat is lost by convection to the cold bottom layer and by radiation to the floor. All these heat transfer processes are indicated in Figure 1.4 by thick red arrows.

It is interesting to note that the oxidizer can be heated, e.g., in 'vitiated conditions' in enclosures (Section 5.2.1), and when igniting a flammable fuel, oxidizer mixture heat is provided to both fuel and oxidizer, but very often the major interaction is between fuel and heat, on the one hand, and between fuel and oxidizer, on the other hand.

At the level of the fuel surface, additional radiative feedback is provided by the hot ceiling and walls as well as the smoke layer. This is indicated in Figure 1.4 by thin red arrows. This radiative feedback enhances the pyrolysis process and causes more rapid fire development in enclosures compared to free fire/flame spread. Re-radiation from the fuel surface also incides onto the walls

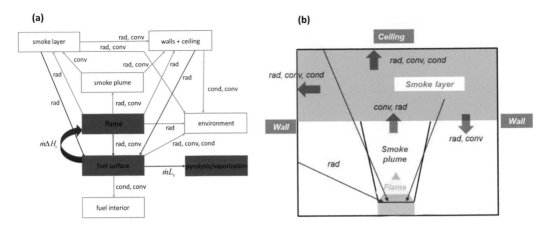

FIGURE 1.4 Enclosure effects on heat transfer processes in a room fire. (a) A schematic (same nomenclature as in Figure 1.3a). (b) The case of a pool fire in a confined room.

and the ceilings and into the smoke layer (but this has not been added in Figure 1.4 in order not to overload it).

There are other scenarios involving heat transfer in enclosure fires which are not sketched in Figure 1.4. For example, in the presence of an opening (e.g., a door or a window), heat is taken away by convection from the smoke layer to the environment. Heat losses by radiation are possible through, e.g., windows, although radiation losses are typically limited for smoke layers (unless the temperature becomes 500°C or higher). This is addressed in more detail in Chapter 5. Another scenario that is not sketched in Figure 1.4 is the use of water sprays that take away heat from the fire and thus allow to control it, if not extinguish it. The effect of water on flames and smoke is discussed in Chapter 7.

1.2.3 FLUID MECHANICS AND TURBULENCE

In Figure 1.1, one can notice the smooth and regular shape of the candle flame which is said to be laminar. Such regular shape is essentially due to the low fuel vapour velocities and the small size of the candle. If a flame spans over a fuel bed with dimensions larger than few centimetres, the flow becomes turbulent. The shape of the flame is not smooth anymore, but it is rather wrinkled and made of 'structures' that are in seemingly chaotic motions (see Figure 1.5).

Turbulence plays an important role in fires. For instance, the diffusion process at the molecular level (as mentioned earlier for the laminar candle flame) becomes overwhelmed by turbulent diffusion. Also, the development of a smoke plume (see Chapter 4) under the effect of buoyancy is strongly dependent on the amount of surrounding 'fresh air' entrained through the column of hot gases. Air entrainment occurs often under turbulent conditions. Another example showing the importance of fluid mechanics and turbulence in fire dynamics is the interaction of water sprays with smoke and flames (see Chapter 7). The spray pattern (i.e., distribution of water droplets and associated velocities) is strongly dependent on the flow field (and thus turbulence) in the gas phase. A similar statement is valid for the heat transfer between the liquid phase (i.e., droplets) and the gas phase.

Generally speaking, mass and heat transfers in enclosure fires take place in turbulent flows and are dominated by fluid mechanics aspects. That is why we choose in our book to particularly focus

FIGURE 1.5   Illustration of turbulent flames, namely a pool fire.

on fluid mechanics. Fundamentals of turbulence are provided in Chapters 2 and 3, and examples of its importance in the context of enclosure fires are highlighted throughout the book.

1.3 COMBUSTION AND FIRE

So far, we have not made a clear distinction between combustion and fire. In fact, as stated by Quintiere [5], *in scientific terms, combustion and fire are synonymous.* Fire is undoubtedly a combustion process. However, it is quite a particular process that could very often grow to levels where it becomes difficult to *control*, leading to life and property loss. In order to illustrate best this notion of *control*, let us compare fire to another combustion system, a combustion engine for example. In a combustion engine, the amounts of fuel and air are to some extent well known and adjusted by the designer. In a car, hitting the gas pedal means that you are providing more fuel (and along more air). When you stop injecting fuel, the process stops. A fire, on the other hand, is a more *autonomous* process. As soon as it is initiated, it will seek fuel and air by itself. This self-sustained and dynamic complex process, as opposed to controlled combustion processes, is very difficult to characterize, for example in terms of the transient profiles of oxygen near the flame. This is another example describing the importance of fluid mechanics in fire.

Is a candle flame a fire? A candle flame surely presents (as described above) the main principles of fire behaviour (i.e., the fire triangle). It is clearly a self-sustained system driven by a chain of chemical and physical processes: (a) solid wax melting, (b) liquid wax absorption by the wick under the capillary force, (c) liquid wax vaporization under the flame heat, (d) diffusion of oxygen molecules near the flame envelope and mixing with fuel molecules, and (e) finally, burning that generates heat and light. The 'controlled' size of a candle makes it, however, prone to an easy extinction, and thus controllable, by simply blowing strong enough in front of it, or by putting a cup on top of it to prevent oxygen diffusion, or by using water. Furthermore, a candle flame is not representative of a 'real' fire, especially from a fluid mechanics standpoint, because it is not turbulent and turbulence is a dominant mechanism for mass and heat transfers in (enclosure) fires.

1.4 FIRE MODELLING

The main purpose of fire modelling is to provide a good representation of the fire development under given conditions (i.e., fire scenarios) in order to understand fire dynamics, and ultimately design 'fire-safe' buildings. The latter implies, for example, early fire detection and efficient fire suppression strategies. A good representation of the fire development is performed by an identification of the key elements of a given fire scenario (e.g., geometry, properties of the fuel package) followed by a corresponding mathematical description. There are several degrees of complexity in such a description, depending on the fire scenario as well as the desired level of detail. For example, an enclosure fire in its early stages could be either solved using a rather 'simple' two-zone model (see Chapter 5) or a more sophisticated Computational Fluid Dynamics (CFD) approach (see Chapter 8), which contains itself several degrees of complexity. The reader is guided throughout the whole book towards a (hopefully) clear and thorough understanding of the mathematical representation of several aspects of fire dynamics.

REFERENCES

1. M. Faraday (October 1988) "The chemical history of a candle", *Chicago Review Pr*, First edition.
2. D. Drysdale (2011) *An Introduction to Fire Dynamics*, 3rd Ed. Wiley.
3. A. Stec and R. Hull (Eds) (2010) *Fire Toxicity*. Woodhead Publishing.
4. B. Karlsson, and J.G. Quintiere (2000) *Enclosure Fire Dynamics*. CRC Press: Boca Raton, FL.
5. J.G. Quintiere (2006) *Fundamentals of Fire Phenomena*, John Wiley and Sons Ltd.

2 Turbulent Flows with Chemical Reaction

2.1 FLUID PROPERTIES – STATE PROPERTIES – MIXTURES

In order to understand fluid *mechanics*, a number of fluid *properties* must be defined first, where a fluid can be a liquid or a gas (vapour). Fluid properties must be clearly distinguished from *flow* properties, which are discussed below. Fluid properties are material properties, which can vary according to the state (e.g., temperature and pressure, aggregation state) of the fluid.

In order to allow for the definition and quantification of fluid properties, an implicit assumption in the classical fluid mechanics is the 'continuum hypothesis'. This implies treating fluids as continuous media, not as an ensemble of individual molecules [1]. This is justified in 'normal' fire-related applications. The continuum hypothesis effectively allows defining *local* fluid properties, which can be interpreted as average values over a very small volume around the position considered (but assuming that this volume is still very large compared to distances between molecules). The continuum hypothesis is adopted throughout this book.

2.1.1 FLUID PROPERTIES

2.1.1.1 Mass Density

The mass density (or 'density'), ρ, is the amount of fluid mass, m, inside a volume, V:

$$\rho = \frac{m}{V}. \tag{2.1}$$

Its unit is kg/m^3.

In a variable density flow, the mass density can vary in space and in time and the *local* density at a certain time is defined as in Eq. (2.1), taking the local limit for a small volume. Noting the position in physical space as $\bar{x} = (x, y, z)$ and time as t, the dependence of density is noted as: $\rho = \rho(x, y, z, t)$.

In an *incompressible* flow, the density does not vary. In general, liquids can be considered 'incompressible'. In gases, the density can vary due to variations in pressure or temperature (see Section 2.1.2.6).

The reciprocal of density is the 'specific volume' (m^3/kg).

2.1.1.2 Viscosity

Fluids can flow. Yet, if no force is exerted, the natural state of a fluid is to stand still. There is a 'natural' resistance against flow. The fluid property *viscosity* quantifies this resistance.

Expressed in a slightly more general way, it can be stated that neighbouring fluid particles with different velocities have the tendency to evolve towards the same common velocity, through exchange of momentum.

Before providing a more general mathematical description, Figure 2.1 visualizes the concept for a simple incompressible unidirectional laminar flow with uniform *velocity gradient* (or 'strain rate') dU/dy, i.e., constant derivative of velocity in the transversal direction. Neighbouring fluid

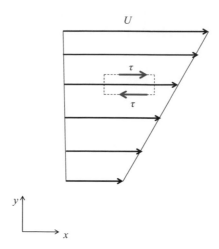

FIGURE 2.1   Illustration of the shear stress in an incompressible unidirectional laminar flow with uniform velocity gradient.

layers have different velocities, and they exert a shear stress τ onto each other: the slower layer tries to decelerate the faster layer above and vice versa. An external driving force is required to maintain the velocity gradient. If the fluid can be considered 'Newtonian', the shear stress increases linearly with the strain rate: $\tau = \mu \dfrac{dU}{dy}$. This is the case in practically all fire-related applications. The unit of τ is Pa (=N/m²). The proportionality factor, relating the velocity gradient to the shear stress, is the *dynamic* molecular viscosity μ (unit: Pa s). In gases, μ typically increases with (approximately the square root of) temperature, whereas in liquids, it decreases with increasing temperature.

The *kinematic* viscosity is related to the dynamic viscosity through the density:

$$v = \frac{\mu}{\rho} \tag{2.2}$$

Its unit is m²/s.

Expressed in more general terms, both the shear stress and the strain rate are *tensors*. The proportionality factor remains the simple scalar dynamic viscosity. The general expression for the shear stress tensor component τ_{ij} in variable density flows reads (Stokes' law):

$$\tau_{ij} = \mu \left[\left(\frac{\partial v_i}{\partial x_j} + \frac{\partial v_j}{\partial x_i} \right) - \frac{2}{3} \delta_{ij} \frac{\partial v_k}{\partial x_k} \right]. \tag{2.3}$$

A few new notations have been introduced in Eq. (2.3):

- The components i, j and k refer to coordinate directions x_i, x_j and x_k, respectively. With the above notation (and for convenience), x_1 corresponds to x, x_2 to y and x_3 to z.
- The Einstein summation convention applies: when the same index appears twice (which is the case for k in the last term in Eq. (2.3)), the sum is made over all components.
- The Kronecker delta δ_{ij} is a short notation: $\delta_{ij} = 0$ if $i \neq j$ and $\delta_{ij} = 1$ if $i = j$.

In general, flows are three-dimensional, so indices i, j and k each take values 1–3.

For the sake of clarity, two examples are provided for expression (2.3):

$$i = 1, j = 2: \tau_{12} = \mu \left(\frac{\partial v_1}{\partial x_2} + \frac{\partial v_2}{\partial x_1} \right), \tag{2.3a}$$

$$i = 1, j = 1: \tau_{11} = \mu \left(2\frac{\partial v_1}{\partial x_1} - \frac{2}{3} \left(\frac{\partial v_1}{\partial x_1} + \frac{\partial v_2}{\partial x_2} + \frac{\partial v_3}{\partial x_3} \right) \right). \tag{2.3b}$$

The final term in Eq. (2.3) can only be non-zero in compressible flows (see Section 2.3.1). It takes into account possible dilatation or compression of the fluid. It is referred to as 'Stokes' law'. Very often this term is relatively small in fire-related applications.

From the definition in Eq. (2.3), it is clear that the shear stress tensor is symmetric. The off-diagonal components cause friction losses in case of flow. The higher the viscosity, the larger the flow losses become for the same velocity gradient. In other words, the resistance of the fluid against (imposed) flow increases with increasing viscosity.

The viscosity of a fluid is never zero. The important implication is that whenever there is a solid boundary, this boundary always exerts an influence on the flow field (e.g., causing the development of a boundary layer). We come back to this point in Section 2.9.

2.1.1.3　Specific Heat

The *specific heat or thermal capacity*, c, is the amount of energy required to cause a temperature rise of 1 K (or 1°C) in 1 kg of the fluid. Its unit is J/(kg K).

In gases, the value of the specific heat depends on the circumstances in which the energy is supplied. If the pressure is kept constant, the notation is c_p. If the volume is kept constant, the notation is c_v. The difference between the two values is called the gas constant R (also in J/(kg K)):

$$c_p = c_v + R \tag{2.4}$$

For liquids and solids, $c_p \approx c_v$. For most fire-related applications, pressure can be assumed constant and the value c_p is to be used for gases.

In general, c_p increases slightly with temperature and polynomials exist in the literature for different gases. In fire-related applications, c_p is typically assumed constant and the value for air is used: $c_p = 1$ kJ/(kg K). Also for the gas constant R, the value for dry air is used: $R = 287$ J/(kg K).

Sometimes use is made of 'averaged specific heat' values. The averaging refers to the mean value over a temperature range. Table 2.1 provides a number of values for relevant gases in the context of fire. Note the high value for CO_2 when expressed in J/(kmol K). Adding CO_2 to a mixture leads to a reduction in temperature (see Section 2.2.2.1), which explains why addition of CO_2 is an effective means of suppression (see Section 3.1).

TABLE 2.1
Averaged Specific Heat Values Over the Range of
[300–2,000 K] for O_2, N_2, CO_2 and H_2O

	O_2	N_2	CO_2	H_2O
$\bar{c}_p$ (J/(kg K))	1,070	1,160	1,190	2,340
$\bar{c}_p$ (J/(kmol K))	34,300	32,500	52,300	42,100

2.1.1.4   Conduction Coefficient

The *conduction coefficient or thermal conductivity* expresses how easily heat flows inside a material. Its value quantifies how the heat flux per unit area (W/m²) relates to the local spatial temperature gradient (K/m):

$$\vec{q} = -k\nabla T = -\lambda\nabla T \tag{2.5}$$

This is known as 'Fourier's law'. The minus sign indicates that the heat flux is always from high temperature to low temperature. Note that Eq. (2.5) is a vector expression, with three components (one per coordinate direction).

The unit of the conduction coefficient (k or λ) is W/(m K).

It is noteworthy that Eq. (2.5) is not the most general expression. It implicitly assumes 'isotropy', i.e., the conductivity is the same in all directions. This assumption allows the use of a single scalar quantity to relate the heat flux vector to the temperature gradient. It also makes the heat flux vector aligned with the temperature gradient vector (and therefore makes the heat flux vector field locally perpendicular to isothermal lines throughout the domain). Some materials, with wood as a well-known example, are intrinsically anisotropic, though. For the example of wood, conductivity is much higher along the grains than perpendicular to the grains. Since this book focuses on fluid mechanics, not on the solid phase, this topic is not discussed in more detail here.

The conduction coefficient, specific heat and density can be combined to obtain the *thermal diffusivity*:

$$\alpha = \frac{k}{\rho c}. \tag{2.6}$$

The unit of α is m²/s. It is noteworthy that this is the same unit as for kinematic viscosity (Section 2.1.1.2). Indeed, α and ν play a similar role: the thermal diffusivity tries to bring temperature gradients to zero through heat flux (the higher α, the higher the heat flux for a given temperature gradient), just like the kinematic viscosity ν tries to relax velocity gradients to zero through exchange of momentum between neighbouring fluid particles (the higher ν, the stronger the momentum exchange for a given velocity gradient). We come back to this point in Section 2.7.

2.1.1.5   Diffusion Coefficient

In a mixture of fluids (see Section 2.1.3), one species can diffuse in the mixture due to concentration gradients of that species in the mixture. An elaborate discussion is found in many text books on reacting flows (e.g. [2–4]).

In fire-related applications, it is common practice to simplify the most general transport equations through application of 'Fick's law' to the diffusion of species k within the mixture:

$$\vec{J}_k = -\rho D_k \nabla Y_k \tag{2.7}$$

Note that the Einstein summation convention does not apply in Eq. (2.7). The *diffusion coefficient or mass diffusivity* D thus provides the relation between the diffusion flux vector J_k (kg/(m²s)) for species k and the spatial gradient of the local mass fraction Y_k (i.e., the local amount of mass of species k per kg mixture) of that species. The minus sign expresses that the diffusion flux is always from higher concentration to lower concentration.

Most species encountered in fire applications have comparable diffusivities (with the important exception of hydrogen, which has a much higher diffusivity). Also, mixtures in fire-related applications typically consist mainly of air and combustion products (H_2O, CO_2, CO, …).

Therefore, Eq. (2.7) is often a justifiable simplification of the most general expressions for species diffusion in mixtures. The diffusion coefficient D_k then relates to the diffusion of species k in air.

The unit of D is m²/s. It is noteworthy that this is again the same unit as for kinematic viscosity (Section 2.1.1.2). Indeed, D and v play a similar role: the mass diffusivity tries to bring concentration gradients to zero through mass transfer (the higher D, the higher the mass flux for a given concentration gradient), just like the kinematic viscosity v tries to relax velocity gradients to zero through exchange of momentum between neighbouring fluid particles (the higher v, the stronger the momentum exchange for a given velocity gradient), and just like the thermal diffusivity tries to bring temperature gradients to zero through heat flux (the higher α, the higher the heat flux for a given temperature gradient). We come back to this point in Section 2.7.

2.1.2 STATE PROPERTIES

State properties describe the state of the fluid. These are not material properties of the fluid.

2.1.2.1 Pressure

The pressure (p) can be defined as the normal force per unit area at a certain point. The unit is Pa. Pressure differences are the driving force for fluid flows.

2.1.2.2 Temperature

The unit of temperature (T) is Kelvin (K). The temperature must not be confused with heat (the unit of which is Joule, J).

2.1.2.3 Internal Energy

The local motion of molecules in a fluid is related to the internal energy (e or u, with unit J/kg). This is a measure for the thermal energy.

2.1.2.4 Enthalpy

The (static) enthalpy (h, with unit J/kg) is related to the internal energy through addition of pressure, divided by mass density:

$$h = u + \frac{p}{\rho} = e + \frac{p}{\rho} \tag{2.8}$$

2.1.2.5 Entropy

Entropy is a measure for the disorder in the fluid. It is related to the second law of thermodynamics. This law relates to, e.g., the fact that heat is transferred from higher temperature to lower temperature. Thus, while it is implicitly important, it does not appear explicitly in fire-related fluid mechanics. Therefore, it is not discussed in detail.

2.1.2.6 Equation of State

The 'equation of state' relates the local mass density of the fluid to its thermodynamic state, typically characterized by the state properties temperature and pressure in fire-related applications:

$$\rho = f(p,T). \tag{2.9}$$

In liquids, the density is essentially constant. Its dependence on pressure and temperature is relatively very weak.

In gases, it is common practice to specify Eq. (2.9) through the 'ideal gas law':

$$p = \rho R T. \tag{2.10}$$

For fire-related flows, this is justified. Most gases behave as air, which behaves as an ideal gas (with the exception of extremely low or high pressure or temperature, but this is not relevant for real-life fire applications). The gas constant R (J/(kg K)) has been introduced in Eq. (2.4), and the temperature T is expressed in Kelvin (K).

Equation (2.10) plays a central role in fire-related fluid mechanics.

Homework:

Consider a volume of $2\,m^3$ ($70\,ft^3$) air at ambient temperature equal to 20°C (68°F) and atmospheric pressure 1,00,000 Pa (1 bar). The gas constant of air is 287.1 J/(kg K).

a. Calculate the mass density of air. (Answer: $1.19\,kg/m^3$)
b. Assuming that the pressure is constant and that no mass is lost from the volume, calculate the density of air if the temperature increases to 200°C (392°F). Calculate the volume occupied by the air. (Answer: $3.23\,m^3$)
c. Assuming that the volume is kept constant and that no mass is lost from the volume, calculate the pressure if the temperature is equal to 200°C (392°F). (Answer: 1,61,400 Pa)
d. Which of the above conditions (b or c) is most typical for fires? (Answer: b; condition c only applies if the volume is sealed very tightly)

2.1.3 MIXTURES

In fire-related applications, the fluid is often a mixture. Yet, mixtures of gases typically do not receive much attention when considering fire-related fluid mechanics. There are a number of reasons for this.

First of all, turbulence is often dominant, overwhelming effects from molecular viscosity and molecular diffusion of heat and species. Turbulence is discussed in more detail in Section 2.8.

Second, except in the flame region, mixtures are mainly composed of air, as explained in Chapter 4. Therefore, the common simplification to treat the mixture of hot gases as hot air, applying the ideal gas law (2.10) with the gas constant for air and using the (temperature dependent) viscosity for hot air, is practically always justified.

Yet, a few definitions are introduced in this section. The mass fraction Y_i of species i is the ratio of the local amount of mass of species i to the local amount of mass of mixture. It is therefore a non-dimensional quantity. Conservation of mass leads to the fact that, at all times and everywhere in physical space, the sum of all mass fractions of all species equals unity: $\sum_{i=1}^{N} Y_i = 1$, with N the number of species in the mixture.

Using the notion of mass fractions, the fluid properties of mixtures can be determined from the fluid properties of their constituent species. For example, the specific heat becomes $c = \sum_{i=1}^{N} Y_i c_i$.

Also state properties can be defined accordingly. For example, static enthalpy becomes $h = \sum_{i=1}^{N} Y_i h_i$.

2.2 COMBUSTION

Entire monographs have been devoted to combustion (e.g. [2–8]). For the book at hand, focusing on fluid mechanics, not all detail is required. First of all, we restrict ourselves to phenomena in the gas phase. Although very important for fire dynamics in general, chemistry inside the solid (or liquid) phase is not considered, because it does not directly affect the fluid mechanics aspects in the gas phase.

Even in the gas phase alone, combustion is a very complex phenomenon. Yet, focusing on fluid mechanics, strong simplifications are possible. As explained in Section 2.3, the primary fluid property in the flow equations is the mass density, as this has a very direct impact on the flow field. As such, this determines to a large extent the mixing/entrainment processes as well, as explained in Chapters 3 and 4. In other words, at the level of fluid mechanics, combustion *physics* is more influential than combustion *chemistry*. Still, some chemical aspects are required as well. This is obvious when toxicity is an issue, but chemical kinetics can also be important for the determination of the mass density, as explained below.

2.2.1 CHEMICAL REACTION

In the context of fire, chemical reactions in the gas phase relate to 'combustion', i.e., the reaction of a 'fuel' with oxygen, where the oxygen typically stems from air. The 'fuel' is not restricted to economically and technically relevant fuels as used in industrial combustion processes, but is to be interpreted broadly as 'anything that can burn'.

In reality, a combustion process involves complex mechanisms of hundreds of elementary reactions with hundreds of intermediate species [2–8]. Yet, in fire applications, this entire complex process is often simplified to a limited set of global reactions of the type:

$$F(uel) + O(xidizer) \rightarrow P(roducts). \tag{2.11}$$

The fuel elements are typically carbon (C) and hydrogen (H) atoms, while the oxidizer is, as stated above, typically oxygen.

The nature of the products strongly depends on the 'completeness' of the reaction. The reaction is called 'complete' when all chemical reactions have come to completion or equilibrium. In fires with pure hydrocarbon fuels, the only reaction products are water vapour (H_2O) and carbon dioxide (CO_2) in such circumstances. A number of conditions need to be fulfilled for a combustion reaction to be complete:

- The fuel/oxidizer mixture must be 'flammable' (see Section 3.1).
- There must be a sufficient amount of oxidizer, so all the fuel has the opportunity to react.
- There must be a sufficient amount of time available for the reactions to take place. This is related to chemical kinetics (Section 2.2.3).

Even when all these circumstances are fulfilled, the combustion can still be 'incomplete' due to, e.g., insufficient mixing of fuel and oxidizer or too low local temperatures (see Section 2.2.3). This can lead to unburnt hydrocarbons and other combustion products, such as carbon monoxide (CO), at the right hand side of Eq. (2.11).

It should also be noted that the formation of soot is typical in fires. Soot essentially consists of carbon and is important with respect to radiation and smoke. Smoke is discussed in more detail in Chapter 4.

From Eq. (2.11), the notion 'stoichiometry' can be introduced. The theoretical notion 'stoichiometric conditions' refers to the situation where there is just enough oxidizer to allow for complete reaction of all the fuel. In other words, there is no lack of oxidizer, nor excess oxidizer. In such circumstances, the reaction can in principle be complete (but in practice it will not be, because the idealized theoretical conditions are not met in a fire). If there is too little oxidizer or, equivalently, too much fuel, the mixture is called '(fuel) rich'. By definition, the reaction cannot be complete in such circumstances. If there is excess oxidizer, the mixture is called '(fuel) lean'. In such circumstances, the reaction can be complete, but the excess oxidizer will also appear at the right hand side of Eq. (2.11). In this context, a few examples can be written for Eq. (2.11) as follows:

- For complete stoichiometric combustion with pure oxygen:

$$C_mH_n + \left(m + \frac{n}{4}\right)O_2 \rightarrow mCO_2 + \frac{n}{2}H_2O. \tag{2.11a}$$

- For complete stoichiometric combustion with air (where the air composition is simplified as 21% O_2 and 79% N_2 by volume):

$$C_mH_n + \left(m + \frac{n}{4}\right)O_2 + \frac{79}{21}\left(m + \frac{n}{4}\right)N_2 \rightarrow mCO_2 + \frac{n}{2}H_2O + \frac{79}{21}\left(m + \frac{n}{4}\right)N_2 \tag{2.11b}$$

- For complete combustion with excess air (where the air composition is simplified as 21% O_2 and 79% N_2 by volume and $\varepsilon \geq 1$):

$$C_mH_n + \varepsilon\left(m + \frac{n}{4}\right)O_2 + \frac{79}{21}\varepsilon\left(m + \frac{n}{4}\right)N_2 \rightarrow mCO_2 + \frac{n}{2}H_2O + (\varepsilon - 1)\left(m + \frac{n}{4}\right)O_2$$

$$+ \frac{79}{21}\varepsilon\left(m + \frac{n}{4}\right)N_2, \varepsilon \geq 1 \tag{2.11c}$$

Equations (2.11) illustrate that the result of the chemical reaction is a rearrangement of the chemical bonds among these elements. In other words, new species are created. The thermodynamics and chemical kinetics involved are discussed in Sections 2.2.2 and 2.2.3, respectively. Note that Eq. (2.11c) is only valid for $\varepsilon \geq 1$: if $\varepsilon < 1$, also products like CO and H_2 are formed.

It is important to appreciate that the number of elements remains constant throughout the reaction process, while the number of species need not remain constant.

An important quantity in combustion is the 'equivalence ratio' Φ, which is defined as the ratio of the amount (e.g., mass m) of fuel to the amount of oxidizer, divided by the same ratio at stoichiometric conditions:

$$\Phi = \frac{m_F/m_O}{\left(m_F/m_O\right)_{st}} \tag{2.12}$$

In other words,

- $\Phi < 1$ corresponds to fuel-lean conditions;
- $\Phi = 1$ corresponds to stoichiometric conditions;
- $\Phi > 1$ corresponds to fuel-rich conditions.

Homework:

Consider a hazard where methane is released through a leakage hole at a rate of 1 kg/s. Calculate the minimum amount of air needed for complete combustion (expressed in kg/s) of the methane. (Answer: 17.4 kg/s)

2.2.2 THERMODYNAMICS

2.2.2.1 Enthalpy

In the context of combustion, it is relevant to interpret the static enthalpy as the sum of the 'standard formation enthalpy' (or 'chemical' enthalpy) and the 'sensible enthalpy':

$$h(T) = h^c + h^s(T) = h_0(T_{\text{ref}}) + \int_{T_{\text{ref}}}^{T} c_p(T)dT. \tag{2.13}$$

The standard formation enthalpy of a species is the change in static enthalpy upon formation of 1 mole of that species in its standard state, from its composing elements in their standard state, at standard 'reference' temperature (298.15 K) and standard pressure (1 atm = 1,01,325 Pa). The 'standard state' is the most stable state at given pressure and temperature. This is a material property.

The standard formation enthalpy by definition equals zero for elements in their standard state at reference pressure and temperature. Examples are N_2 and O_2. The standard formation enthalpy is positive when heat needs to be supplied to create the species at standard pressure and temperature. Examples are NO and NO_2. The standard formation enthalpy is negative when heat needs to be removed when creating the species at standard pressure and temperature. Examples are water vapour (H_2O) and CO_2.

However, it is essential to realize that enthalpy *differences* are crucial, not absolute enthalpy values.

In a closed adiabatic system without mechanical work, the first law of thermodynamics states that there is no change in static enthalpy during the chemical reaction. However, for combustion reactions (Eq. 2.11), it is typical that the chemical enthalpy of the mixture of products is substantially lower than the chemical enthalpy of the mixture of fuel and oxidizer. Such reactions are 'exothermic'. The difference in chemical enthalpy is then transformed into sensible enthalpy (or 'heat'). This transformation typically causes a rise in temperature, as explained in Section 2.2.2.2.

Figure 2.2 visualizes the transformation of chemical enthalpy into sensible enthalpy. The notion 'activation energy' is explained in Section 2.2.3. The 'heat of reaction' of reaction (2.11) is defined as follows:

$$\Delta h_{c,\text{mixt}} = h^{c,u}(T_{\text{ref}}) - h^{c,b}(T_{\text{ref}}) \tag{2.14}$$

In Eq. (2.14), the superscript 'u' refers to 'unburnt mixture', while 'b' refers to 'burnt' conditions. The argument T_{ref} illustrates that the formation enthalpies are calculated at temperature T_{ref}.

Decomposing this into the different species, Eq. (2.14) reads:

$$\Delta h_{c,\text{mixt}} = \sum_{i=1}^{N} \left(Y_i^u - Y_i^b\right) h_{i,0}(T_{\text{ref}}) \tag{2.15}$$

The unit of $\Delta h_{c,\text{mixt}}$ is J/kg mixture. It can also be expressed in terms of J/mole mixture (in which case volume fractions, rather than mass fractions, need to be used in the summation).

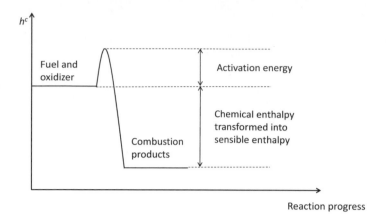

FIGURE 2.2   Sketch of the evolution of chemical enthalpy during a global combustion reaction (fuel + oxidizer → combustion products) in a closed adiabatic system without mechanical energy. The activation energy is indicated, as well as the amount of chemical enthalpy that is transformed into sensible enthalpy.

From Eq. (2.15), the 'heat of combustion' of a fuel can be calculated as well:

$$\Delta h_c = \Delta h_{c,\mathrm{mixt}} \frac{m_{\mathrm{mixt}}}{m_F} \tag{2.16}$$

The unit of Δh_c is then J/kg fuel. Complete combustion is assumed in the definition of the heat of combustion. Note that the heat of combustion can also be expressed in J/mole fuel (replacing then also the mass ratio in Eq. (2.16) by a molar ratio).

It is interesting to note that Eq. (2.16) can be rewritten in terms of J/kg air consumed (for combustion with air):

$$\Delta h_{c,\mathrm{air}} = \Delta h_{c,\mathrm{mixt}} \frac{m_{\mathrm{mixt}}}{m_{\mathrm{air}}}. \tag{2.16a}$$

In Eq. (2.16a), m_{air} is the amount of mass air consumed for complete combustion of the amount of mass fuel, m_F. For many organic fuels, the value is $\Delta h_{c,\mathrm{air}} \approx 3 \; \mathrm{MJ/kg\,air}$ [10].

Equation (2.16) can also be rewritten in terms of *J/kg* oxygen consumed:

$$\Delta h_{c,ox} = \Delta h_{c,\mathrm{mixt}} \frac{m_{\mathrm{mixt}}}{m_{ox}} \tag{2.16b}$$

In Eq. (2.16b), m_{ox} is the amount of mass oxygen consumed for complete combustion of the amount of mass fuel, m_F.

Assuming combustion of fuel with air and labelling the 'air-to-fuel' ratio at stoichiometric conditions S (in kg air per kg fuel), Eq. (2.17) can be rewritten as follows:

$$\Delta h_c = \Delta h_{c,\mathrm{mixt}} \left(1 + S\right). \tag{2.17}$$

Table 2.2 provides a few values for formation enthalpies $h_0\left(T_{\mathrm{ref}}\right)$ at $T_{\mathrm{ref}} = 298$ K. The formation enthalpy of O_2 and N_2 equals zero. From this table, the heat of combustion of a fuel can be calculated, as illustrated in the example.

TABLE 2.2

Formation Enthalpy of a Few Gases at T_{ref} = 298 K

Gas	$h_o(T_{ref})$(kJ/mol)
CH_4	−74.75
C_3H_8	−103.6
CO_2	−393.5
H_2O (vapour)	−241.8

Example 2.1

Use Table 2.2 to estimate the heat of combustion of methane.

The chemical reaction for complete combustion reads: $CH_4 + 2O_2 + 7.52N_2 \rightarrow CO_2 + 2H_2O + 7.52N_2$. Note that, in terms of enthalpy balances, addition of nitrogen or excess air does not alter the calculations, because the formation enthalpy of O_2 and N_2 equals zero. Thus, the calculation is performed here for stoichiometric conditions. In the unburnt conditions, the mass fractions read: $Y_{CH_4} = 0.055, Y_{O_2} = 0.22, Y_{N_2} = 0.725, Y_{CO_2} = 0, Y_{H_2O} = 0$. After complete combustion, the mass fractions read: $Y_{CH_4} = 0, Y_{O_2} = 0, Y_{N_2} = 0.725, Y_{CO_2} = 0.151, Y_{H_2O} = 0.124$. Thus, Eq. (2.15) yields:

$$\Delta h_{c,mixt} = 0.055 \times \frac{(-74.75)}{0.016} + 0.22 \times 0 + 0.725 \times 0 - 0.151 \times \frac{(-393.5)}{0.044}$$

$$- 0.124 \times \frac{(-241.8)}{0.018} - 0.725 \times 0 = 2760 \text{ kJ/kg}.$$

The air-to-fuel ratio reads: $S = \dfrac{2 \times 32 + 7.52 \times 28}{16} = 17.2$. Thus, Eq. (2.17) yields the heat of combustion of methane: $\Delta h_c = \Delta h_{c,mixt}(1+S) = 2,760 \times 18.2 = 50.1 \text{ MJ/kg}$.

Exercise: Use Table 2.2 to estimate the heat of combustion of propane. (Answer: 2,044 kJ/mol or 46.5 MJ/kg)

2.2.2.2 Temperature

Temperature is an important quantity in fire. For obvious reasons, it is important for the sake of heat transfer (conduction, convection and radiation), but it is equally important with respect to fluid mechanics. Indeed, through the equation of state (Eq. 2.9 or 2.10), it directly affects the mass density, which in turn directly affects the fluid mechanics, as explained in Section 2.3.

The local temperature in a fire is determined by the complex interplay of combustion reactions, convection, diffusion and radiation and can be determined from the transport equations (Section 2.3), but it is interesting to first discuss the notion 'adiabatic flame temperature'. This is in fact the temperature of the products in Eq. (2.11) in the following conditions:

- The system is closed and adiabatic;
- There is no work, nor changes in kinetic or potential energy;
- The combustion reaction (Eq. 2.11) is complete.

Unless if heat were supplied, the adiabatic flame temperature is the highest temperature possible. Indeed, any incomplete combustion or heat losses would lower the temperature.

The determination of the adiabatic flame temperature is through application of the first law in thermodynamics on an adiabatic closed system at constant pressure, stating that under the conditions mentioned, there is no change in static enthalpy: $dh = 0$. Using the superscripts 'u' for 'unburnt' and 'b' for burnt and for a mixture of N species (where N is the total number of species in the fuel, oxidizer and products in Eq. 2.11), this reads:

$$h^u = h(T_u) = \sum_{i=1}^{N} Y_i^u h_i^u = \sum_{i=1}^{N} Y_i^b h_i^b = h(T_b) = h^b. \tag{2.18}$$

The left hand side refers to the enthalpy of the reactants, at temperature T_u, whereas the right hand side concerns the enthalpy of the mixture of combustion products, at temperature T_b. The mass fraction Y_i of species i is zero if that species is absent.

For each component, the following holds:

$$h_i^b = h_i(T_b) = h_{i,0}(T_{\text{ref}}) + \int_{T_{\text{ref}}}^{T_b} c_{p,i}(T)dT = h_{i,0}(T_{\text{ref}}) + \int_{T_{\text{ref}}}^{T_u} c_{p,i}(T)dT + \int_{T_u}^{T_b} c_{p,i}(T)dT$$

$$= h_i(T_u) + \int_{T_u}^{T_b} c_{p,i}(T)dT = h_i^u + \int_{T_u}^{T_b} c_{p,i}(T)dT \tag{2.19}$$

Using Eqs. (2.13), (2.18) and (2.19), the adiabatic temperature $T_{b,ad}$ can be computed:

$$\sum_{i=1}^{N} Y_i^b \left(h_i(T_u) + \int_{T_u}^{T_{b,ad}} c_{p,i}(T)dT \right) = \sum_{i=1}^{N} Y_i^u h_i(T_u)$$

$$\rightarrow \sum_{i=1}^{N} Y_i^b \int_{T_u}^{T_{b,ad}} c_{p,i}(T)dT = \sum_{i=1}^{N} \left(Y_i^u - Y_i^b \right) h_i(T_u) \tag{2.20}$$

The determination of the adiabatic temperature is iterative due to the temperature dependence of the specific heats of the species. Averaged values $\left(\overline{c}_{p,i} \right)$ can be defined for the specific heat the interval $\left[T_u ; T_{b,ad} \right]$, but as $T_{b,ad}$ is not known a priori, this averaged value needs to be updated in an iterative manner, leading to the end result:

$$\sum_{i=1}^{N} Y_i^b \overline{c}_{p,i} \left(T_{b,ad} - T_u \right) = \sum_{i=1}^{N} \left(Y_i^u - Y_i^b \right) h_i(T_u) \tag{2.21}$$

If the unburnt mixture is at $T_u = T_{\text{ref}}$, the right hand side equals the difference in chemical formation enthalpy:

$$\sum_{i=1}^{N} Y_i^b \overline{c}_{p,i} \left(T_{b,ad} - T_u \right) = \sum_{i=1}^{N} \left(Y_i^u - Y_i^b \right) h_{i,0}(T_{\text{ref}}) \tag{2.22}$$

Thus, using Eq. (2.15), this yields:

$$\sum_{i=1}^{N} Y_i^b \overline{c}_{p,i} \left(T_{b,ad} - T_u \right) = \Delta h_{c,\text{mixt}} \tag{2.23}$$

Thus, rephrasing this in terms of heat of combustion of a fuel (in J/kg fuel), Eq. (2.17) finally leads to:

$$\sum_{i=1}^{N} Y_i^b \overline{c}_{p,i} \left(T_{b,ad} - T_u \right) = \frac{\Delta h_c}{1 + S} \tag{2.24}$$

As stated above, S is the 'air-to-fuel' ratio (in kg air per kg fuel) at stoichiometric conditions.

The adiabatic flame temperature is the highest for stoichiometric conditions. For lean mixtures, the excess oxidizer does not participate in the chemical reactions. Thermodynamically, the process can then be interpreted as the mixing of the cold excess air with the hot products (generated in stoichiometric conditions), so that the temperature of the end mixture is lower than the temperature of the pure combustion products. For rich mixtures, not all fuel can react due to lack of oxidizer. Thermodynamically, the process can then be interpreted as the mixing of the cold excess fuel with the hot products (generated in stoichiometric conditions), so that the temperature of the end mixture is lower than the temperature of the pure combustion products. (Note that reality will be more complex due to the presence of products as generated by incomplete combustion.)

Finally, it is noted that adiabatic flame temperatures are much higher for combustion with pure oxygen than for combustion with air. Indeed, the nitrogen in air is inert. In other words, it does not participate in the chemical reactions. Thermodynamically, the process can then be interpreted as the mixing of the cold nitrogen with the hot products (generated in stoichiometric conditions through reaction with the oxygen), so that the temperature of the end mixture is lower than the temperature of the pure combustion products. This illustrates that the adiabatic flame temperature is not a fuel property, because it depends on the reaction conditions.

A few examples, indicating the order of magnitude of the adiabatic flame temperature for combustion with air at stoichiometric conditions, are: 2,250 K (for methane), 2,380 K (for hydrogen), 1,975 K (for methanol), 2,155 K (for ethanol) and 2,145 K (for dry wood). Real temperatures in fires are much lower, primarily due to heat losses (convection and radiation), incomplete combustion and soot formation.

Example 2.2

Estimate the adiabatic flame temperature for stoichiometric complete combustion of methane in air, with initial temperature equal to 25°C. Repeat the same exercise for stoichiometric complete combustion of methane in pure oxygen.

The heat of combustion of methane is 50.1 MJ/kg (see example in Section 2.2.2.1). After complete combustion in air, the mass fractions read (see example in Section 2.2.2.1): $Y_{CH_4} = 0, Y_{O_2} = 0, Y_{N_2} = 0.725, Y_{CO_2} = 0.151, Y_{H_2O} = 0.124$. Thus, using the values in Table 2.1 (avoiding an iterative procedure here by taking these averaged specific heat values as fixed), Eq. (2.24) yields: $(0.725 \times 1,160 + 0.151 \times 1,190 + 0.124 \times 2,340)\left(T_{b,ad} - 298 \right) = \dfrac{50.1 \times 10^6}{18.2}$, so that the adiabatic temperature reads: $T_{b,ad} = 2,400$ K. This value is higher than what is reported in the literature (see above), but the deviation is not surprising, as no iterative procedure has been performed and possible dissociation reactions have been ignored.

For complete combustion in pure oxygen, the mass fractions become $Y_{CH_4} = 0, Y_{O_2} = 0, Y_{N_2} = 0, Y_{CO_2} = 0.55, Y_{H_2O} = 0.45$ in the burnt mixture. The oxygen to fuel ratio reads: $S = 2 \times 32 / 16 = 4$.

Thus, Eq. (2.25) yields $(0.55 \times 1,190 + 0.45 \times 2,340)\left(T_{b,ad} - 298 \right) = \dfrac{50.1 \times 10^6}{4}$, so that the adiabatic temperature becomes $T_{b,ad} = 5,870$ K. This is completely unrealistic, because, e.g., dissociation reactions are ignored. Yet, it does indicate that adiabatic flame temperatures are much higher in case of combustion with pure oxygen, compared to combustion in air. Indeed, the inert nitrogen in air acts as thermal ballast and reduces the temperature.

2.2.3 Chemical Kinetics

In Section 2.2.2, thermodynamic aspects of combustion chemistry have been discussed. However, thermodynamics does not provide information on the progress of reactions like Eq. (2.11). This progress is related to chemical kinetics.

As mentioned in Section 2.2.1, Eq. (2.11) is a severe simplification of the very complex reality of chemistry schemes, involving hundreds of elementary reactions with short-lived intermediate radicals. Accordingly, the *reaction rate* of Eq. (2.11) is typically severely simplified in the context of fire, stating that the rate at which the fuel and oxidizer react (expressed in, e.g., kg/s or mole/s), is proportional to:

$$k = A \, \exp\left(-\frac{E_a}{RT} \right) \qquad (2.25)$$

Equation (2.25) is an 'Arrhenius' expression for the reaction rate constant. The 'pre-exponential constant' A has unit s^{-1}. More important is the exponential expression, in which R denotes the gas constant (see Section 2.1.1.3), which has the universal value $R = 8.314$ J/(mol K). This expression reveals a strong dependence on two factors. The numerator E_a is the activation energy, i.e., the amount of energy that needs to be overcome for the reaction (Eq. 2.11) to take place. This is visualized in Figure 2.2. The higher E_a, the more difficult it becomes to 'ignite' the mixture. Yet, the most important observation from Eq. (2.25) is the strong dependence on temperature T. The higher the temperature, the faster the chemical reactions take place. If the temperature becomes too low, the reactions become so slow that the mixture is no longer considered as 'flammable' (see Section 3.1). As explained in Section 2.2.2.2, such conditions will occur if the mixture is too rich or too lean. Figure 2.3 illustrates the strong temperature dependence of the exponential factor in k for a value $E_a/R = 15,000$ K, which is typical for hydrocarbon gas fuels burning in air [9]. In Section 2.2.2.2, it was mentioned that the adiabatic flame temperature for combustion of hydrocarbon gas with air at atmospheric pressures is about 2,200 K at stoichiometric conditions.

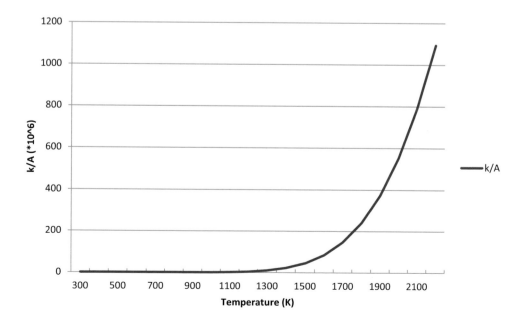

FIGURE 2.3 Illustration of the strong temperature dependence of k/A (Eq. 2.25) for $E_a/R = 15,000$ K.

Figure 2.3 reveals that if the mixture hydrocarbon gas/air is so lean or so rich that the adiabatic flame temperature becomes as low as 1,500 K, the reaction rate is at least 2 orders of magnitude less than in stoichiometric conditions. In such circumstances, the mixture becomes 'non-flammable' [10], as discussed in Section 3.1.

2.3 TRANSPORT EQUATIONS

Figure 2.4 visualizes a streamline through a surface of a (control) volume. This concept will be used to develop the conservation equations in the integral formulation. A streamline is defined such that at any point on the streamline the local velocity vector is tangent to the streamline. A collection of streamlines is called a stream tube.

2.3.1 CONSERVATION OF MASS

Conservation of mass reads:

The amount of mass flowing into a stationary volume per unit time equals the outflow of mass per unit time out of that same volume plus the amount of mass accumulation per unit time in that same volume.

Mathematically, this is expressed as follows (with notations as in Figure 2.4):

$$\frac{\partial}{\partial t} \iiint_V \rho \, dV + \iint_{\partial V} \rho \vec{v} \cdot \vec{n} \, dA = 0. \tag{2.26}$$

The first term is the accumulation of mass per unit time in the control volume. The second term is the net outflow per unit time, i.e., a closed surface integral over the entire area of the manifold ∂V. Note that the inner product $\vec{v} \cdot \vec{n} > 0$ for outflow, while $\vec{v} \cdot \vec{n} < 0$ for inflow.

Equation (2.26) is also called the *continuity equation*.

An important simplification is found in the case of *permanent* (or 'steady') *motion*. In that case, the time derivative disappears in Eq. (2.26):

$$\iint_{\partial V} \rho \vec{v} \cdot \vec{n} \, dA = 0. \tag{2.27}$$

A further simplification concerns *incompressible fluids* (e.g., water in a pipe under normal conditions). In that case, density does not change, so that not only Eq. (2.27) applies, but it further simplifies to read:

$$\iint_{\partial V} \vec{v} \cdot \vec{n} \, dA = 0 \tag{2.28}$$

FIGURE 2.4 Streamline through a surface. Notation: dA is the area of an infinitesimal part of the surface; $\vec{n}$ is the local normal vector on the surface (i.e., the vector with length equal to 1, locally perpendicular to dA and pointing outwards); $\vec{v}$ is the local flow velocity vector at position dA; θ is the angle between vectors $\vec{n}$ and $\vec{v}$.

Example 2.3

A simple illustration of Eq. (2.18) is provided on the basis of Figure 2.5. As there is no flow through the solid boundaries (solid lines in Figure 2.5), the only contributions to $\iint_{\partial V} \rho \vec{v} \cdot \vec{n} \, dA$ stem from surfaces 1 and 2. In surface 1, the velocity vector is pointing inwards, while the normal vector is by definition pointing outwards, so the contribution (under the simplified assumption of uniform flow through the cross section) becomes: $-\rho_1 v_1 A_1$. On surface 2, the velocity and the normal vectors are both pointing outwards, leading to: $+\rho_2 v_2 A_2$. Equation (2.27) thus provides: $-\rho_1 v_1 A_1 + \rho_2 v_2 A_2 = 0 \rightarrow \rho_1 v_1 A_1 = \rho_2 v_2 A_2$. In case of incompressible flow (Eq. 2.28), this further simplifies to: $v_1 A_1 = v_2 A_2$.

The surface integral in Eq. (2.26), in fact, refers to the total net mass flow rate (kg/s) through a surface with area A:

$$\dot{m} = \iint_A \rho \vec{v} \cdot \vec{n} \, dA. \tag{2.29}$$

If the mass density is not included, the total net volume flow rate (m³/s) through a surface with area A is found:

$$\dot{V} = \iint_A \vec{v} \cdot \vec{n} \, dA. \tag{2.30}$$

Expression (2.26) can also be formulated in differential form, applying Green's theorem:

$$\frac{\partial \rho}{\partial t} + \nabla \cdot \left(\rho \vec{v} \right) = 0 \tag{2.31}$$

The symbol ∇ is the divergence operator:

$$\nabla \cdot \vec{v} = \left(\frac{\partial}{\partial x_1} \vec{1}_{x_1} + \frac{\partial}{\partial x_2} \vec{1}_{x_2} + \frac{\partial}{\partial x_3} \vec{1}_{x_3} \right) \cdot \left(v_1 \vec{1}_{x_1} + v_2 \vec{1}_{x_2} + v_3 \vec{1}_{x_3} \right)$$

$$= \frac{\partial v_1}{\partial x_1} + \frac{\partial v_2}{\partial x_2} + \frac{\partial v_3}{\partial x_3} \tag{2.32}$$

In Eq. (2.32), $\vec{1}_{x_i}$ is the notation for the unity vector, i.e. a vector with length equal to unity, in the x_i-direction.

Expression (2.27), for steady flow, reads in differential form:

$$\nabla \cdot \left(\rho \vec{v} \right) = 0, \tag{2.33}$$

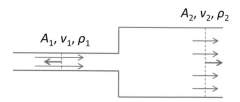

FIGURE 2.5 Illustration of conservation of mass for steady flow (Eq. 2.27) through a sudden pipe expansion. Dashed lines: boundary of control volume. Bold arrows: normal vectors (unity length, perpendicular to surface and pointing outwards). The other vectors indicate velocity vectors.

while expression (2.28), for incompressible fluids, becomes:

$$\nabla \cdot \vec{v} = 0. \tag{2.34}$$

This shows that the velocity field for any flow of an incompressible fluid is 'divergence free', or 'solenoidal'.

2.3.2 MOMENTUM EQUATIONS

Figure 2.4 serves as the basic sketch for the discussion of the integral formulation of conservation of total momentum. This boils down to the expression of Newton's second law, applied to flows: the net change in momentum of a system per unit time in a certain sense and direction equals the net force exerted on that system in that sense and direction. Expressed for a stationary volume, this reads:

The total force exerted onto a stationary volume equals the sum of the net outflow of momentum per unit time out of that same volume plus the accumulation of momentum per unit time in that same volume.

The local amount of momentum per unit volume is $\rho \vec{v}$ (kg/(m²s)), so Newton's second law reads, in vector notation:

$$\frac{\partial}{\partial t} \iiint_V \rho \vec{v} \, dV + \iint_{\partial V} \rho \vec{v} \left(\vec{v} \cdot \vec{n} \right) dA = \vec{F}_{\text{tot}}. \tag{2.35}$$

Equation (2.35) is valid for each component/direction individually.

For a permanent (or 'steady') motion, Eq. (2.35) simplifies to:

$$\iint_{\partial V} \rho \vec{v} \left(\vec{v} \cdot \vec{n} \right) dA = \vec{F}_{\text{tot}}. \tag{2.36}$$

The total force consists of:

- Surface forces:
  - Pressure p;
  - Viscous stresses;
- Body forces:
  - Gravity;
  - Others (not relevant for fire-related flows).

These forces are discussed now, in differential formulation:

$$F_{\text{tot},1} = -\frac{\partial p}{\partial x_1} + \frac{\partial \tau_{11}}{\partial x_1} + \frac{\partial \tau_{12}}{\partial x_2} + \frac{\partial \tau_{13}}{\partial x_3} + \rho g_1 \tag{2.37a}$$

$$F_{\text{tot},2} = -\frac{\partial p}{\partial x_2} + \frac{\partial \tau_{21}}{\partial x_1} + \frac{\partial \tau_{22}}{\partial x_2} + \frac{\partial \tau_{23}}{\partial x_3} + \rho g_2 \tag{2.37b}$$

$$F_{\text{tot},3} = -\frac{\partial p}{\partial x_3} + \frac{\partial \tau_{31}}{\partial x_1} + \frac{\partial \tau_{32}}{\partial x_2} + \frac{\partial \tau_{33}}{\partial x_3} + \rho g_3 \tag{2.37c}$$

The final terms in Eq. (2.37) refer to the gravity acceleration vector, multiplied with the local mass density. If y (or x_2) is chosen vertically upwards, then $g_1 = g_3 = 0$ and $g_2 = -9.81$ m/s².

Figure 2.6 shows how the normal stresses and shear stresses are defined.

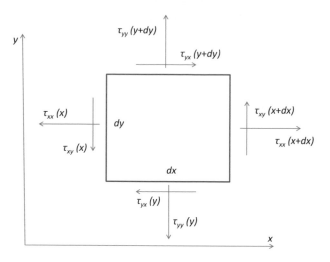

FIGURE 2.6   Definition of normal stresses and shear stresses (sketched in two dimensions only).

The shear stresses are assumed proportional to the dynamic viscosity and the local velocity gradients (Stokes' law), Eq. (2.3).

The above results in the *Navier–Stokes equations:*

$$\frac{\partial}{\partial t}(\rho v_i) + \rho v_1 \frac{\partial v_i}{\partial x_1} + \rho v_2 \frac{\partial v_i}{\partial x_2} + \rho v_3 \frac{\partial v_i}{\partial x_3} = -\frac{\partial p}{\partial x_i} + \frac{\partial \tau_{1i}}{\partial x_1} + \frac{\partial \tau_{2i}}{\partial x_2} + \frac{\partial \tau_{3i}}{\partial x_3} + \rho g_i, \quad i = 1,2,3 \quad (2.38)$$

It is noteworthy that the presence of the gravity force is essential to account for the Archimedes force. This is essential for buoyancy-driven forces, which is important in the context of fire. This is discussed further in Sections 2.5 and 2.6.

It is also noteworthy that pressure *gradients* (or pressure *differences*) are the driving force for flows, not the absolute pressure level.

2.3.3   CONSERVATION OF ENERGY

Conservation of energy concerns the first law of thermodynamics:

The change (per unit time) of the total internal energy of a system equals the sum of the heat added (per unit time) to the system and the work (per unit time) exerted onto that system.

The total internal energy consists of:

*   Static internal energy e (J/kg) or ρe (J/m³);
*   Kinetic energy $\rho v^2/2$ (J/m³).

Using Figure 2.4 as the basic sketch, the mathematical formulation of the first law of thermodynamics for a stationary open system reads:

$$\frac{\partial}{\partial t}\iiint_v \rho\left(e + \frac{1}{2}v^2\right)dV + \iint_{\partial V} \rho\left(e + \frac{1}{2}v^2\right)(\vec{v}\cdot\vec{n})dA = -\iint_{\partial V} p(\vec{v}\cdot\vec{n})dA + \iint_{\partial V}\left(\bar{\bar{\tau}}\cdot\vec{v}\right)\cdot\vec{n}\,dA$$

$$+ \iiint_v \rho\vec{g}\cdot\vec{v}\,dV + \iiint_v \rho S_h\,dV - \iint_{\partial V}\vec{q}\cdot\vec{n}\,dA. \quad (2.39)$$

The double arrow on top of τ reflects the fact that this is a tensor, Eq. (2.3).

The terms on the right hand side are as follows:

- *First term*: net inflow of total internal energy into the control volume ('convection'); the minus sign is necessary to comply with the sign convection (see previous sections: the normal vector is pointing outwards).
- *Second term*: work of the flow against pressure. This is work from a force (pressure), exerted onto the surface. The work by the pressure onto the flow is positive for inflow and negative for outflow, which explains the minus sign.
- *Third term*: work by the viscous stresses. This is work from a force (viscous stresses, Eq. 2.3), exerted onto the surface of the control volume.
- *Fourth term*: work by gravity. This is work by a volume force, exerted inside the volume. This work is positive for a downward flow, so that no minus sign is required in this term (if the y-direction is positive vertically upwards).
- *Fifth term*: volumetric source term of heat/internal energy (e.g., radiation). This term can be positive or negative.
- *Final term*: net incoming flux of heat/internal energy (e.g., conduction). The flux with the flow cannot be added to this term (as it is already included in the convection term).

The energy equation can also be formulated, using enthalpy (Eq. 2.8):

$$\frac{\partial}{\partial t}\iiint_v \rho\left(h + \frac{1}{2}v^2\right)dV = \frac{\partial}{\partial t}\iiint_v p\,dV - \iint_{\partial V}\rho\left(h + \frac{1}{2}v^2\right)(\vec{v}\cdot\vec{n})dA - \iint_{\partial V}p(\vec{v}\cdot\vec{n})dA$$

$$+\iint_{\partial V}(\overset{\leftrightarrow}{\tau}\cdot\vec{v})\cdot\vec{n}\,dA + \iiint_v \rho\vec{g}\cdot\vec{v}\,dV + \iiint_v \rho S_h\,dV - \iint_{\partial V}\vec{q}\cdot\vec{n}\,dA. \quad (2.40)$$

In differential formulation, and generalizing for mixtures of n species, this reads:

$$\frac{\partial}{\partial t}\left(\rho h + \frac{1}{2}\rho v^2\right) + \nabla\cdot\left(\rho\left(h + \frac{1}{2}v^2\right)\vec{v}\right) = \frac{\partial p}{\partial t} + \nabla\cdot(\overset{\leftrightarrow}{\tau}\cdot\vec{v}) + \rho S_h - \nabla\cdot\vec{q}. \quad (2.41)$$

As mentioned in Section 2.1.3, the (static) enthalpy is the mass-weighted sum of the enthalpies of species i:

$$h = \sum_{i=1}^{N} Y_i h_i, \quad (2.42)$$

where the enthalpy h_i for each species consists of chemical and sensible enthalpy, Eq. (2.13).

It is noteworthy that in Eq. (2.41), expressed in terms of enthalpy, the source term ρS_h contains, e.g., radiation, but not a heat release rate due to combustion. As mentioned in Section 2.2.2.1, combustion reactions transform chemical enthalpy into sensible enthalpy, but the sum of sensible and chemical enthalpy does not change locally. If the energy equation is expressed in terms of temperature (or sensible enthalpy), a source term due to the combustion heat release rate does appear.

The final term in Eq. (2.41) reads:

$$\nabla\cdot\vec{q} = -\nabla\cdot(k\nabla T) - \nabla\cdot\left(\rho\sum_{i=1}^{N}h_i D_i\nabla Y_i\right) + \text{D.E.} \quad (2.43)$$

The abbreviation 'D.E.' refers to the 'Dufour effect', i.e., an additional enthalpy flux due to species concentration differences. This effect is ignored in fire-related flows. The first term in Eq. (2.43) refers to Fourier's law for heat conduction. The middle term denotes an enthalpy flux due to diffusion, using Fick's law, Eq. (2.7).

Many fire-induced flows are low-Mach number flows, i.e., the velocities are much lower than the speed of sound (note: this is not true for explosions). Also, temporal changes in pressure can often be ignored in Eq. (2.41), as well as the work done by gravity, by the viscous shear stresses and by the normal stresses. Typically also the kinetic energy is negligible. Using the Prandtl and Schmidt numbers, as defined in Eqs. (2.88) and (2.89), respectively, the energy equation simplifies to:

$$\frac{\partial}{\partial t}(\rho h) + \nabla \cdot (\rho h \vec{v}) = \nabla \cdot \left(\frac{\mu}{Pr} \nabla h + \mu \sum_{i=1}^{N} \left(\frac{1}{Sc_i} - \frac{1}{Pr} \right) h_i \nabla Y_i \right) + \rho S_h \qquad (2.44)$$

For flows of mixtures with unity Lewis number for all species ($Le_i = 1$ for all i, Eq. 2.90), this further simplifies to:

$$\frac{\partial}{\partial t}(\rho h) + \nabla \cdot (\rho h \vec{v}) = \nabla \cdot \left(\frac{\mu}{Pr} \nabla h \right) + \rho S_h. \qquad (2.45)$$

Equation (2.45) contains the three types of heat transfer, which are briefly discussed first now.

2.3.3.1   Convection

Convection refers to heat exchange between a fluid in motion and a surface. It concerns the second term on the left hand side of Eq. (2.45). The fact that the fluid is in motion implies that also the conservation of mass, Eq. (2.31), as well as the Navier–Stokes equations, Eq. (2.38), need to be solved in order to determine the temperature field. However, very often convection is expressed in a macroscopic manner, introducing the convection coefficient h (W/(m^{2}K)):

$$\dot{q}_{conv} = h \Delta T. \qquad (2.46)$$

The convection coefficient must not be confused with the static enthalpy. In Eq. (2.46), $\dot{q}_{conv}$ is the convective heat transfer per unit time and per unit area available for the convective heat transfer (in W/m^2) and ΔT is a characteristic temperature difference between the surface and the fluid.

Correlations exist in the literature for the convection coefficient. They are expressed in a non-dimensional manner, in the form of a Nusselt number:

$$Nu = \frac{hL}{k}. \qquad (2.47)$$

In Eq. (2.47), L is a characteristic length scale for the problem and k is the conductivity of the fluid (Section 2.1.1.4).

Two types of convection exist: forced convection and natural convection. Forced convection refers to conditions where the motion in the fluid is not caused by the heat transfer itself, but rather by a device such as a fan or a pump. In such conditions, a characteristic velocity, v, can be defined, and thus also a characteristic Reynolds number can be defined:

$$Re = \frac{\rho v L}{\mu} = \frac{vL}{\nu}. \qquad (2.48)$$

A more detailed discussion on the Reynolds number is postponed until Sections 2.7 and 2.8. The correlations in the context of forced convection read:

$$Nu = f(Re, Pr). \tag{2.49}$$

In Eq. (2.49), the Reynolds number provides information on the flow and the geometry, whereas the Prandtl number (see below, Eq. 2.88) provides information on the fluid.

In case of natural convection, the fluid motion is caused by the heat transfer itself and disappears if the heat transfer process stops. In such circumstances, it is less clear how to define a characteristic velocity, so that the use of Eq. (2.48) is less straightforward. On the other hand, the temperature difference, causing an Archimedes force, is now a logical choice as characteristic quantity. Thus, in the context of natural convection, the Reynolds number is replaced by the Rayleigh or Grashof number (see below, Eqs. (2.96) and (2.97)) in the correlations:

$$Nu = f(Ra, Pr). \tag{2.50}$$

The Prandtl number (see below, Eq. (2.88)) still provides information on the fluid, while the Rayleigh (or Grashof) number provides information on the configuration (geometry and temperature difference).

The reader is referred to specialized literature for explicit formulations of correlations of the type (2.49) and (2.50).

In the context of fire, the convection coefficient of air flows in natural convection is typically within the range of 5–25 W/(m² K), with 10–15 W/(m² K) the most common range.

2.3.3.2   Conduction

Conduction refers to Eq. (2.45) in the absence of fluid motion. Conduction also takes place if the fluid is in motion, but in that case, it is overwhelmed by convection (Section 2.3.3.1). In the absence of motion, Eq. (2.45) becomes in terms of temperature:

$$\frac{\partial}{\partial t}(\rho c_p T) = \nabla \cdot (k \nabla T) + \rho S_h. \tag{2.51}$$

Recall that Fourier's law, Eq. (2.5), has been introduced in the right hand side of Eq. (2.51).

In the absence of a source term $(\rho S_h = 0)$ and in steady-state conditions (left hand side equal to zero in Eq. 2.51), this reveals that in materials with uniform conduction coefficient, the temperature field is linear.

In the case of fire, often transient conduction is important. Indeed, conduction in the solid phase is typically much slower than convection in the gas phase. Moreover, if radiation (see Section 2.3.3.3) incides onto the surface of a solid, this also leads to unsteady conduction inside the solid.

Consider now the situation where there is no source (nor sink) term for heat inside the solid and uniform material properties. Equation (2.51) then simplifies to:

$$\frac{\partial T(x, y, z, t)}{\partial t} = \alpha \nabla \cdot \big(\nabla T(x, y, z, t)\big). \tag{2.52}$$

In Eq. (2.52), the dependence on location and time has been made explicit and α denotes the thermal diffusivity (see Section 2.1.1.4). From Eq. (2.52), with given initial conditions and boundary conditions, the temperature field inside the solid can be calculated.

It can be found in numerous text books that the Fourier number represents time in a non-dimensional manner:

$$Fo = \frac{\alpha t}{L^2}. \tag{2.53}$$

The key question is what the characteristic length scale is. To that purpose, it is most instructive to consider the 'thermal penetration depth':

$$L_{\text{th,pen}} = C\sqrt{\alpha t}. \tag{2.54}$$

If $C = 4$ is chosen, temperature changes at the surface are not felt at depth $L_{\text{th,pen}}$ from the surface (i.e., the temperature change, defined as the deviation from the initial conditions, at depth $L_{\text{th,pen}}$ is <1% of the temperature change at the surface). If the geometrical characteristic dimension of the solid in which the temperature field is computed is larger than $L_{\text{th,pen}}$, the solid is 'thermally thick' and $L_{\text{th,pen}}$ is the characteristic length scale to be used in Eq. (2.53). An example of 'geometrical characteristic dimension of the solid' is the thickness of a wall where the temperature changes at one side (e.g., as a consequence of a fire at that side). If the geometrical characteristic dimension of the solid, L_{geom}, is smaller than $L_{\text{th,pen}}$, then L_{geom} is to be used in Eq. (2.53). The solid then behaves as 'thermally thin'. The notions thermally thin and thermally thick play a central role in the phenomenon of flame spread (Section 3.6). Note that, implicitly, a one-dimensional approximation has been introduced in the discussion of the wall as example. The assumption of one-dimensional conduction is often valid in the context of fire, although two- or three-dimensional effects can sometimes be observed in the context of flame spread (Section 3.6).

It is noted that in Refs. [9,10], a material is called 'thermally thick' as long as the back surface does not affect the temperature field near the surface, exposed to the fire. This is in line with the above. However, in Refs. [9,10], the link is also made between 'thermal thickness' and temperature gradients inside the solid. It is stated in Refs. [9,10] that as long as temperature gradients inside the solid are small, a material can be labelled 'thermally thin'. Indeed, the limit of zero temperature gradient corresponds to the limit of an infinite thermal penetration depth, so the criterion 'small temperature gradients' corresponds to $L_{\text{geom}} \ll L_{\text{th,pen}}$, which is consistent with the above. It is interesting to make the connection to the Biot number, which is defined as follows:

$$Bi = \frac{hL}{k}. \tag{2.55}$$

In Eq. (2.55), h is the convection coefficient in the fluid that is in contact with the surface and, in contrast to k in the Nusselt number (Eq. 2.47), k is the conduction coefficient of the solid. The Biot number can be interpreted as a ratio of thermal resistances, namely the ratio of the thermal resistance against conduction $\left(R_{\text{th,cond}} = \frac{L}{kA} \right.$, with A the area for the heat flux by conduction) to the thermal resistance against convection $\left(R_{\text{th,conv}} = \frac{1}{hA} \right)$. This interpretation of the Biot number reveals that low values ($Bi \ll 1$) correspond to very small temperature gradients inside the solid, given the relatively very low resistance against conduction. This corresponds to thermally thin behaviour indeed, in line with the criterion that the back surface boundary condition affects the temperature evolution: the temperature at the back surface quickly follows temperature changes at the front surface. In other words, the criterion $Bi \ll 1$ can also be used to label a material 'thermally thin', as it corresponds to the situation $L_{\text{geom}} \ll L_{\text{th,pen}}$, as explained above. The latter criterion is more general, though, and remains in place regardless of the convection (e.g., in a radiation-dominated heat transfer problem).

To conclude this section, it is mentioned that for a thermally thick material, the heat flux, absorbed per unit time by conduction into the solid material, experiencing a temperature rise at its surface from the initial temperature T_0 to T_s, is proportional to:

$$\dot{q}_{\text{cond}} \propto \frac{k|T_s - T_0|}{L_{\text{th,pen}}} \propto \sqrt{\rho k c} \, \frac{|T_s - T_0|}{\sqrt{t}}. \tag{2.56}$$

This illustrates the importance of the combination $(\rho k c)$, which is called the 'thermal inertia' (see also Section 3.6). Materials with low values for $\rho k c$, such as insulation materials, follow temperature changes easily (near the surface) and thus result in small temperature differences ΔT in Eq. (2.46). In other words, relatively little heat is taken out of a compartment (with, e.g., a fire) by convection. The opposite is true for materials with high $\rho k c$ value, such as steel or concrete. As a consequence, thermal feedback to the fire will be higher in a compartment with low $\rho k c$ value boundaries (walls and ceilings) than in a compartment with high $\rho k c$ value boundaries. Consequently, more rapid fire development can be expected in the former (see Chapter 5).

2.3.3.3 Radiation

Radiation concerns heat transfer over a distance. No medium is required. A basic law for radiation is the Stefan–Boltzmann law, quantifying the radiative power emitted per unit area by a black body at absolute temperature T (in K):

$$E_b = \sigma T^4. \tag{2.57}$$

The constant $\sigma = 5.67 \ 10^{-8}$ W/$(\text{m}^2 \text{K}^4)$ is the Stefan–Boltzmann constant. Expression (2.57) reveals a very strong dependence on absolute temperature. The fact that σ is so small also implies that radiation remains relatively unimportant at low temperatures. Table 2.3 provides an overview of values for $\sigma(T^4 - T_{\text{amb}}^4)$, compared to the convection heat transfer (per unit area) $h\Delta T$ (Eq. 2.37), with $h = 10$ W/$(\text{m}^2 \text{K})$ and $T_{\text{amb}} = 300$ K. Clearly radiation becomes dominant from temperatures equal to 700 K and higher, but also for temperatures equal to, e.g., 500 K, radiation is not necessarily negligible (depending on the view factor and emissivity, as explained next).

It must be appreciated that Eq. (2.57) is an upper boundary for the true radiative power of a body at temperature T. Indeed, a black body is the perfect emitter, i.e., Eq. (2.57) is the theoretically maximum possible radiative power (per unit area) to be emitted by a body at temperature T. A real body will emit less power, and a correction to Eq. (2.57) is made, introducing the emissivity ε:

$$E = \varepsilon \sigma T^4. \tag{2.58}$$

The emissivity ε is a non-dimensional number, $0 \le \varepsilon \le 1$. In general, the emissivity depends on the direction (i.e., the emissivity need not be the same in all directions) and on the wavelength of the radiation. In particular for gases, the latter can be quite important. If a 'grey' model is used, the spectral dependence is ignored. If a diffuse approximation is adopted, the directional dependence is ignored.

TABLE 2.3

Values (in kW/m²) for $\sigma(T^4 - T_{\text{amb}}^4)$ and $h\Delta T$(Eq. 2.46), with $h = 10$ W/(m²K) and $T_{\text{amb}} = 300$ K

T (K)	300	400	500	600	700	800	900	1,000	1,100	1,200	1,400	1,600
$\sigma(T^4 - T_{\text{amb}}^4)$	0	1	3	6.9	13.2	22.8	36.7	56.2	82.6	117.1	217.4	371.1
$h\Delta T$	0	1	2	3	4	5	6	7	8	9	11	13

Equation (2.58) provides the radiative power (per unit area) emitted by an object, or a volume of gas, at uniform temperature T. This radiative power incides onto surrounding objects and surfaces. However, these objects or surfaces receive only a fraction of the emitted power. This fraction is determined by the distance from the source of radiation and the relative orientation and is quantified by a 'view factor' or 'configuration factor'. This can be understood as follows. Consider a surface, with area A_1, at temperature T_1, exchanging heat by radiation with another surface, with area A_2, at temperature T_2. In order to calculate the total amount of radiative heat transfer (per unit time), both surfaces are subdivided into infinitesimal surfaces first. Then, the radiative heat transfer (per unit time) between these infinitesimal surfaces is computed and an integration is made over both surfaces. If the source of radiation can be modelled as a point source, only one integration is performed, namely over the receiving surface. It can be illustrated that if both surfaces are black bodies, the net radiative heat exchange between the infinitesimal surfaces reads:

$$\dot{Q}_{dA_1-dA_2} = \sigma\left(T_1^4 - T_2^4\right)\frac{\cos\varphi_1\cos\varphi_2}{\pi r^2}\,dA_1 dA_2. \tag{2.59}$$

In Eq. (2.59), φ_1 and φ_2 denote the angle between the ray of radiation, exchanged between the infinitesimal surfaces dA_1 and dA_2, and the normal direction, perpendicular to dA_1, resp. dA_2. The distance between dA_1 and dA_2 is r. Equation (2.59) reveals, for given temperatures, an increase in radiative heat exchange as surfaces dA_1 and dA_2 are more aligned (i.e., φ_1 and φ_2 being closer to zero) and, more importantly, as the distance r between dA_1 and dA_2 decreases.

The total amount of net radiative heat exchange between the surfaces is obtained through integration:

$$\dot{Q}_{1-2} = \sigma\left(T_1^4 - T_2^4\right)\int\int_{A_1 A_2}\frac{\cos\varphi_1\cos\varphi_2}{\pi r^2}\,dA_1 dA_2 = \sigma\left(T_1^4 - T_2^4\right)A_1 F_{12}. \tag{2.60}$$

The latter expression introduces the view factor:

$$F_{12} = \frac{1}{A_1}\int\int_{A_1 A_2}\frac{\cos\varphi_1\cos\varphi_2}{\pi r^2}\,dA_1 dA_2. \tag{2.61}$$

Note that this is a purely geometrical quantity. Tables, formulas and graphs exist in the literature for view factors.

If the emitter is a point source, one integration disappears and Eq. (2.60) can be rewritten as follows:

$$\dot{Q}_{1-2} = \sigma\left(T_1^4 - T_2^4\right)\int_{A_2}\frac{\cos\varphi_2}{\pi r^2}\,dA_2 = \sigma\varphi\left(T_1^4 - T_2^4\right). \tag{2.62}$$

In Eq. (2.62), φ is called a 'configuration factor'.

If the emissivities are taken into account, the equations become:

$$\dot{Q}_{1-2} = \sigma\left(\varepsilon_1 T_1^4 - \varepsilon_2 T_2^4\right)A_1 F_{12}, \tag{2.63}$$

$$\dot{Q}_{1-2} = \sigma\varphi\left(\varepsilon_1 T_1^4 - \varepsilon_2 T_2^4\right). \tag{2.64}$$

Note that 'surface' 2 need not really be a surface. It can also be ambient. In that case, expressions (2.63) and (2.64) quantify the amount of radiative power going to ambient.

An object or gas (e.g., smoke) that receives incident radiation does not absorb this incoming heat flux completely. If the total irradiation (in W/m²) is noted as G, a fraction αG is absorbed (with α the absorptivity), a fraction ρG is reflected (with ρ the reflectivity) and a fraction τG is transmitted (with τ the transmissivity). The absorptivity, reflectivity and transmissivity are non-dimensional numbers within the range of [0–1] and their sum equals 1 at all times:

$$\alpha + \rho + \tau = 1. \tag{2.65}$$

If $\tau = 0$, the body is called opaque.

Gases with symmetric molecules, such as O_2 and N_2, are completely transparent for radiation: $\tau = 1, \alpha = \rho = 0$. Asymmetric molecules, such as CO_2, H_2O and CH_4, absorb and emit radiation within certain wavelength bands. As such, they constitute a 'participating medium' for radiation. In smoke, which contains soot particles (see Chapters 4 and 8), the absorption is continuous over the radiative wavelength spectrum.

The flame or gas emissivity is typically approximated as follows:

$$\varepsilon_g = 1 - \exp\left(-\kappa_g L\right). \tag{2.66}$$

In Eq. (2.66), κ_g is called the 'absorption coefficient' in Ref. [9] and the 'effective emission coefficient' in Ref. [10]. This might seem strange at first sight, but this is not in conflict with each other. Indeed, Kirchhoff's law states that for all wavelengths, the monochromatic absorptivity equals the monochromatic emissivity:

$$\alpha_\lambda = \varepsilon_\lambda. \tag{2.67}$$

In Eq. (2.67), the subscript λ refers to the wavelength. Note that Kirchhoff's law is only valid in the case of sufficiently small temperature differences. It must not be applied to, e.g., solar radiation. In the context of fire, expression (2.67) is valid.

Note that values for κ_g are in the order of $1\,m^{-1}$ [9,10]. L is a characteristic length scale (e.g., the average flame height or the smoke layer thickness). Thus, Eq. (2.66) reveals that flames of 2 m high approach black-body emitters.

Finally, note that with Eq. (2.67), Eq. (2.66) also allows to calculate how much radiation is absorbed in a smoke layer of thickness L, if κ_g is known. Similarly, it can be computed from Eq. (2.66) how much radiation is emitted by the smoke layer, if a uniform smoke layer temperature T_s is assumed. Using the view factor or configuration factor, the incident radiative power on the floor (important for fire dynamics, see Chapter 5) or on people (important for life safety) can be calculated.

2.3.4 TRANSPORT OF SPECIES

The transport equation for mass fraction Y_i of species i, using Fick's law (Eq. 2.7), reads:

$$\frac{\partial}{\partial t}\left(\rho Y_i\right) + \nabla \cdot \left(\rho Y_i \vec{v}\right) = \nabla \cdot \left(\rho D_i \nabla Y_i\right) + \rho S_i, \ i = 1, \ldots, N \tag{2.68}$$

In Eq. (2.68), ρS_i is the local chemical source term of species i (unit: kg/(m³s)). This source term can be positive (production of species i) or negative (consumption of species i). In general, as mentioned before, chemistry mechanisms are very complex. The expressions for the chemical source terms are strongly non-linear, dependent on density and on concentrations of the different species, and with reaction rate constants that strongly depend on temperature (and pressure) [2–8].

For the sake of completeness, it is noted that the 'Soret' effect (i.e., an additional diffusion flux due to temperature gradients) is neglected. This is common practice in fire-related applications.

Whereas Eq. (2.68) is valid for all N species, only $(N-1)$ equations need to be solved, because the sum of all mass fractions must be equal to 1, everywhere and at all times:

$$\sum_{i=1}^{N} Y_i = 1. \tag{2.69}$$

This algebraic expression replaces the N-th transport equation.

2.3.5 MIXTURE FRACTION

Very often, in case of fire, the combustion process is of the 'non-premixed' or 'diffusion' type: fuel and oxidizer are initially separate, and combustion reactions only take place as fuel and oxidize mix. An important quantity in this type of combustion is the *mixture fraction*. This quantity is defined on the basis of *elements* (atoms), rather than species (molecules). A major advantage is that throughout the chemical combustion process, the number of elements does not change, as mentioned in Section 2.2.1. This is an essential difference, compared to species, which are created or destroyed during the process (Section 2.2.1).

Consequently, in contrast to Eq. (2.68), there is no chemical source term in the transport equations for mass fractions of elements. This can be formalized as follows. The mass fraction of element i reads: $Z_i = \sum_{j=1}^{N} \mu_{ij} Y_j, i = 1, \ldots, M$, where μ_{ij} is the amount of mass of element i per kg species j. The number of elements in the mixture is denoted as M. Expressing conservation of element i during the reactions reads:

$$\sum_{j=1}^{N} \mu_{ij} \rho S_j = 0, i = 1, \ldots, M. \tag{2.70}$$

Consequently, expressing Eq. (2.68) for species j, multiplying this equation by μ_{ij}, summing over all species, and using property (2.70), one obtains:

$$\frac{\partial}{\partial t}(\rho Z_i) + \nabla.(\rho Z_i \vec{v}) = \nabla. \left(\rho \sum_{j=1}^{N} \mu_{ij} D_j \nabla Y_j \right), i = 1, \ldots, M. \tag{2.71}$$

The right hand side remains complex, despite application of Fick's law (Eq. 2.7). This is due to multi-component diffusion. However, things simplify substantially if all diffusion coefficients can be assumed equal, i.e., $D_j = D$. For most components in fire-related flows, this is a valid assumption. (However, an important exception concerns hydrogen (H_2), which diffuses much faster.) With this assumption, Eq. (2.71) simplifies to:

$$\frac{\partial}{\partial t}(\rho Z_i) + \nabla \cdot (\rho Z_i \vec{v}) = \nabla \cdot (\rho D \nabla Z_i), i = 1, \ldots, M. \tag{2.72}$$

One more step can be taken, by normalizing all element mass fractions, so that they all take on values between 0 and 1. This can be done as follows:

$$\xi_i = \frac{Z_i - Z_{i,O}}{Z_{i,F} - Z_{i,O}}. \tag{2.73}$$

The subscripts F and O refer to 'fuel' and 'oxidizer', respectively. Equation (2.73) reveals that ξ_i equals 0 in pure oxidizer and equals 1 in pure fuel. It takes a value in between for mixtures of fuel and oxidizer, hence the name 'mixture fraction'. The local value of a mixture fraction expresses what fraction of the total mass of the local mixture stems from the fuel. It is thus a measure for the amount of mixing of fuel and oxidizer.

It is immediately clear that Eq. (2.72) can be rewritten in terms of ξ_i, rather than Z_i. This implies that all mixture fractions ξ_i must obey the same transport equation, with the same initial conditions and the same boundary conditions. This implies that all mixture fractions are equal. In other words, there is a single *mixture fraction*, with the following transport equation:

$$\frac{\partial}{\partial t}(\rho\xi) + \nabla \cdot (\rho\xi\vec{v}) = \nabla \cdot (\rho D \nabla \xi). \tag{2.74}$$

It has already been noted that there is no chemical source term in Eq. (2.74). The mixture fraction is thus a 'conserved scalar'.

In Chapter 8, the value of the concept 'mixture fraction' will become clear in the context of CFD ('Computational Fluid Dynamics') calculations. Here, the discussion is closed by noting that the original transport equations, Eq. (2.68), for N species, have been simplified to a single transport equation, Eq. (2.74), for the mixture fraction. As such, not only has the number of equations been drastically reduced, but also the chemical source terms have been removed. While this is intentional, the information from the chemical process has been lost. This information is brought back into the CFD calculations through a chemistry model, relating the temperature and species mass fractions to the value of the mixture fraction. This is explained next on the basis of a simple example. Before elaborating on the example, it is noted that the maximum adiabatic flame temperature (Section 2.2.2.2) is obtained at the 'stoichiometric mixture fraction' ξ_{st}, i.e., the mixture fraction that corresponds to stoichiometric conditions. Lean mixtures correspond to the region $\xi < \xi_{st}$, whereas rich mixtures correspond to $\xi > \xi_{st}$. As explained in Section 2.2.2.2, the adiabatic flame temperatures for such conditions are inevitably lower.

Example 2.4

Consider the reaction of methane with air at atmospheric conditions, assuming complete combustion and ignoring intermediate species. In that case, there are $N=5$ species: CH_4, O_2, N_2, CO_2 and H_2O. We label them as $j=1, ..., 5$ in this order. There are $M=4$ elements: C, H, O and N. We label them as $i=1, ..., 4$ in this order.

With the notation introduced above, we have, e.g., μ_{11} (i.e., the amount of mass of element C per kg species CH_4) $= 12/(12+4) = 0.75$; μ_{12} (i.e., the amount of mass of element C per kg species O_2) $= 0$; $\mu_{13} = 0$; $\mu_{14} = 12/44$; $\mu_{15} = 0$; $\mu_{21} = 4/16$; $\mu_{22} = 0$; $\mu_{23} = 0$; $\mu_{24} = 0$; $\mu_{25} = 2/18$; $\mu_{31} = 0$; $\mu_{32} = 1$; $\mu_{33} = 0$; $\mu_{34} = 32/44$; $\mu_{35} = 16/18$; $\mu_{41} = 0$; $\mu_{42} = 0$; $\mu_{43} = 1$; $\mu_{44} = 0$; and $\mu_{45} = 0$.

Then element mass fraction Z_1 reads:

$$Z_1 = Z_C = \sum_{j=1}^{5} \mu_{1j} Y_j = 0.75 Y_{CH_4} + 0.273 Y_{CO_2};$$

$$Z_2 = Z_H = 0.25 Y_{CH_4} + 0.111 Y_{H_2O};$$

$$Z_3 = Z_O = Y_{O_2} + 0.727 Y_{CO_2} + 0.889 Y_{H_2O};$$

$$Z_4 = Z_N = Y_{N_2}.$$

For the normalization, it is recalled that the fuel F is CH_4 and the oxidizer air is a mixture of 23.3% O_2 and 76.7% N_2 (by mass). Thus,

$$Z_{1,F} = Z_{C,F} = 0.75,$$

$$Z_{2,F} = Z_{H,F} = 0.25,$$

$$Z_{3,F} = Z_{O,F} = 0,$$

$$Z_{4,F} = Z_{N,F} = 0,$$

$$Z_{1,O} = Z_{C,O} = 0,$$

$$Z_{2,O} = Z_{H,O} = 0,$$

$$Z_{3,O} = Z_{O,O} = 0.233,$$

$$Z_{4,O} = Z_{N,O} = 0.767.$$

This leads to (Eq. 2.71):

$$\xi_1 = \frac{0.75Y_{CH_4} + 0.273Y_{CO_2}}{0.75} = Y_{CH_4} + 0.364Y_{CO_2},$$

$$\xi_2 = \frac{0.25Y_{CH_4} + 0.111Y_{H_2O}}{0.25} = Y_{CH_4} + 0.444Y_{H_2O},$$

$$\xi_3 = \frac{Y_{O_2} + 0.727Y_{CO_2} + 0.889Y_{H_2O} - 0.233}{-0.233}$$
$$= 1 - 4.29Y_{O_2} - 3.12Y_{CO_2} - 3.815Y_{H_2O},$$

$$\xi_4 = \frac{Y_{N_2} - 0.767}{-0.767} = 1 - 1.304Y_{N_2}.$$

Consider now three situations, illustrating that all mixture fractions are equal:

- *Pure fuel:* $Y_{CH_4} = 1, Y_{O_2} = 0, Y_{N_2} = 0, Y_{CO_2} = 0, Y_{H_2O} = 0 \rightarrow \xi_1 = \xi_2 = \xi_3 = \xi_4 = 1,$
- *Pure oxidizer:* $Y_{CH_4} = 0, Y_{O_2} = 0.23, Y_{N_2} = 0.77, Y_{CO_2} = 0, Y_{H_2O} = 0 \rightarrow \xi_1 = \xi_2 = \xi_3 = \xi_4 = 0,$
- *Stoichiometric conditions:* these can be derived from the reaction, Eq. (2.11b): $CH_4 + 2O_2 + \frac{79}{21}2N_2 \rightarrow CO_2 + 2H_2O + \frac{79}{21}2N_2$. Before the reaction, such a mixture has the following composition: $Y_{CH_4} = 0.055, Y_{O_2} = 0.22, Y_{N_2} = 0.725, Y_{CO_2} = 0, Y_{H_2O} = 0$. This leads to $\xi_1 = \xi_2 = \xi_3 = \xi_4 = 0.055$, which is the 'stoichiometric mixture fraction'. After complete reaction, the composition of the mixture reads: $Y_{CH_4} = 0, Y_{O_2} = 0, Y_{N_2} = 0.725, Y_{CO_2} = 0.151, Y_{H_2O} = 0.124.$. This leads again to $\xi_1 = \xi_2 = \xi_3 = \xi_4 = 0.055$. This illustrates that the mixture fraction value does not provide any information on the progress of the reaction.

Recalling from Section 2.2.2.2 that the adiabatic temperature at stoichiometric conditions is 2,250 K for methane–air combustion at atmospheric conditions and that the temperature of pure fuel and pure air is 300 K, it is easily seen that there is a much steeper temperature increase at the lean side ($0 < \xi < \xi_{st} = 0.055$) than on the rich side ($0.055 < \xi < 1$). This point will be recalled in Section 3.1 while discussing flammability limits.

Homework:

Consider the reaction of propane with air at atmospheric conditions, assuming complete combustion and ignoring intermediate species.

a. Calculate the stoichiometric mixture fraction. (Answer: 0.06)
b. Determine the amount of air (in kg) required to consume 1 kg of propane by complete combustion in stoichiometric conditions. (Answer: 15.6 kg)
c. Assuming that pure propane and pure air are at 300 K and that the adiabatic flame temperature is 2,265 K, sketch piecewise linear temperature profiles as function of mixture fraction.
d. Determine the mixture fraction values for which $T = 1,500$ K. (Answer: 0.037; 0.426)
e. Assuming that 1,500 K would be a threshold value for 'flammability' due to too slow chemical kinetics below this value (see Figure 2.3), calculate the mixture composition (before the onset of any reactions) in concentrations (volume percent) for the lowest mixture fraction value of question d. [This is called the 'lower flammability limit'; see Section 3.1.1.] (Answer: 2.4% C_3H_8 by vol., 97.6% air)

2.4 BERNOULLI

The Bernoulli equation is valid on any streamline (as defined in Section 2.3). It is important to note that the equation has been developed for steady incompressible flows (liquids). Yet, to a very good approximation, it can also be applied to low-Mach number flows where the density does not change rapidly along the streamline.

Taking the z-direction as vertically upwards, Bernoulli's equation reads:

$$p + \frac{1}{2}\rho v^2 + \rho g z = \text{const.} \tag{2.75}$$

Stated in another manner, if points 1 and 2 are on the same streamline (and the density does not change strongly between points 1 and 2):

$$p_1 + \frac{1}{2}\rho v_1^2 + \rho g z_1 = p_2 + \frac{1}{2}\rho v_2^2 + \rho g z_2. \tag{2.76}$$

This is sometimes referred to as 'conservation of total mechanical energy'. Indeed, the first term refers to static pressure, the second term is kinetic energy and the third term is potential energy.

A few application examples of application of Bernoulli's equation are briefly mentioned here. In Chapters 5 and 6, this equation will be applied repeatedly.

Example 2.5: Velocity Measurement with a Pitot Tube

A Pitot tube is a simple device to measure flow velocity by creating flow stagnation inside the tube. As the velocity becomes zero by stagnation, the pressure increases from the static pressure in the flow to the 'stagnation' pressure. Aligning the tube with the streamline and assuming horizontal positioning (or ignoring height differences along the measurement zone), Bernoulli's equation illustrates that the flow velocity can be computed from the measurement of pressure difference:

$$p_{\text{stag}} + \frac{1}{2}\rho 0^2 = p + \frac{1}{2}\rho v^2 \rightarrow v = \sqrt{\frac{2(p_{\text{stag}} - p)}{\rho}} = \sqrt{\frac{2\Delta p}{\rho}} \tag{2.77}$$

Example 2.6: Flow Rate Measurement with a Venturi Tube

A Venturi meter essentially consists of a converging cone (from cross-sectional area A_1 to a smaller cross section A_2), from which the flow rate through a pipe can be calculated. Indeed, applying Bernoulli's at constant height z yields:

$$p_1 + \frac{1}{2}\rho v_1^2 = p_2 + \frac{1}{2}\rho v_2^2. \tag{2.78}$$

Conservation of mass for a steady incompressible flow yields: $v_1 = v_2 \dfrac{A_2}{A_1}$. Noting the volume flow rate as $\dot{V} = v_2 A_2$, it can be computed from the measurement of the pressure difference and the Venturi geometry:

$$\dot{V} = A_2 \sqrt{\frac{2A_1^2 (p_1 - p_2)}{\rho(A_1^2 - A_2^2)}} = A_2 \sqrt{\frac{2\Delta p}{\rho} \frac{1}{1 - \left(\dfrac{A_2}{A_1}\right)^2}}. \tag{2.79}$$

It is noteworthy that both examples clearly illustrate that *pressure differences* are the driving force for flows. Bernoulli's equation is the basic equation for flows through openings, as discussed extensively in Chapters 5 and 6. A 'discharge coefficient' is introduced then, as briefly discussed in the next example.

Example 2.7: Flow through an Orifice

As just mentioned, the pressure difference (Δp) over an opening determines the flow through that opening. From the (mean) velocity $\left(v = \sqrt{\dfrac{2\Delta p}{\rho}} \right)$ and the cross-sectional area (A) of the opening, the volume flow rate through that opening can be computed:

$$\dot{V} = C_d A \sqrt{\frac{2\Delta p}{\rho}}. \tag{2.80}$$

The discharge coefficient C_d, introduced in Eq. (2.80), can be interpreted as a 'correction' to the cross-section area that is effectively used for the outflow (or inflow) through the opening ('vena contracta' effect). The combination $C_d A$ is then called the 'aerodynamic area', to be distinguished from the geometric cross-sectional area A. For an orifice, the value of C_d is around 0.6 for, e.g., open doors or windows, going up to about 0.7 for flows through small gaps [11] (Figure 2.7).

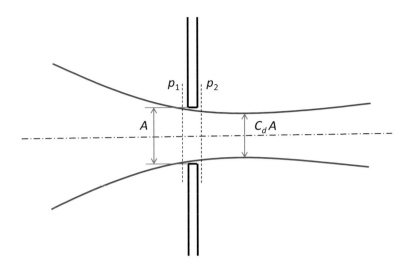

FIGURE 2.7   Illustration of the 'vena contracta' effect in the flow through an opening. The pressure difference $\Delta p = p_1 - p_2$ is the driving force. The discharge coefficient C_d can be interpreted as the ratio of the 'effective' stream tube cross-section area to the geometric opening area A.

2.5   HYDROSTATICS

Hydrostatics refers to the state where there is no flow. In other words, all velocities in a certain environment are equal to zero. Introducing this into the general Navier–Stokes equations (2.38), the basic law for hydrostatics reads:

$$\nabla p = \rho_{\text{amb}} \vec{g} \tag{2.81}$$

Equation (2.81) is valid at any time (in the absence of motion). It indicates that the pressure field, and therefore pressure differences, are directly proportional to the density of the medium considered and to the gravity field in place.

Restricting ourselves to applications on earth and defining z as the vertical upward direction, the gravity vector reads: $\vec{g} = -g\vec{1}_{x_3}$, with $g = 9.81$ m/s^2.

The vector equation (2.81) reads in the z-direction:

$$\frac{dp}{dz}\left(= \frac{dp}{dx_3} \right) = -\rho_{\text{amb}} g \tag{2.82}$$

Given that the gravity vector in Eq. (2.81) only has a vertical component, the vector equation (2.81) implies that pressure does not vary in the horizontal directions in the absence of flow.

Finally, choosing a reference pressure p_{ref} at reference height z_{ref}, Eq. (2.82) can be integrated over height to read:

$$p = p_{\text{ref}} - \rho_{\text{amb}} g \left(z - z_{\text{ref}} \right). \tag{2.83}$$

This equation illustrates that pressure decreases linearly with height. This is illustrated in Figure 2.8: pressure evolves with height according to a straight line. The slope of the line is proportional to the density of the medium, ρ_{amb}, and the gravity field acceleration constant, g.

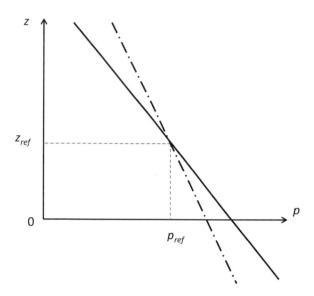

FIGURE 2.8   Linear evolution of pressure with height: the law of hydrostatics (Eq. 2.82). The slope depends on the density of the medium. The solid line refers to a medium with higher density than the dash-dotted line.

Example 2.8

A person takes an elevator in a high-rise building from floor level up to the 80th floor. Each floor is 2.5 m (8.2 ft) high. The average pressure inside the building is 1 00,000 Pa (6.89 10^8 psi, 1 bar). The temperature inside the building is 20°C (68°F). What pressure difference does the person experience after having completed the elevator ride?

Answer:

The temperature inside the building is 20°C = 293 K.
    Using the ideal gas law, Eq. (2.10), the average air density inside the building is calculated to be: $\rho_{amb} = 1,00,000/(287.1 \times 293) = 1.19$ kg/m³.
    The height difference between the ground floor ('gf') and the 80th floor ('80f') is 80×2.5 m = 200 m (656.2 ft).
    Using Eq. (2.51), $p_{gf} - p_{80f} = 1.19 \times 9.81 \times 200 = 2,332$ Pa (16.1 × 10^8 psi).

Homework:

Consider an airtight 50-storey high-rise building in summer conditions. Each floor is 2.5 m (8.2 ft) high. The atmospheric pressure at ground level is 1,05,000 Pa (7.24 × 10^8 psi), and the outside temperature is 35°C (95°F). Inside the building, the temperature is kept at 20°C (68°F). Consider a 6 m wide staircase, completely surrounded by glass windows at the outside (i.e., by glass windows of 6 m × 2.5 m (19.685 ft × 8.202 ft) each). Assuming that at mid-height of the building the pressure inside the building equals the atmospheric pressure:

a. Calculate the total force on each of the windows at floor level. (Answer: 548 N (55.9 kgf))
b. Is the force inwards or outwards? (Answer: outwards)
c. Calculate the total force on each of the windows on the 50th floor. (Answer: 548 N (55.9 kgf))
d. Is the force inwards or outwards? (Answer: inwards)

2.6 BUOYANCY

The main relevance of the fundamental law of hydrostatics, Eq. (2.81), lies in the fact that in many fire-related flows, buoyancy plays a dominant role. This can be learnt from the Navier–Stokes equations, Eq. (2.38), combining the forces due to pressure gradients and gravity. In the vertical direction (still with $\vec{g} = -g\vec{1}_{x_3}$), using Eq. (2.82), the resulting force per unit area (i.e., the right hand side of Eq. 2.38), reads:

$$-\frac{dp}{dz} - \rho g = \left(\rho_{\text{amb}} - \rho\right)g. \tag{2.84}$$

In the process of getting to expression (2.84), pressure differences in the horizontal directions are implicitly assumed small. As explained in the previous section, the pressure differences are zero in the special case of hydrostatics. As long as the horizontal velocity components remain relatively small in comparison to the vertical flow, the assumption of negligible horizontal pressure gradients remains valid. In case of forced horizontal ventilation (see Chapter 6), this assumption may no longer be valid. Yet, the buoyancy term, Eq. (2.84), remains in place.

Equation (2.84) reveals that the driving force in situations where buoyancy dominates, stems from density differences in the presence of a gravity field. This is known as *Archimedes'* law. Note that as gravity acts in the vertical direction only, buoyancy forces by definition also act in the vertical direction only.

Equation (2.84) reveals that:

- the net force due to buoyancy is upwards in situations where the density of the ambient medium, ρ_{amb}, is higher than the density of the fluid considered, ρ.
- the net force due to buoyancy is downwards in situations where the density of the ambient medium, ρ_{amb}, is lower than the density of the fluid considered, ρ.

For small density differences, the additional approximation $\rho \approx \rho_{\text{amb}}$ is typically made in the Navier–Stokes equations (2.38). Only the difference $\left(\rho_{\text{amb}} - \rho\right)$ is accounted for in combination with gravity (Eq. 2.83). This is called *Boussinesq*'s approximation.

In the context of small density differences *in a given fluid*, expression (2.84) can be developed by means of a Taylor series expansion, in terms of the independent variables temperature and pressure:

$$\rho = \rho(T, p) \rightarrow \rho \approx \rho_{\text{amb}} + \left(\frac{\partial \rho}{\partial T}\right)_p (T - T_{\text{amb}}) + \left(\frac{\partial \rho}{\partial p}\right)_T (p - p_{\text{amb}}).$$

Usually the pressure term is much smaller than the temperature term. Using the thermal volumetric expansion coefficient:

$$\beta - -\frac{1}{\rho}\left(\frac{\partial \rho}{\partial T}\right)_p, \tag{2.85}$$

The right hand side of Eq. (2.84) becomes:

$$\left(\rho_{\text{amb}} - \rho\right)g = \rho_{\text{amb}}\beta(T - T_{\text{amb}})g, \quad \text{if } \beta(T - T_{\text{amb}}) \ll 1. \tag{2.86}$$

Equation (2.86) reveals that:

- the net force due to buoyancy is upwards in situations where the temperature of the ambient medium, T_{amb}, is lower than the temperature of the fluid considered, T.
- the net force due to buoyancy is downwards in situations where the temperature of the ambient medium, T_{amb}, is higher than the temperature of the fluid considered, T.

It is recalled that the basic expression is Eq. (2.84), based on density differences, while Eq. (2.86) is only valid for sufficiently small temperature differences inside a given fluid.

Finally, for ideal gases, the following approximation can be made to calculate the thermal volumetric expansion coefficient, Eq. (2.85), at a given temperature T, using the ideal gas law, Eq. (2.10):

$$p = \rho RT \rightarrow \beta = -\frac{1}{\rho}\left(\frac{\partial \rho}{\partial T}\right)_p = -\frac{RT}{p}\left(-\frac{p}{RT^2}\right) = \frac{1}{T} \tag{2.87}$$

Equation (2.86) then reads:

$$\left(\rho_{amb} - \rho\right)g = \rho_{amb}\left(1 - \frac{T_{amb}}{T}\right)g, \; if \; \left(1 - \frac{T_{amb}}{T}\right) \ll 1 \tag{2.88}$$

It is important to note that in Eq. (2.88), temperatures must be expressed in Kelvin.

Example 2.9

A spherical balloon of $2\,m^3$ ($70.6\,ft^3$) is filled with methane. The temperature inside the balloon is 15°C (59°F), equal to the temperature of the surrounding air. The average pressure inside the balloon is 1,00,000 Pa (6.89×10^8 psi, 1 bar), equal to the atmospheric pressure surrounding the balloon. Calculate the buoyancy force exerted on the balloon. Ignoring the weight of the balloon itself, will the balloon rise, or drop to the floor? What if the balloon had been filled with propane?

Answer:

The mass density of methane (CH_4) at 15°C and 1,00,000 Pa is: $\rho_{CH_4} = \dfrac{1,00,000}{518.3 \times 288} = 0.67\,kg/m^3$.

The mass density of air at 15°C and 1,00,000 Pa is: $\rho_{air} = \dfrac{1,00,000}{287.1 \times 288} = 1.21\,kg/m^3$.

The buoyancy force on the $2\,m^3$ large balloon is found from Eq. (2.53). The force per unit volume reads: $\left(\rho_{amb} - \rho\right)g = (1.21 - 0.67) \times 9.81 = 5.3\,N/m^3$. Thus, the total force on the $2\,m^3$ large balloon is: $2 \times 5.3 = 10.6 N$ (1.08 kgf).

Because $\rho_{CH_4} < \rho_{air}$, the balloon will rise.

For propane (C_3H_8), the mass density is $\rho_{C_3H_8} = \dfrac{1,00,000}{189 \times 288} = 1.84\,kg/m^3$. The total force on the $2\,m^3$ large balloon is: $2 \times 6.16 = 12.3 N$ (1.25 kgf).

Because $\rho_{C_3H_8} > \rho_{air}$, the balloon will drop.

Homework:

1. Due to a fire, hot smoke (with assumed uniform temperature, T_{smoke}) is generated in still ambient air (with temperature, T_{amb}). Assuming that the fluid properties of the smoke can be determined as if the smoke were hot air, calculate the buoyancy force exerted onto $1\,m^3$ ($35.3\,ft^3$) smoke in air, assuming the following temperatures (and assuming that the smoke temperature does not vary):

    a. $T_{amb} = 20°C$ (68°F) and $T_{smoke} = 120°C$ (248°F); (Answer: 3 N, 0.306 kgf)

    b. $T_{amb} = 20°C$ (68°F) and $T_{smoke} = 220°C$ (428°F); (Answer: 4.8 N, 0.49 kgf)

    c. $T_{amb} = 20°C$ (68°F) and $T_{smoke} = 320°C$ (608°F); (Answer: 6 N, 0.612 kgf)

    d. Note that the temperature difference $T_{smoke} - T_{amb}$ increases linearly from a to c. Does the buoyancy force increase linearly? (Answer: no)

    e. $T_{amb} = 0°C$ (32°F) and $T_{smoke} = 120°C$ (248°F); (Answer: 3.9 N, 0.4 kgf)

    f. $T_{amb} = -20°C$ (−4°F) and $T_{smoke} = 120°C$ (248°F); (Answer: 4.9 N, 0.5 kgf)

    g. Why does the buoyancy force increase, comparing a to e to f? (Answer: due to stronger density differences)

2. In a basement, methane is leaking through the pipeline supply. Will the methane rise into the building on top? Is it safe to turn on the lights (mounted on the ceiling) in the basement? (Answers: yes; no, a spark when switching on the light can act as ignition source in a potentially flammable mixture meets the source (see Section 3.1, on 'flammability').)

3. On a day without wind, propane is leaking from a tank, standing on the ground. Describe what will happen. Is this a danger for the surrounding? (Answers: propane stays close to the ground and starts spreading horizontally; yes, this is dangerous, since an ignition source may lead to a very hazardous situation when a flammable mixture meets the source (see Section 3.1, on 'flammability').)

2.7 NON-DIMENSIONAL NUMBERS

2.7.1 FLUID PROPERTIES

By combining fluid properties, dimensionless groups can be constructed. Indeed, it has already been mentioned that the units of kinematic viscosity (v), thermal diffusivity (α) and molecular diffusivity (D) are identical (m²/s). Physically, the interpretation is that the kinematic viscosity tries to make the velocity field uniform inside a fluid (through exchange of momentum), the thermal diffusivity tries to make the temperature field uniform (through heat exchange by conduction) and the molecular diffusivity tries to make the concentration field homogeneous in a mixture (through concentration gradient-driven diffusion).

The resulting non-dimensional groups read:

- The Prandtl number, comparing kinematic viscosity (exchange of momentum) to thermal diffusivity (heat exchange by conduction):

$$\text{Pr} = \frac{v}{\alpha} = \frac{\mu c_p}{\lambda} = \frac{\mu c_p}{k}. \tag{2.89}$$

- The Schmidt number, comparing kinematic viscosity (exchange of momentum) to molecular diffusivity (mass diffusion):

$$\text{Sc} = \frac{v}{D}. \tag{2.90}$$

- The Lewis number, comparing thermal diffusivity (heat exchange by conduction) to molecular diffusivity (mass diffusion):

$$\text{Le} = \frac{\alpha}{D}. \tag{2.91}$$

Clearly, the non-dimensional numbers are not all independent of each other:

$$\text{Le} = \text{Sc} \cdot \text{Pr}^{-1}. \tag{2.92}$$

It is important to note that the non-dimensional groups are still fluid properties, not flow properties.

2.7.2 FLOW PROPERTIES

Non-dimensional flow numbers are interesting to characterize the flow, given a certain fluid. The characteristic length scale is noted as L, and the characteristic velocity as v.

Examination of the terms in the Navier–Stokes equations, Eq. (2.38), combined with Eq. (2.3), leads to the following proportionalities: $\dfrac{\rho v}{t} \propto \dfrac{\rho v^2}{L} \propto \dfrac{\Delta p}{L} \propto \Delta \rho g \propto \mu \dfrac{v}{L^2}$.

Several non-dimensional numbers, characterizing the flow, can be derived now. The importance of each of the numbers mentioned below depends on the importance of the corresponding terms in the Navier–Stokes equations (2.38).

The convection term/inertia term is always important in the flow. Depending on the flow configuration, one or more terms are in competition with (or determine) the inertia term (or thus the flow).

If the viscous stresses prevail, the proportionality $\dfrac{\rho v^2}{L} \propto \mu \dfrac{v}{L^2}$ leads to the Reynolds number which is the ratio of inertial forces to viscous forces (2.48):

$$\mathrm{Re} = \frac{\rho v L}{\mu} = \frac{v L}{\nu}. \tag{2.93}$$

The viscous forces tend to damp the inherent instabilities in the non-linear convection terms in the Navier–Stokes equations, while these instabilities can evolve towards fully developed turbulence for a large-enough Reynolds number. This is addressed in Section 2.8.

If buoyancy is dominant, the proportionality $\dfrac{\rho v^2}{L} \propto \Delta \rho g$ leads to the Froude number, which is the ratio of inertial forces to the Archimedes force:

$$\mathrm{Fr} = \frac{\rho v^2}{\Delta \rho g L}. \tag{2.94}$$

In the fire community, this is often simplified to:

$$\mathrm{Fr} = \frac{v^2}{g L}. \tag{2.95}$$

Expression (2.94) resembles the underlying physics more correctly than Eq. (2.95). On the other hand, the difference between expressions (2.94) and (2.95) is no more than a numerical factor, depending on the densities at hand. Moreover, in many experiments, it is much more straightforward to measure velocities than mass densities, so that it is easier to characterize the experimental set-up through formulation (2.95). This explains why the use of (2.95) is popular in diagrams and correlations.

If large (imposed) pressure differences occur, sometimes the Euler number becomes relevant, through $\dfrac{\rho v^2}{L} \propto \dfrac{\Delta p}{L}$:

$$\mathrm{Eu} = \frac{\Delta p}{v^2}. \tag{2.96}$$

In fire-related flows, the Euler number is often not important.

In buoyancy-driven flows, applying Boussinesq's hypothesis, the driving force is written in terms of temperature differences (Eq. 2.85). This explains the physics behind the Rayleigh number:

$$\mathrm{Ra} = \frac{L^3 g \beta \Delta T}{\alpha \nu} \tag{2.97}$$

Alternatively, the Grashof number can be used:

$$\mathrm{Gr} = \frac{L^3 g \beta \Delta T}{\nu^2}. \tag{2.98}$$

The relation between the two is:

$$\mathrm{Ra} = \mathrm{Gr} \cdot \mathrm{Pr}, \tag{2.99}$$

with the Prandtl number as defined in Eq. (2.89). The Grashof number can be interpreted as a ratio of buoyancy forces (with Boussinesq's approximation) to the viscous forces. This is relevant in boundary layer flows (Section 2.9).

2.7.3 SCALING LAWS

For an extensive discussion on scaling, the reader is referred to, e.g. [9]. Here scaling is only briefly discussed in the context of fluid mechanics. Consequently, only the momentum equation is considered (although, at the end of this section, some remarks are formulated on the fire heat release rate using the energy equation, and on unsteady phenomena, using the mass conservation equation).

The main non-dimensional numbers in low-Mach number flows are the Reynolds number (2.93) and the Froude number (2.94 or 2.95). The only way to preserve both numbers when scaling (up or down) a flow in a certain geometry, is through the use of different fluids. Indeed, if the fluid does not change, preservation of Re (2.93) states: $\mathrm{Re}_1 = \mathrm{Re}_2 \rightarrow \frac{v_1 L_1}{\nu} = \frac{v_2 L_2}{\nu} \rightarrow v_2 = v_1 \frac{L_1}{L_2}$.

Preservation of Fr (2.95) yields: $\mathrm{Fr}_1 = \mathrm{Fr}_2 \rightarrow \frac{v_1^2}{g L_1} = \frac{v_2^2}{g L_2} \rightarrow v_2 = v_1 \sqrt{\frac{L_2}{L_1}}$. Clearly, these two requirements cannot be fulfilled simultaneously. Both numbers can be preserved if, starting from the requirement for preservation of the Froude number, the fluid's viscosity is modified such that also the Reynolds number is preserved. This is not straightforward.

Fortunately, the Reynolds number has the property that, as soon as it is large enough, its actual value becomes irrelevant. In other words, as soon as it is sufficiently high, the qualification 'high' is sufficient, not the exact number. This is due to turbulence, overwhelming molecular phenomena (see Section 2.8). This can also be understood intuitively. As stated above, the Reynolds number is the ratio of inertia to viscous damping forces. Either the damping force is strong enough to overcome the inherent instabilities in the non-linear convection terms in the Navier–Stokes equations (laminar flow), almost strong enough (transitional flow) or not strong enough (turbulent flow). When turbulence is fully developed, the strength of the viscous stress becomes irrelevant, i.e., the true value of the Reynolds number becomes irrelevant.

Knowing this, it is instructive to examine the order or magnitude of Reynolds number and Froude number in fire-related flows, in order to appreciate which number needs to be preserved in scaling. Typical dimensions are in the order of 1 m: $L = O(\mathrm{m})$. Typical velocities are in the order of 1 m/s:

$v = O(m/s)$. Densities are in the order of $1\,\mathrm{kg/m^3}$: $\rho = O(\mathrm{kg/m^3})$. The dynamic viscosity in gases is in the order of 10^{-6} Pa s: $\mu = O(10^{-6}\,\mathrm{Pa\ s})$. Using these numbers, the Reynolds number (2.93) is: $\mathrm{Re} = O\!\left(\dfrac{1 \times 1 \times 1}{10^{-6}}\right) = O(10^6)$, while the Froude number (2.94) is: $\mathrm{Fr} = O\!\left(\dfrac{1 \times 1}{1 \times 10 \times 1}\right) = O(0.1)$. Obviously, these are rough order-of-magnitude analyses, but it is clear that in fire-related flows, the choice will be made to preserve the Froude number, not the Reynolds number, when scaling is applied.

The energy equation also provides information regarding scaling laws. The simplified formulation, Eq. (2.44), can be used for fire-related flows. Yet, temperatures are very important in fire-related flows, so the energy equation should be interpreted in terms of sensible enthalpy, in which case the fire heat release rate ($\dot{Q}$, in W) comes into play. Knowing that, in terms of dimensions, (sensible) enthalpy differences can be rewritten as the product of specific heat and temperature differences, Eq. (2.45) leads to the following proportionalities: $\dfrac{\rho c_p \Delta T}{t} \propto \dfrac{\rho c_p \Delta T v}{L} \propto \dfrac{\dot{Q}}{L^3} \propto \dfrac{k \Delta T}{L^2}$.

This reveals that:

$$\dot{Q} \propto \rho c_p \Delta T v L^2. \tag{2.100}$$

It is common practice to scale configurations such that the temperatures remain the same. This also implies that densities do not change (if the same fluid is applied). As has just been explained, the Froude number (2.94) is preserved, so that the velocity scales as $v \propto \sqrt{L}$. As a consequence, the fire heat release rate scales as:

$$\frac{\dot{Q}_1}{\dot{Q}_2} = \frac{\sqrt{L_1}\,L_1^2}{\sqrt{L_2}\,L_2^2} \to \dot{Q} \propto L^{5/2}. \tag{2.101}$$

Finally, it is noteworthy that the conservation of mass, Eq. (2.31), reveals that:

$$t \propto \frac{L}{v} \propto \sqrt{L}. \tag{2.102}$$

Froude scaling has been applied in the second proportionality in Eq. (2.102). Expression (2.102) shows that the temporal evolution of quantities (e.g., temperature) depends on the dimensions of the configuration as $t \propto \sqrt{L}$. This is relevant when unsteady phenomena are studied.

2.8   TURBULENCE

There are numerous text books on turbulence and turbulent flows, e.g., [12,13]. The discussion provided in this section does not supersede the level of a basic introduction on the phenomenology of turbulence. In Chapter 8, the modelling of turbulence in CFD (Computational Fluid Dynamics) simulations is discussed more extensively.

2.8.1   REYNOLDS NUMBER

In the previous section, it has been mentioned that the Reynolds number, Eq. (2.93), is the ratio on inertia to viscous forces. It is well known that the convection terms in the Navier–Stokes equations (2.38) are intrinsically unstable: any flow becomes turbulent if the viscous forces are not strong enough to damp the instabilities. In other words, if the Reynolds number is below a certain threshold number, the flow remains 'laminar'. If it is higher than a 'critical value', the flow is turbulent. The change from 'laminar' to 'turbulent' is not sudden: there is a 'transition' zone in between.

Care must be taken in the definition of this 'critical' Reynolds number, Re_c, in the sense that the length scale must be defined. In flows over flat plates, it is common practice to use the distance from the leading edge of the plate. In that case, Re_c is in the order of 500.000. In pipe flows, it is common practice to use the pipe diameter as characteristic length scale and Re_c is in the order of 2.000.

It is recalled that the Reynolds number is a flow property, not a fluid property.

Turbulence is typically defined on the basis of a number of properties [13]:

- Randomness: there are fluctuations in the flow.
- Three-dimensionality: even if the mean flow is 2D or axisymmetric, the vortices or 'eddies' are always three-dimensional.
- There is a wide range of length scales and time scales in the flow. The largest scales are determined by the geometry at hand, while the smallest scales are determined by the Reynolds number. The smallest scales can easily be 10.000 times smaller than the largest scales.
- Turbulent mixing is very effective.
- There is strong diffusion and dissipation. Turbulence dies out quickly if not sustained by velocity gradients in the mean flow.
- There is an energy cascade, transferring energy from the mean flow (large scales) to turbulent fluctuations (to smaller and smaller scales). At the smallest scales, the 'turbulence kinetic energy' is dissipated into heat due to the viscous forces.

It is instructive to briefly discuss the randomness in the flow. Indeed, given that the Navier–Stokes equations (2.38) are deterministic, the question arises how it is possible that randomness occurs when applying deterministic boundary and initial conditions. The reason is that there are always small fluctuations, i.e., the boundary and initial conditions are never known with infinite precision. Due to the unstable convection terms in the Navier–Stokes equations, turbulent flows are extremely sensitive to 'details', and this creates randomness in the instantaneous flow fields. This makes it impossible to make long-term predictions of instantaneous turbulent flow fields and explains why turbulent flows are tackled in simulations through statistical approaches (see Chapter 8). Obviously, the mean flow can still be deterministic, as explained next.

2.8.2 REYNOLDS AVERAGING

As mentioned in the previous section, the fluctuations in a turbulent flow make a direct analysis through the Navier–Stokes equations (2.38) practically impossible. Therefore, a statistical approach is adopted. The concept of Reynolds averaging is explained first.

Consider a turbulent flow. Measuring a velocity component (or, e.g., a temperature) at a certain location will then yield a fluctuating signal, as explained. One can now determine the 'average' of that signal. The true definition of a Reynolds average [12,13] is that many realizations of the 'same' turbulent flow are made, repetitive measurements of the quantity are made at the same location, and the average value of the measurements is determined. In a simplified manner, though, one can think of this procedure as a time averaging, where the averaging period Δt is sufficiently long, compared to the largest turbulent time scales, but sufficiently short compared to time scales associated with possible unsteadiness in the mean flow:

$$\bar{v}_i(t) = \frac{1}{\Delta t} \int_{t-\Delta t}^{t} v_i(t')dt'; \ \bar{T}(t) = \frac{1}{\Delta t} \int_{t-\Delta t}^{t} T(t')dt'. \qquad (2.103)$$

Clearly, this is only possible if the turbulent time scales are short, compared to time scales in the mean flow. The 'integral' turbulent time scale is typically <1 s, so in many fire-related flows, this concept of Reynolds averaging is possible. Using Eq. (2.102), the instantaneous value can be expressed as the sum of the (Reynolds) averaged value and the instantaneous 'turbulent' fluctuation around that value:

$$v_i(t) = \bar{v}_i(t) + v_i'(t); T(t) = \bar{T}(t) + T'(t). \tag{2.104}$$

Note that, by definition:

$$\overline{v_i'(t)} = 0; \overline{T'(t)} = 0; \overline{\bar{v}_i(t)} = \bar{v}_i(t); \overline{\bar{T}(t)} = \bar{T}(t). \tag{2.105}$$

For variable density flows, use is made of 'mass-weighted' or 'Favre' averages. The Favre average value of a quantity φ is defined as follows:

$$\tilde{\varphi} = \frac{\overline{\rho\varphi}}{\bar{\rho}}. \tag{2.106}$$

The equivalent of Eq. (2.104) is:

$$v_i(t) = \tilde{v}_i(t) + v_i''(t); T(t) = \tilde{T}(t) + T''(t). \tag{2.107}$$

The equivalent of Eq. (2.105) is:

$$\widetilde{v_i''(t)} = 0; \ \widetilde{T''(t)} = 0; \ \widetilde{\tilde{v}_i(t)} = \tilde{v}_i(t); \ \widetilde{\tilde{T}(t)} = \tilde{T}(t). \tag{2.108}$$

Applying this averaging technique to the conservation equations (2.31), (2.38) and (2.45), the following equations are obtained:

$$\frac{\partial \bar{\rho}}{\partial t} + \nabla \cdot \left(\bar{\rho}\tilde{v} \right) = 0, \tag{2.109}$$

$$\frac{\partial}{\partial t}\left(\bar{\rho}\tilde{v} \right) + \nabla \cdot \left(\bar{\rho}\tilde{v}\tilde{v} \right) = -\nabla\bar{p} + \nabla \cdot \left(\bar{\bar{\tau}} + \bar{\bar{\tau}}_{\text{turb}} \right) + \bar{\rho}\tilde{g}, \tag{2.110}$$

$$\frac{\partial}{\partial t}\left(\bar{\rho}\tilde{h} \right) + \nabla \cdot \left(\bar{\rho}\tilde{h}\tilde{v} \right) = \nabla \cdot \left(\frac{\bar{\mu}}{\text{Pr}}\nabla\tilde{h} + \bar{\bar{q}}_{\text{turb}} \right) + \bar{\rho}\tilde{S}_h. \tag{2.111}$$

They are very similar to the instantaneous equations, but some additional terms appear:

- Reynolds stresses (turbulent stress tensor) in the momentum equations, Eq. (2.110);
- Turbulent heat fluxes (vector) in the energy equation, Eq. (2.111).

The additional terms appear as a consequence of the presence of products in the convection terms in the instantaneous equations (left hand side). Indeed, the mean value of the product is not equal to the product of the mean values:

$$\widetilde{v_iv_j} = \tilde{v}_i\tilde{v}_j + \widetilde{v_i''v_j''}; \ \widetilde{v_ih} = \tilde{v}_i\tilde{h} + \widetilde{v_i''h''}. \tag{2.112}$$

In other words, the turbulent stress tensor (or 'Reynolds stress tensor') and turbulent heat flux vector (or 'Reynolds flux vector') components read:

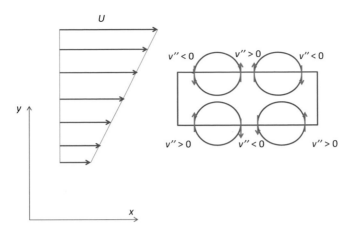

FIGURE 2.9 Sketch of turbulent fluctuations (vortices) in a flow with a mean velocity gradient.

$$\tilde{\tau}_{\text{turb},ij} = -\bar{\rho}\widetilde{v_i'' v_j''}, i = 1,\ldots,3, j = 1,\ldots,3; \tilde{q}_{\text{turb},i} = -\bar{\rho}\widetilde{v_i'' h''}, i = 1,\ldots,3. \tag{2.113}$$

Modelling of Eq. (2.113) is discussed in Chapter 8. Yet, it is noted already here that the terms, stemming from turbulence, are similar in nature to the molecular viscous stresses and the molecular thermal diffusion terms.

The main question is now what the correlations between the turbulent fluctuations look like. Indeed, terms like $\widetilde{v_i'' v_j''}$ are only non-zero if the turbulent fluctuations of the velocity components in the different directions are statistically correlated. This is the case, as can be understood from Figure 2.9, showing a situation in a flow with a mean velocity gradient. The discussion is given here for the top right vortex, but it prevails for all vortices. At the left side of the top right vortex, the instantaneous motion is upwards, as indicated by the arrow. Knowing that the mean velocity in the vertical direction equals zero, this implies that $v'' > 0$. In its upward motion, the eddy brings along fluid with a lower (mean) velocity in the horizontal direction into a region with higher (mean) velocity. Thus, the impact is a local decrease in horizontal velocity; in other words $u'' < 0$. Clearly, from a statistical point of view, the velocity fluctuations in both directions are correlated, in such a manner that $\widetilde{u'' v''} < 0$. At the right side of the top right eddy, the instantaneous motion is downwards ($v'' < 0$) and (in the mean) higher horizontal velocity is brought into a region with (in the mean) lower horizontal velocity, causing $u'' > 0$. Thus, again $\widetilde{u'' v''} < 0$.

A similar reasoning can be made for the temperature (or enthalpy) fluxes for flows with a temperature (enthalpy) gradient.

The most important observation is the fact that there is a non-zero correlation. This explains the importance of turbulence at the level of the mean flow and temperature field.

Additionally, it is clear that the fluctuations, caused by the turbulent motion, are larger as the mean velocity and temperature gradients are larger. The above led to the following 'eddy viscosity' modelling concept. This concept has been introduced by Boussinesq. For the turbulent stress tensor, this boils down to using Eq. (2.3), replacing the instantaneous velocities by their mean values and by replacing the molecular viscosity by an artificial 'turbulent viscosity' (or 'eddy viscosity'):

$$\tilde{\tau}_{\text{turb},ij} = \bar{\mu}_t \left[\left(\frac{\partial \tilde{v}_i}{\partial x_j} + \frac{\partial \tilde{v}_j}{\partial x_i} \right) - \frac{2}{3} \delta_{ij} \frac{\partial \tilde{v}_k}{\partial x_k} \right]. \tag{2.114}$$

In other words, in Eq. (2.110), a 'turbulent' or 'eddy' viscosity is simply added to the molecular viscosity. This reflects the physical observation that momentum transfer increases in turbulent flows through the turbulent motion of eddies. These cause 'large-scale' momentum transfer. Yet, it must be appreciated that the turbulent viscosity is not a physical quantity (let alone a fluid property). It is a parameter that indicates the effect of the phenomenon turbulence at the level of the mean flow field.

Similarly, this concept can be applied to the heat fluxes:

$$\tilde{q}_{\text{turb},i} = \frac{\bar{\mu}_t}{\text{Pr}_t} \frac{\partial \tilde{h}}{\partial x_i}.$$ (2.115)

In other words, the addition of a turbulent thermal diffusivity to the molecular thermal diffusivity reflects the physical observation that heat transfer increases in turbulent flows through the turbulent motion of eddies. These cause 'large-scale' heat transfer. Again, it must be appreciated that the turbulent thermal diffusivity is not a physical quantity (let alone a fluid property). It is a parameter that indicates the effect of the phenomenon turbulence at the level of the mean temperature (enthalpy) field.

2.8.3   TURBULENCE MODELLING

Only some global notions are mentioned here. More concrete details on turbulence modelling are discussed in Chapter 8. The reader is also referred to [13] as an excellent book on turbulence for more details.

2.8.3.1   Energy Cascade

As mentioned in Section 2.8.1, there is always a wide range of length scales and time scales in turbulent flows. These scales are related to the turbulent vortices. The higher the Reynolds number, the wider this range becomes, because the smallest scales become smaller and smaller. This is discussed in Section 2.8.3.2.

The largest turbulence scales are called the 'integral' scales. The smallest ones are called the 'Kolmogorov' scales. A detailed discussion of the spectrum is outside the scope of this section, but it is important to appreciate that most of the turbulent kinetic energy is in the integral scale range ('energy containing range'), while turbulence is dissipated at scales around the Kolmogorov scales. Indeed, at those scales, damping of the fluctuations by viscous forces 'kills' turbulence, i.e., dissipates the turbulent kinetic energy into heat.

An important notion is the 'energy cascade', as introduced by Richardson [12,13]. The basic mechanism is as follows:

- Energy is taken from the mean flow and transferred to kinetic energy of turbulent vortices; this occurs around the integral scales;
- The turbulent vortices break up, transferring their energy to vortices of smaller scale; only little energy is dissipated in this break-up process;
- The break-up process of vortices continues ('cascade process') until the vortices become so small that they cannot survive the damping action of viscosity anymore;
- The dissipation takes place at the smallest turbulence scales.

It is important to note that whereas the dissipation takes place at the smallest scales, the dissipation *rate* is determined by the production rate of turbulence from the mean flow in the energy containing range (in equilibrium conditions). Figure 2.10 provides a sketch.

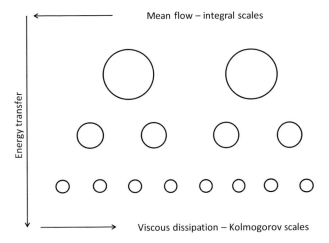

Mean flow – integral scales

Energy transfer

Viscous dissipation – Kolmogorov scales

FIGURE 2.10 Sketch of the Richardson energy cascade in turbulence. Kinetic energy, taken from the mean flow (integral scales), is transferred to smaller turbulent eddies due to eddy break-up and is eventually dissipated by viscosity at the smallest scales (Kolmogorov scales).

2.8.3.2 Turbulent Scales

As mentioned in Section 2.8.2, there is a degree of statistical correlation within the apparent randomness in turbulence. At a local position in physical space, for sufficiently small variation in time s, there is correlation between, e.g., the velocity fluctuation at times t and $(t + s)$:

$$\rho(s) = \frac{\overline{v''(t)v''(t+s)}}{\widehat{v''^2(t)}}. \tag{2.116}$$

This is an 'autocorrelation' in time, which expresses that if, e.g., an instantaneous velocity fluctuation is positive (in a certain direction) at a given location and at a certain moment in time, it is likely that 'a little bit later' in time (and at the same position), the velocity fluctuation is still positive. The notion 'a little bit later' depends on how long the velocity fluctuations are correlated in time. This can be formalized through the following integral:

$$\tau_{\text{int}} = \int_0^\infty \rho(s)\,ds. \tag{2.117}$$

This is the 'integral time scale'. It only exists if the integral converges, which happens to be the case in turbulence. The larger the integral time scales, the longer fluctuations of the variable (in this example: velocity fluctuations) are correlated in time.

In a similar way, correlations in space can be defined. Taking the same example of velocity fluctuation, looking at instantaneous velocity fluctuations at a certain moment in time and at two positions that are 'not too far away from each other', a statistical correlation can be assumed (i.e., if an instantaneous velocity fluctuation is positive (in a certain direction) at a given location and at a certain moment in time, it is likely that 'not too far away' in physical space (and at the same time), the velocity fluctuation is still positive). The notion 'not too far away' depends on how far the velocity fluctuations are correlated in physical space. This can be formalized through an integral like (Eq. 2.116), but in physical space, rather than in time. This leads to 'integral length scales'. This determines the size of the largest 'eddies' in the turbulent flow.

In, e.g., [13], it is explained that the integral scales are of the same order of magnitude as the geometry (length scales) and mean flow (velocity and time scales). The smallest turbulent scales are called the 'Kolmogorov scales'. As illustrated in Figure 2.10, the dominant parameters in this range are the viscosity (v, dimension m²/s, see Section 2.1.1.2) and the dissipation rate of the turbulent kinetic energy (ε, dimension m²/s³). The following 'Kolmogorov' scales can be constructed:

$$\eta = \left(\frac{v^3}{\varepsilon}\right)^{1/4} ; v_\eta = (\varepsilon v)^{1/4} ; \tau_\eta = \left(\frac{v}{\varepsilon}\right)^{1/2} . \tag{2.118}$$

Note that these scales are independent of the geometry and the mean flow. They are related to the integral scales through the Reynolds number (2.93) [13]:

$$\frac{\eta}{l_{\text{int}}} \propto \text{Re}^{-\frac{3}{4}}; \frac{v_\eta}{v_{\text{int}}} \propto \text{Re}^{-\frac{1}{4}}; \frac{\tau_\eta}{\tau_{\text{int}}} \propto \text{Re}^{-\frac{1}{2}}. \tag{2.119}$$

Accepting that the integral scales are determined by geometry and mean flow, Eq. (2.119) shows that the smallest scales in turbulence become smaller as the Reynolds number increases.

2.8.3.3   Turbulence Modelling

This phenomenology is reflected in the choice made for turbulence modelling in CFD simulations. One extreme approach is not to model turbulence at all. In other words, all turbulent motions, down to the smallest scales, are resolved. This is called DNS: Direct Numerical Simulations of the instantaneous Navier–Stokes equations (2.29). Knowing that these small scales can easily be in the order of 0.1 mm or less, and realizing that the computational mesh needs to be sufficiently fine to resolve the smallest eddies (see Chapter 8), it is immediately clear that this approach is not feasible for typical fire-related flow simulations, where geometry dimensions are in the order of 1 m or larger. Worse than that, in addition to unacceptable computing time and memory requirements, most of the time and memory would be devoted to simulating the smallest scales [13], whereas the primary interest is typically in the large-scale flow phenomena (or in the mean flow).

The other extreme is the RANS (Reynolds-Averaged Navier–Stokes) turbulence modelling approach. In this approach, Reynolds (or Favre) averaging is applied as discussed in Section 2.8.2. All turbulent motions, i.e., the entire turbulent spectrum, are modelled, and only the mean flow is resolved. Note that the mean flow can be unsteady. The $k - \varepsilon$ model belongs to this class of models. The advantages of the RANS approach are clear:

- the computational mesh only needs to be fine enough to resolve the mean flows;
- the time step (in transient calculations) can be chosen on the basis of mean flow phenomena;
- one immediately gets a solution for the mean flow.

There are also major disadvantages, though. Firstly, all turbulence is modelled. Knowing that the largest turbulent scales are configuration dependent, it cannot be expected that a single RANS model can deal with arbitrary configurations in a reliable manner. To be more precise, two-equation RANS models are relatively accurate for shear stress-driven flows, but are less reliable for impingement flows, where turbulence is typically over-predicted. Although adjustments can be made to these turbulence models (see, e.g., [14]), these are often not in place in software packages used for fire simulations.

Second, the effect of buoyancy on turbulence production requires special attention. At the level of the mean flow, buoyancy is present through the gravity term in Eq. (2.79), but often the effect on turbulence itself (in, e.g., the determination of the turbulent viscosity μ_t) is not (properly) taken into account. Again, although adjustments can be made to these turbulence models (see, e.g., [15]), these are often not in place in software packages used for fire simulations.

Third, in fire-related flows, large-scale flow unsteadiness often plays an important role, e.g., in the entrainment process of air into flames or smoke (see Chapters 3–5). Such unsteadiness is not captured in (unsteady) RANS and must be modelled. Again, being configuration dependent, RANS models cannot be expected to be as accurate as approaches where large-scale turbulent unsteadiness is resolved.

This explains the popularity of the LES (Large-Eddy Simulations) technique in CFD for fire-related flows. In this technique, the large-scale vortices (or 'eddies') are resolved and the small eddies are filtered out. Only the effect of the small-scale eddies on the flow field then needs to be modelled. It is common practice to use the computational mesh as filter in the LES approach, i.e., the size of the computational mesh cells determines the size of the eddies that are still resolved. Note that the use of the mesh as filter implies that no grid-independent results can be expected from LES. Yet, it makes sense to examine how sensitive the simulation results are to the choice of the computational mesh size.

The LES technique offers the advantage of resolving the large-scale flow unsteadiness (and buoyancy effects). Also, the unacceptable fineness of the computational mesh as required in DNS is avoided. Yet, in order to guarantee the quality of LES results, 80% of the turbulent kinetic energy must be resolved [13]. A more extensive discussion is postponed until Chapter 8.

2.9   BOUNDARY LAYER FLOW

In Section 2.7.3, it was mentioned that the absolute value of the Reynolds number (2.93) becomes irrelevant as soon as Re is high enough. In the absence of solid boundaries, the flow can be considered 'inviscid' ($\mu = 0$). However, as viscosity is never really equal to zero, there is always a 'no-slip' boundary condition at any solid boundary: due to the viscous forces, the fluid locally takes the velocity of the solid boundary. In fire-related flows, the solid boundaries typically stand still, so that the no-slip boundary condition implies that the fluid velocity equals zero.

In fire-related flows, boundary layers appear as 'external' flows or in 'internal flows'. Examples of fire-related boundary layers in external flows are: the flow over surfaces (horizontal or vertical) with, e.g., flame spread, flow of smoke underneath a ceiling, atmospheric boundary layers in forest fires, etc. Examples of fire-related boundary layers in internal flows are: flow of water through a piping system for sprinklers or water mist, flow of smoke through a duct channel, etc. 'Internal' flows are discussed in Section 2.10.

Consider first the situation of a flow with free stream velocity U_∞ over a smooth flat plate, without external pressure gradient. At the flat plate, $U = 0$ (no-slip boundary condition), whereas 'sufficiently far away' from the plate $U = U_\infty$. The notion 'sufficiently far away' is related to the thickness of the boundary layer, which can be defined as follows:

$$y = \delta : v_x = 0.99 U_\infty. \qquad (2.120)$$

In words, the boundary layer thickness δ is the distance from the plate where the velocity equals 99% of the free stream velocity. Other measures, such as displacement thickness and momentum thickness, can also be used to characterize the boundary layer thickness, but this is not essential for the present discussion. Two flow regions can be defined now:

- $y < \delta$ strong velocity gradients and large viscous shear stresses;
- $y > \delta$ negligible velocity gradients and small viscous shear stresses.

From an order-of-magnitude analysis, in the assumption that $\delta \ll x$, with x the distance from the leading edge of the flat plate, and in the assumption of laminar flow in the boundary layer, it can be shown that the boundary layer thickness grows as follows:

$$\delta_{\text{lam}} \propto \left(\frac{\mu x}{\rho U_\infty} \right)^{1/2}. \tag{2.121}$$

In words, the laminar boundary layer thickness grows with the square root of the distance from the leading edge. It is thicker as the kinematic viscosity is higher. The latter shows that the influence region of the flat plate is larger for fluids with higher viscosity.

Using x as characteristic distance, the following Reynolds number can be defined:

$$\text{Re}_x = \frac{U_\infty x}{\nu}. \tag{2.122}$$

The viscous shear stress at the plate:

$$\tau_w = \mu \left. \frac{\partial v_x}{\partial y} \right|_{y=0} \propto \rho \nu \frac{U_\infty}{\delta} \propto \rho U_\infty^2 \left(\frac{U_\infty x}{\nu} \right)^{-1/2} \tag{2.123}$$

can be expressed in a non-dimensional manner, by introducing the friction coefficient:

$$C_{f,x} = \frac{\tau_w}{\frac{1}{2} \rho U_\infty^2} \propto \text{Re}_x^{-1/2}. \tag{2.124}$$

The 'Blasius' solution for laminar boundary layers over smooth flat plates indeed yields:

$$\delta_{\text{lam}} = 4.92 x \, \text{Re}_x^{-1/2}; \; C_{f,x,\text{lam}} = 0.664 \, \text{Re}_x^{-1/2}. \tag{2.125}$$

However, as mentioned, there are inherent instabilities in the convection terms in the Navier–Stokes equations. These instabilities are damped near the flat plate, primarily due to the blocking effect and the viscous forces, so that turbulent vortices (eddies) cannot develop. However, the laminar boundary layer thickness grows with the distance from the leading edge (Eq. 2.120), so that turbulence can start to develop. There is a critical Reynolds number $\text{Re}_{x,c}$ beyond which there is transition from laminar to turbulent flow. For a smooth flat plate, $\text{Re}_{x,c}$ is in the order of 500.000. Roughness stimulates the transition to turbulence.

It is instructive to interpret the transition from laminar flow to turbulent flow in terms of the possibility for eddies to develop inside the boundary layer. Indeed, as long as the boundary layer is too thin for eddies to develop, i.e., the eddies are blocked/damped by the presence of the wall, turbulence cannot develop. The qualitative notion 'thin' can be quantified from Eq. (2.125), using the value :

$$\text{Re}_{x,c} = 500000 : \frac{\delta_{\text{lam},c}}{x_c} = 4.92 \times 5,00,000^{-\frac{1}{2}} = 0.007 \rightarrow \text{Re}_{\delta,c} = 3,500.$$ In other words, expressed

in terms of the boundary layer thickness, the critical Reynolds number is two orders of magnitude lower than when the distance from the leading edge is used as characteristic length scale.

As mentioned in Section 2.8, the momentum and heat transfers strongly increase in turbulent motions, as compared to laminar flow conditions, because momentum and heat are transferred on a larger scale through the turbulent eddies. As a result, the surface friction and heat transfer increase and the boundary layer becomes thicker. It can be shown that:

$$\delta_{\text{turb}} = 0.37 x \, \text{Re}_x^{-1/5}; \; C_{f,x,\text{turb}} = 0.0592 \, \text{Re}_x^{-1/5}. \tag{2.126}$$

Thus, a turbulent boundary layer grows more rapidly than a laminar boundary layer.

It is worth mentioning that regardless of how turbulent the flow or the boundary layer is, there is always a 'laminar' 'viscous sublayer' sufficiently close to the wall, as a consequence of viscosity (no-slip boundary condition) and the blocking of the eddies. This can affect near-wall modelling in CFD simulations, since different types of boundary conditions must be imposed, depending on whether or not the first grid cell at the wall is inside this viscous sublayer or not. The reader is referred to Chapter 8 and to specialized literature on this topic for more detail.

Thermal boundary layers can be defined in a very similar way as velocity boundary layers:

$$y = \delta_T : T - T_s = 0.99(T_\infty - T_s), \tag{2.127}$$

with T_s the surface temperature at the flat plate.

Given the physical similarity of the kinematic viscosity and the thermal diffusivity, it is not surprising that the Prandtl number (2.89) determines whether the thermal boundary layer is thicker or not than the flow boundary layer:

- Pr = 1: $\delta = \delta_T$
- Pr < 1: $\delta < \delta_T$ (n example of such a fluid is air);
- Pr > 1: $\delta > \delta_T$ (an example of such a fluid is water).

The above discussion has been presented in the context of 'forced convection' (see Section 2.3.3.1): the free stream velocity is provided by a force that is independent of possible heat exchange between fluid and plate. However, boundary layer flows can also develop due to local density differences (read: temperature differences) near the flat plate, due to the buoyancy force (Eq. (2.84) or (2.86)). This is called 'natural convection'. Care must be taken, though, since the terminology can be confusing, as illustrated next with a few examples.

First example (no confusion possible): consider a fire at floor level in a small compartment. Assume that within a few minutes a stable situation is created where there are essentially two zones: a hot upper layer, filled with smoke, and a cold bottom layer, consisting of fresh air. The hot smoke macroscopically 'rests' upon the cold air due to buoyancy. Consider now a vertical wall of that compartment:

- The bottom part of the wall is heated by the fire source through radiation. The air is transparent for radiation. Consequence: the wall is hotter than the neighbouring air. Consequently, due to heat exchange, a boundary layer of higher temperature (lower density) develops at the wall and an upward flow of air is generated at the bottom part of the wall due to natural convection (buoyancy term (2.84) or (2.86) in the Navier–Stokes equations (2.38)). As soon as the upward moving air enters the hot upper layer, the buoyancy force becomes downwards again, pushing the air back down into the cold bottom layer.
- The top part of the wall (which might be heated by the fire source through radiation) is in contact with hot smoke. The temperature of the smoke is higher than the wall temperature, because gas phase phenomena are much faster than the heating processes in solids. Consequence: the wall is cooler than the neighbouring smoke. Consequently, due to heat exchange, a boundary layer of lower temperature (higher density) develops at the wall and a downward flow of smoke is generated at the top part of the wall due to natural convection (buoyancy term (2.84) or (2.86) in the Navier–Stokes equations (2.38)). As soon as the hot smoke enters the cold bottom air layer, the buoyancy force becomes upwards again, pushing the smoke back up into the hot upper layer.

Second example (confusion possible): consider a fire source at the bottom of a non-combustible vertical wall. The fire generates a flow, due to buoyancy. This is discussed in Chapter 3. This phenomenon is 'natural convection': it is the temperature rise due to the fire itself that generates the flow. However, from the viewpoint of the vertical wall, the flow is 'forced': the fire is the 'external' source, driving the flow. Note that, in fact, the convection at the wall is rather 'mixed': as the hot combustion products (smoke) from the fire are hotter than the wall, locally near the wall the gas temperatures are lower than further away from the wall (but still in the smoke plume above the fire). As such, natural convection as triggered by heat exchange between the smoke and the vertical wall, would cause a downward motion. Yet, the 'forced' upward motion (due to 'natural convection' as generated by the fire source) typically overwhelms this downward motion.

2.10   INTERNAL FLOWS – PRESSURE LOSSES

A major difference from external flows (Section 2.9) is that in internal flows, the notion 'free stream velocity' does not exist. In fully developed flow conditions, the flow is entirely affected by the presence of the solid boundary and, consequently, by the fluid's viscosity. The discussion is based here on pipe flows, because pipes are a common configuration (e.g., water through pipes for sprinklers or water mist systems, or water hoses in fire service intervention). Some comments are formulated for flows through ducts in the end.

In the entrance region of a pipe, a boundary layer develops from the solid boundary, very similar to what has been described in Section 2.9. However, this boundary layer grows on the entire surface, so there is a point where the entire cross section is covered by a 'boundary layer'. This point determines the 'entrance length'. From that point onwards, the boundary layers do not evolve anymore and the flow becomes fully developed.

Depending on the Reynolds number, the flow is again laminar or turbulent. The distance from the entrance (which would be the equivalent of the distance from the leading edge of a flat plate, Section 2.9) is not useful as characteristic length: in fully developed flow conditions, the velocity profiles are independent of that distance. Clearly, the pipe diameter is a useful quantity. Indeed, the boundary layer growth as discussed in Section 2.9 is limited by the pipe diameter.

At the same time, there is no free stream velocity. A mean velocity U_m can be computed from the volume flow rate and the cross-sectional area. Therefore, the Reynolds number is now defined as follows:

$$\text{Re} = \frac{U_m D}{\nu}. \tag{2.128}$$

The critical Reynolds number, beyond which the flow becomes turbulent, is around $\text{Re}_c = 2,300$. Note that this value is of the same order of magnitude as the critical Reynolds number in Section 2.9, if the boundary layer thickness is used as characteristic length scale.

If the cross section is not round, the diameter D is replaced by the hydraulic diameter D_h, defined as four times the cross-sectional area divided by the cross-section perimeter:

$$D_h = \frac{4A}{P}. \tag{2.129}$$

It is straightforward to show that for fully developed laminar flows, the following expressions hold (with $R = D/2$ the radius of the pipe and r the radial distance from the pipe symmetry axis):

$$\text{Parabolic velocity profile}: \frac{u(r)}{U_m} = 2\left(1 - \frac{r}{R}\right)^2; \tag{2.130}$$

$$\text{Wall shear stress (friction)}: \tau_w = 4\mu \frac{U_m}{R}; \tag{2.131}$$

$$\text{Friction factor}: f = \frac{\tau_w}{\frac{1}{2}\rho U_m^2} = \frac{16}{\text{Re}}; \tag{2.132}$$

$$\text{Pressure loss over a distance } L \text{ in the pipe}: \Delta p = f\frac{1}{2}\rho U_m^2 \frac{4L}{D}. \tag{2.133}$$

Note that the pressure loss increases linearly with the length of the pipe and, with Eq. (2.133), linearly with the mean velocity and inversely proportional with the square of the pipe's diameter. Recall that this is only true for laminar flows.

For turbulent flows, the expressions become more complex. The velocity profile can be approximated as follows:

$$\frac{u(r)}{U_{\max}} = \left(1 - \frac{r}{R}\right)^{1/7}; U_{\max} = \frac{U_m}{0.817}. \tag{2.134}$$

Expression (2.133) still holds for the pressure losses, but the friction factor is no longer obtained from (2.132). Rather, a Moody diagram [16] is used. Such a diagram reveals that for large-enough Reynolds number, the friction factor is determined by the relative roughness of the pipe, independent of the Reynolds number. As a consequence, for turbulent flows, the pressure loss is proportional to the length of the pipe in a linear manner, but proportional to the mean velocity squared and inversely proportional to the pipe diameter.

In duct flows, essentially the same reasoning holds. The major difference is in the value for the friction factor f, important to estimate the pressure losses. Secondary flows appear in the corners of ducts, transporting momentum from the centre to the corners and leading to a relative increase in velocity near the corners.

It is instructive to quantify pressure losses for internal flows as follows:

$$\Delta p_L = C_L \frac{1}{2}\rho U_m^2. \tag{2.135}$$

The loss coefficient C_L must be defined, depending on the situation (geometry and flow type – laminar/turbulent). All pressure losses must be accounted for in the design. This holds for, e.g., the design of the piping system for sprinklers or water mist systems (i.e., what pump must be chosen) or the design of a smoke extraction system (i.e., what extraction fans are required to overcome all pressure losses, including the ones in the exhaust system).

Some examples are briefly mentioned here:

- *Straight sections*: C_L is determined from a Moody diagram [16].
- *Curves/bends*: C_L is determined by the total angle and the radius of the bend (e.g., $C_L = 0.14$ for an angle of 90° with radius 2D, but it is about 1.2 for the same angle of 90° but with radius $= 0$, i.e. a sharp bend). Curves and bends are always important to consider in calculations of pressure losses.
- *Sudden pipe expansion*: $C_L - \left(1 - \frac{A_1}{A_2}\right)^2$. A special case concerns the flow into a large space, i.e., $A_2 \rightarrow \infty$. Then, $C_L = 1$.
- *Sudden pipe constriction*: The flow is constricted and then widens again behind the constriction. A good estimate for a sudden constriction is $C_L = 0.5$, while C_L goes down to 0 for a very gentle constriction.

- *Flows through openings*: C_L primarily depends on the edges of the opening. The most typical situation is that the edges are sharp. In that case, C_L typically varies between the values $C_L = 0.4$ and $C_L = 0.7$.

2.11 ENTRAINMENT

Entrainment into buoyancy-driven flows is shown to be very important for fires in Sections 3.5.1 (for fire plumes) and 4.2 or 4.3 (for smoke plumes). In the present section, entrainment into momentum-driven flows is discussed. The reader is referred to, e.g., [13] for more details.

Consider first a round jet, issued as a top hat profile with velocity U_{jet} from a nozzle with diameter D_n. In Ref. [13], it is illustrated that the velocity on the central axis decays with the distance x from the nozzle according to:

$$\frac{U_o(x)}{U_{jet}} = \frac{B}{\dfrac{x - x_o}{D_n}}. \tag{2.136}$$

In Eq. (2.136), the virtual origin, x_o, relates to a 'developing' or 'core' region in the jet, before reaching a self-similar region. The value for B is approximately 5.9.

The jet half-width, $r_{1/2}(x)$, is defined as the radius where the axial velocity component is half the value on the axis:

$$U(x)\big|_{r=r_{1/2}(x)} = \frac{1}{2}U_o(x). \tag{2.137}$$

The jet spreading rate is then defined as follows:

$$SR(x) = \frac{dr_{1/2}(x)}{dx}. \tag{2.138}$$

For a fully developed round jet, the constant value equals $SR = 0.095$, essentially independent of the Reynolds number [13]. Entrainment occurs into the widening jet. Indeed, the mass flow rate can be shown to increase linearly with x [13].

For the plane jet, i.e., a jet issued from a narrow long slot with height H, the jet half-width can be defined as in Eq. (2.137), replacing the word 'radius' by the 'distance' from the symmetry plane. The spreading rate value, Eq. (2.138), for a plane jet is $SR = 0.10$ [13]. However, the velocity decay is not proportional to x^{-1}, like in Eq. (2.136). For a plane jet, it is proportional to $x^{-1/2}$ in the self-similar region. As for the round jet, entrainment occurs into the widening jet: the mass flow rate increases proportional to $x^{1/2}$.

2.12 IMPINGING FLOW

Impinging flows can be of importance in fire, particularly during its early stages, in the context of fire detection. This is discussed more extensively in Section 4.5. It is also possible for flames to impinge onto a wall or a ceiling (see Section 3.5.3). Only some brief notions are mentioned here, assuming that the flow is not reacting when it impinges onto a solid boundary.

The most typical impinging flow, related to fire, is the 'ceiling jet' flow (Section 4.5). The basic situation is sketched in Figure 2.11. A rising flow exists above a fire source, due to natural convection (buoyancy force, Eq. 2.84): the hot combustion products rise into the colder surrounding air. Note that no wind or forced ventilation is considered in the discussion. As the smoke rises, air is entrained into the 'smoke plume' (Section 4.2). This smoke plume keeps on rising, either

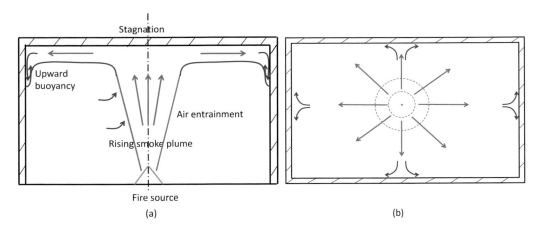

FIGURE 2.11 Schematic sketch of a ceiling jet flow in a compartment. (a) Side view (vertical symmetry plane through fire source). (b) Top view (just below the ceiling). Arrows indicate flow.

until the buoyancy force becomes too weak (in open atmosphere) or until an obstacle is met. A typical 'obstacle' is a ceiling of a compartment. The rising flow stagnates to form a 'ceiling jet'. Bernoulli's equation (2.75) reveals that the highest static pressure is obtained at the 'stagnation point', i.e., the point where the idealized symmetry line of the smoke plume meets the ceiling (Figure 2.11). In reality, the flow will be turbulent and there is no clear stagnation 'point', but there will be a stagnation region around the centre of the smoke plume (still in the absence of forced ventilation). The pressure increase is very low, though. Given that typical velocities are below 5m/s and mass densities are in the order of unity, the pressure rise ($\Delta p = \dfrac{1}{2} \rho v^2$, Eq. 2.75) is in the order of 10 Pa or lower. This is small. The most important effect of the obstacle is its impermeability, forcing the flow from a vertical motion into a horizontal motion. Still in the absence of ventilation and side walls, the flow is radially outwards. In other words, the vertical momentum of the smoke plume is diverted into horizontal momentum of the flow underneath the ceiling. This momentum is reduced by friction losses at the ceiling. There is relatively little entrainment of cold air into the hot smoke (compared to the entrainment into the vertically rising smoke plume), because the buoyancy force acts such that the smoke is pushed upwards towards the ceiling. In other words, there is no 'natural' tendency for air to mix with the smoke. In a compartment, this horizontal flow is typically blocked by the presence of a vertical obstacle (i.e., a side wall). Then the horizontal flow stagnates again, and the flow is pushed either sidewards horizontally (if possible) or downwards. Such downward motion adds to the downward motion due to buoyancy in the hot upper layer next to a side wall (Section 2.9). Still, as mentioned in Section 2.9, as soon as the hot smoke enters the cold bottom layer of fresh air, the buoyancy force pushes the smoke back upwards.

In the context of Section 2.9, from the viewpoint of the ceiling, the convection is 'forced' with the fire source-driven rising smoke plume as 'externally imposed' flow. Due to heat exchange of the hot smoke with the cooler ceiling, close to the ceiling a region of relatively less hot smoke is formed and buoyancy will act downwards on this smoke (as it sees hotter smoke below). Yet, this 'natural' convection is usually ignored compared to the 'forced' convection by the ceiling jet.

2.13   EVAPORATION

So far, only single-phase phenomena have been discussed. However, in case of liquid fuels and in the context of the interaction of water with smoke and fire (Chapter 7), evaporation will also play an important role.

Evaporation is an endothermic process, i.e., it requires energy to occur. The amount of energy required to transform 1 kg of fluid from the liquid state into the gaseous state (vapour) at a given pressure and temperature is called the latent heat of vaporization of the liquid, L_v (unit: J/kg).

An important expression is the Clausius–Clapeyron law, which expresses the relation between the partial vapour pressure p_v of a species and the temperature:

$$\ln(p_v) = -\frac{L_v}{RT} + C. \tag{2.139}$$

Equation (2.139) is valid at pressures that are typical for natural fires (e.g., atmospheric pressure), and under the assumption of temperature independent L_v. R is the gas constant (J/(kg K)), T is the absolute temperature (in K) and C depends on the liquid.

The ratio of the partial vapour pressure to the total pressure yields the mole fraction of the vapour:

$$X_v = \frac{p_v}{p}. \tag{2.140}$$

The mass fraction (see Section 2.1.1.5) can be determined from the mole fraction through multiplication with the molecular weight.

Note that if $p_{v,1}$ is known at a certain temperature T_1, the vapour pressure at another temperature is found as follows:

$$\ln(p_{v,2}) - \ln(p_{v,1}) = \ln\left(\frac{p_{v,2}}{p_{v,1}}\right) = \frac{L_v}{R}\left(\frac{1}{T_1} - \frac{1}{T_2}\right). \tag{2.141}$$

2.14 PYROLYSIS

Pyrolysis refers to the degradation of solid material. In the context of fire, combustible gases are released during this degradation process (Section 1.1).

In contrast to evaporation, pyrolysis is by definition irreversible: once degraded, the solid material cannot be restored in its original form, bringing back the gases released. In the context of fire, though, also evaporation is irreversible once the combustible gases released have reacted with the oxygen. Therefore, in that sense, pyrolysis and evaporation are not dissimilar in the context of fire.

Focusing on the physics of the problem, the main process is similar to evaporation of liquid fuel. Indeed, pyrolysis is an endothermic process, i.e., it requires energy to occur. This endothermic character can be expressed in a simplified manner by introducing the quantity 'heat of pyrolysis', ΔQ_{pyr} : the amount of energy, required to transform 1 kg of solid material ('fuel') into the gaseous state (vapour) at a given pressure and temperature. The 'heat of pyrolysis' is similar to the latent heat of vaporization, L_v (Section 2.13), and has the same unit (J/kg). Note, however, that ΔQ_{pyr} is not a material property, but a model parameter, since it is a simplification of reality, where pyrolysis does not take place at a well-defined unique temperature and where chemistry also takes place. This simplified representation of pyrolysis can prove valuable for modelling in numerical simulations (e.g., [17]).

Still, pyrolysis is much more complex than evaporation, as chemistry inside the solid material is involved. The reader is referred to specialized literature. A possible starting point to learn about state-of-the-art pyrolysis modelling is Ref. [18], the github repository for the Condensed Phase sub-group of the IAFSS MaCFP ('Measurements and Computations of Fire Phenomena') working group.

REFERENCES

1. G.K. Batchelor (1967) *An Introduction to Fluid Dynamics*. Cambridge University Press: Cambridge.
2. P.A. Libby and F.A. Williams (1994) *Turbulent Reacting Flows*. Academic Press: Cambridge, MA.
3. K.K. Kuo (2005) *Principles of Combustion*, 2nd Edition. John Wiley & Sons, Inc: Hoboken, NJ.
4. C.K. Law (2006) *Combustion Physics*. Cambridge University Press: Cambridge.
5. F.A. Williams (1985) *Combustion Theory*, 2nd Edition. Addison-Wesley Publishing Company: Boston, MA.
6. T. Poinsot and D. Veynante (2005) *Theoretical and Numerical Combustion*. R T Edwards, Inc.: Morningside, Australia.
7. N. Peters (2000) *Turbulent Combustion*. Cambridge University Press: Cambridge.
8. J. Warnatz, U. Maas and R.W. Dibble (1996) *Combustion*. Springer Verlag: Berlin, Germany.
9. J.G. Quintiere (2006) *Fundamentals of Fire Phenomena*. John Wiley & Sons Ltd: Hoboken, NJ.
10. D. Drysdale (2011) *An Introduction to Fire Dynamics*, 3rd Edition. John Wiley & Sons, Ltd: Hoboken, NJ.
11. B. Karlsson and J.G. Quintiere (2000) *Enclosure Fire Dynamics*. CRC Press: Boca Raton, FL.
12. H. Tennekes and J.L. Lumley (1972) *A First Course in Turbulence*. MIT Press: Cambridge, MA.
13. S.B. Pope (2000) *Turbulent Flows*. Cambridge University Press: Cambridge.
14. B. Merci and E. Dick (2003) "Heat transfer predictions with a cubic k-ε model for axisymmetric turbulent jets impinging onto a flat plate", *International Journal of Heat and Mass Transfer*, Vol. 46(3), pp. 469–480.
15. K. Van Maele and B. Merci (2006) "Application of two buoyancy-modified k-ε turbulence models to different types of buoyant plumes", *Fire Safety Journal*, Vol. 41(2), pp. 122–138.
16. L.F. Moody (1944) "Friction factors for pipe flow", *Transactions of the ASME*, Vol. 66(8), pp. 671–684.
17. S. Wasan, P. Rauwoens, J. Vierendeels and B. Merci (2010) "An enthalpy-based pyrolysis model for charring and non-charring materials in case of fire", *Combustion and Flame*, Vol. 157, pp. 715–734.
18. https://github.com/MaCFP/matl-db.

3 Turbulent Flames and Fire Plumes

3.1 FLAMMABILITY

As explained in Section 1.1, an essential feature in fires concerns the combustion reactions of combustible gases with oxygen. These reactions must be sufficiently fast to provide the heat required for the evaporation or pyrolysis of the liquid, resp. solid, fuel (Section 1.3). For this to be possible, the mixture of combustible gases and oxygen must be 'flammable'. In case the oxygen is taken from air, the mixture of combustible gases and air must be 'flammable'. Thus, the concept of flammability is essential in the context of fire dynamics. Before entering into a more technical discussion, a few examples are given first.

As the first example, consider the release of a very small amount of a combustible gas in a room, filled with air. In such conditions, the fuel-air mixture will be extremely lean (Section 2.2.1), and whereas the fuel could perhaps be ignited, the mixture would not be 'flammable' because the combustion reactions would stop very quickly (due to the absence of fuel).

As the second example, consider the release of fuel in a room filled with nitrogen or carbon dioxide. In such conditions, the fuel will not find oxygen to react, so the mixture will not be 'flammable'.

From these two extreme examples, it can intuitively be understood that the first example can evolve towards flammable conditions by adding more fuel, whereas the second example can become flammable if a sufficient amount of oxygen is supplied. This is discussed in more quantitative detail in the present section. Numerous textbooks (e.g., [1–3]) provide extensive discussions on flammability and flammability limits. An extensive review on flammability of gas mixtures is provided in Ref. [4]. The reader is referred to those textbooks for in-depth study on this topic (including, e.g., flammability diagrams). The focus in the present book is on the thermodynamics involved.

In reality, flammability is determined by both chemistry (i.e., the presence of chemical reactive species or radicals) and physics (essentially temperature, which determines the reaction rates). Here, as mentioned, only a thermodynamic discussion is provided, using reactivity and the corresponding 'threshold temperature' for a mixture to be flammable, because temperature is often the first-order effect. Yet, it must be appreciated that chemistry can be essential, as mentioned below in, e.g., the context of addition of chemical inhibitors.

Figure 2.2 is a basic figure that illustrates the exothermic nature of combustion. Consider a homogeneous flammable mixture where an initial flame is created through 'ignition' (e.g., by a spark as pilot). The sensible enthalpy generated by the combustion in the flame zone provides the heat required to overcome the 'activation' energy in the neighbouring mixture. This mixture will then react, generating sensible enthalpy that provides the heat required to overcome the 'activation' energy in the neighbouring mixture, which will then react, etc. In other words, a self-sustained combustion process may occur, once initiated (or 'ignited'), and the flame will move inside the mixture, consuming unburnt fuel. This is discussed in more detail in Section 3.2 on premixed flames.

However, it was explained in Section 2.2.3 that a sufficiently high temperature is required for combustion reactions to take place at a notable rate. This relates to the 'activation energy'. Indeed,

DOI: 10.1201/9781003204374-3

as explained in Section 3.1.1, a 'threshold' temperature can be estimated below which the reactions are 'quenched'. The higher the activation energy, the higher the temperature needs to be for reactions to take place at a notable rate. This is addressed in more detail in Sections 3.1.1 and 3.1.2.

It is noteworthy that 'ignitability' need not imply that a mixture is 'flammable', as illustrated in the first example given above. Indeed, 'flammability' refers to sustained combustion. In practice, 'flammability' of a mixture is determined in an apparatus developed at the US Bureau of Mines [5]. The reader is referred to Refs. [1–5] for more details, but essentially a homogeneous mixture is ignited and the flame runs upwards inside a tube with an inner diameter equal to 5 cm. If the flame is still sustained 75 cm away from the ignition source, the mixture is called 'flammable'. While there is a certain level of arbitrariness involved, this standard procedure does allow for categorization of mixtures in terms of risk for hazards. Yet, it is noteworthy that the flammability of a mixture does not only depend on the mixture itself, but also on heat losses. In fact, this is the reason why the diameter of the flammability apparatus is not chosen too small. Indeed, the heat losses become relatively larger as the diameter decreases [1–4], so a mixture would seem less flammable than in a device with a larger diameter. For sufficiently small diameters, the flame is effectively quenched for any possible mixture. This happens for diameters smaller than the 'quenching diameter', the absolute value of which is fuel dependent (in between 1 and 2 mm for most hydrocarbon fuels, but as low as 0.5 mm for hydrogen [3]). In fact, this concept is effectively used in flame arresters: heat is taken away from the reaction zone so quickly that the temperature rise involved becomes too low for the reactions to sustain, by forcing the mixture through a device with sufficiently small holes and taking away the heat. This is discussed in more detail in Section 3.4.

The above makes clear that 'flammability' is not a fundamental thermodynamic or chemical property. Rather, it is a concept of practical importance. 'Flammability' depends on:

- The mixture considered, i.e., fuel, oxidizer and ratio of the amount of fuel to the amount of oxidizer;
- Heat losses from the system;
- Initial temperature;
- Initial pressure.

If nothing is specified, 'flammability' refers to the device mentioned above [5] for a mixture of fuel with air at atmospheric pressure and initial temperature equal to 25°C (298 K).

3.1.1 FLAMMABILITY LIMITS – THRESHOLD TEMPERATURE

Figure 2.3 illustrates the chemical reactivity, using the Arrhenius expression, Eq. (2.25):

$$k = A \exp\left(-\frac{E_a}{RT}\right). \tag{3.1}$$

The value $E_a/R = 15,000$ K is used in Figure 2.3, as it is typical for many hydrocarbon gas fuels burning in air at atmospheric conditions [2]. From this figure, the concept of a 'threshold' temperature T_{th} can be understood: for $T < T_{th}$, the chemical reactions are so slow that they can be ignored. In other words, for $T < T_{th}$, the mixture is considered 'non-flammable'. For many hydrocarbon fuels, the order of magnitude of T_{th} is taken around 1,600 K in air [1]. Hydrogen (980 K) and carbon monoxide (1,300 K) are fuels with strongly different values of T_{th} [1].

For lower values of E_a, the chemical reactivity is stronger, whereas for higher values of E_a, the mixture becomes less reactive. Accepting the concept that flammability relates to exceeding

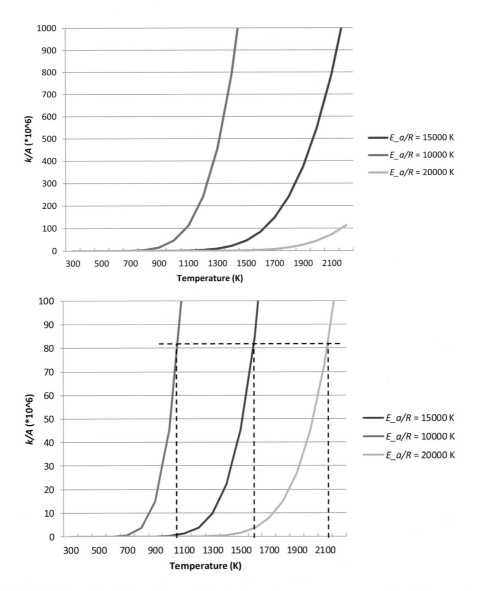

FIGURE 3.1 Illustration of dependence of chemical reaction rate on temperature for different values of activation energy (Eq. 3.1). Lower activation energy denotes more reactive fuel. Bottom: illustration of the concept of threshold temperature: lower T_{th} values correspond to more reactive fuels.

a minimum level of chemical activity, Figure 3.1 illustrates the impact on the value of T_{th}. Using the value $T_{th} = 1,600$ K for $E_a/R = 15,000$ K, T_{th} drops to about $T_{th} = 1,050$ K for $E_a/R = 10,000$ K, while it increases to $T_{th} = 2,120$ K for $E_a/R = 20,000$ K. The threshold temperature is lower for more reactive fuels, since a lower temperature is required to guarantee the minimum level of chemical activity required to sustain the flame (i.e., to make the mixture 'flammable').

The concept of the threshold value can now be used to understand the basic influence parameters on the flammability of a mixture. For the time being, a single fuel is considered, so the term 'mixture' refers to a mixture of pure fuel and oxidizer. Consider adiabatic conditions at atmospheric pressure, with initial temperature (prior to ignition) equal to T_u (where 'u' refers to 'unburnt'), set equal to 298 K. Consider an oxidizer, consisting of a mixture x O_2 and $(1-x)$ N_2, where 'x' is the volume concentration of oxygen. In air, $x = 0.21$. Consider a fuel concentration in the mixture X_F (in volume percent). Using the short notation, $\alpha = \dfrac{X_F}{X_{F,st}}$, with $X_{F,st}$ the fuel concentration (in volume percent) in a mixture at stoichiometric conditions, and focusing on the situation $\alpha \leq 1$, the following global chemical reaction can be written for 1 mole of a fuel that consists of C, H and O:

$$C_m H_n O_p + \frac{1}{\alpha}\left(m + \frac{n}{4} - \frac{p}{2}\right)\left(O_2 + \frac{1-x}{x} N_2\right) \rightarrow m CO_2 + \frac{n}{2} H_2O$$

$$+ \frac{1-\alpha}{\alpha}\left(m + \frac{n}{4} - \frac{p}{2}\right)O_2 + \frac{1}{\alpha}\frac{1-x}{x}\left(m + \frac{n}{4} - \frac{p}{2}\right)N_2; \alpha = \frac{X_F}{X_{F,st}} \leq 1 \qquad (3.2)$$

From an energy point of view, Figure 2.2 illustrates that the chemical enthalpy, contained in the fuel, is transformed into sensible enthalpy. The chemical enthalpy, or 'heat of combustion', of the fuel is noted as Δh_c, expressed in J/mole (fuel). This value is tabulated at 298 K. The chemical enthalpy contents in the mixture of fuel and oxidizer are then $X_F \Delta h_c$ J/mole (mixture). These enthalpy contents are, after completion of combustion and assuming adiabatic conditions, transformed into sensible enthalpy, i.e., into a temperature rise. Using 'averaged' values (where 'averaged' refers to averaging over the temperature interval $[T_u - T_b]$, where 'u' refers to 'unburnt' and 'b' refers to 'burnt') for specific heats (Section 2.1.1.3), the following equation can be used to estimate the temperature of the mixture, expressing the heat balance per mole fuel burnt:

$$\Delta h_c = \left(m c_{pm,CO_2} + \frac{n}{2} c_{pm,H_2O} + \frac{1-\alpha}{\alpha}\left(m + \frac{n}{4} - \frac{p}{2}\right)c_{pm,O_2} + \frac{1}{\alpha}\frac{1-x}{x}\left(m + \frac{n}{4} - \frac{p}{2}\right)c_{pm,N_2}\right)(T_b - T_u)$$

$$\rightarrow T_b = T_u + \left(m c_{pm,CO_2} + \frac{n}{2} c_{pm,H_2O} + \frac{1-\alpha}{\alpha}\left(m + \frac{n}{4} - \frac{p}{2}\right)c_{pm,O_2} + \frac{1}{\alpha}\frac{1-x}{x}\left(m + \frac{n}{4} - \frac{p}{2}\right)c_{pm,N_2}\right)^{-1}\Delta h_c. \quad (3.3)$$

Table 3.1 provides some orders of magnitude for the averaged specific heats, averaged over the interval of 0°C–1,500°C. For more precise calculations, the reader is referred to databases for specific heats. The most important observation in Table 3.1 is that the specific heat of CO_2 is substantially higher than the value for H_2O, which leads to lower temperatures for the same heat of combustion. The high value of specific heat for CO_2 makes it interesting as gaseous fire suppressant (see also Section 3.4.2). Also note that the specific heats of oxygen and nitrogen are very similar.

TABLE 3.1

Order of Magnitude for the Averaged Specific Heats, Averaged over the Interval of 0°C–1,500°C

Species	CO_2	H_2O	O_2	N_2
c_{pm} (J/mol K)	52.3	41.5	34.3	32.5

For more precise calculations, the reader is referred to databases for specific heats.

Equation (3.3) reveals many fundamental issues. Recall that we focus on the lean side of stoichiometry ($\alpha < 1$). In other words, we consider the 'lower flammability limit' (LFL). A similar reasoning can be built for the 'upper flammability limit' (UFL) at the rich side of stoichiometry, but then the chemical reaction becomes more complex than Eq. (3.2) and only a fraction of Δh_c is effectively released, because not all the fuel can find oxygen to react. This is not discussed further here.

First of all, consider a fixed fuel and consider combustion in air. This means that m, n, p and Δh_c are fixed, and $x = 0.21$. At stoichiometry ($\alpha = 1$), Eq. (3.3) reads:

$$\Delta h_c = \left(mc_{pm,CO_2} + \frac{n}{2} c_{pm,H_2O} + \frac{79}{21}\left(m + \frac{n}{4} - \frac{p}{2} \right) c_{pm,N_2} \right)(T_{b,ad} - T_u). \tag{3.4}$$

This yields the adiabatic flame temperature (Section 2.2.2.2) in stoichiometric circumstances (noted here as $T_{ad,st} = T_{b,ad}$). The more α deviates from 1 (i.e., the closer α gets to 0, as we restrict ourselves to the situation $\alpha < 1$), the lower the temperature of the mixture becomes, as can be understood from Eq. (3.3). Now, using the adiabatic flame temperature at stoichiometry (which is around 2,300 K for many common fuels) and a threshold value (e.g., 1,600 K) as value for $T_{th} = T_{ad,LFL}$ at the lower flammability limit, the LFL value of the mixture for combustion in air can be estimated from:

$$\frac{T_{ad,st} - T_u}{T_{ad,LFL} - T_u}$$

$$= \frac{mc_{pm,CO_2} + \frac{n}{2} c_{pm,H_2O} + \frac{1 - \alpha_{LFL}}{\alpha_{LFL}}\left(m + \frac{n}{4} - \frac{p}{2} \right) c_{pm,O_2} + \frac{1}{\alpha_{LFL}} \frac{79}{21}\left(m + \frac{n}{4} - \frac{p}{2} \right) c_{pm,N_2}}{mc_{pm,CO_2} + \frac{n}{2} c_{pm,H_2O} + \frac{79}{21}\left(m + \frac{n}{4} - \frac{p}{2} \right) c_{pm,N_2}}. \tag{3.5}$$

Interestingly [3], the value of α_{LFL} is typically around 0.55. Lower values are observed for hydrogen (0.13 [2]) and carbon monoxide (0.42 [2]), which is in line with the fact that these fuels have lower values for T_{th}, as mentioned above. Figure 3.1 provides the interpretation that these fuels are more reactive. Also acetylene (0.32 [2]) has a substantially lower value, again in line with a higher reactivity (as can be noted from the high laminar burning velocity values, Section 3.2.1) and with the relatively high adiabatic flame temperature at stoichiometric conditions.

Second, fixing the fuel, Eq. (3.3) indicates that the number of moles in the mixture of combustion products decreases (and therefore, its temperature increases), as α approaches 1 (i.e., for mixtures closer to stoichiometry) and as x approaches 1 (i.e., as nitrogen is replaced by oxygen). Interestingly, though, the LFL in pure oxygen is practically the same as in air, because for this limit, lack of fuel is the bottleneck, not lack of oxygen. The excess oxygen only acts as thermal ballast, and as the specific heats of oxygen and nitrogen are very similar (see Table 3.1), this hardly affects the temperature. Obviously, this argument does not hold for the UFL: as nitrogen is replaced by oxygen, more fuel will find oxygen to react and therefore a larger fraction of Δh_c can be released, compared to combustion in air in rich conditions. This implies that the UFL increases as the oxygen concentration in the oxidizer increases.

Equation (3.3) also reveals that it is easier to reach the threshold value for a given mixture of fuel and oxidizer as the initial temperature T_u increases. To put it another way, the flammability limits become wider as the temperature increases. Actually, this can also be expressed as a heat balance per mole mixture, rather than per mole fuel (as was done in Eq. 3.3). In that case, the flammability limit can be interpreted as the minimum concentration (by volume) of fuel in the mixture, required to reach the 'threshold' temperature T_{th}, starting from an initial temperature, T_u. Noting the minimum fuel concentration at the lower flammability as $X_{F,LFL}$ (unit: mol fuel/mol mixture) and the average specific heat of the mixture in the range of $[T_u - T_{th}]$ as c_{pm} (unit: J/(K.mol mixture)), the heat balance reads:

$$X_{F,LFL}(T_u)\Delta h_c = c_{pm}(T_{th} - T_u). \tag{3.6}$$

From Eq. (3.6), it is immediately clear that for a given fuel (i.e., for given Δh_c and T_{th}), $X_{F,LFL}(T_u)$ decreases as T_u increases. It is noted that pressure also affects the flammability limits [3], but this is not discussed here.

Finally, from Eq. (3.3), also the effect of heat losses can be understood. In that case, the equation must be generalized in that not all the chemical enthalpy released is kept as sensible enthalpy in the mixture, because part of the enthalpy is lost as heat loss. This leads to a reduction in temperature, making the mixture (seemingly) less flammable.

To close this section, Le Chatelier's law is mentioned for the determination of the LFL of mixtures of fuels [1–3]:

$$X_{LFL, mixt} = \frac{1}{\sum_{i=1}^{N} \dfrac{X_i}{X_{LFL,i}}}. \tag{3.7}$$

In Eq. (3.7), X_i is the volume percentage of fuel i in the mixture of fuels and $X_{LFL,i}$ is the LFL of fuel i in air alone.

Exercises:

1. The LFL of propane at 20°C is 2.1% (by volume). Estimate the LFL at 100°C. (Answer: 2.0%)
2. Calculate the LFL of a mixture, consisting of 85% methane, 10% ethane and 5% propane. (Answer: 4.4%)
3. What is the adiabatic flame temperature at stoichiometric conditions for propane in air, starting from 20°C? What is the adiabatic flame temperature at the LFL? (Answer: 2,208°C; 1,358°C)
4. Assuming a threshold adiabatic flame temperature of 1,600 K, calculate the LFL of propane in a mixture of 21% (by volume) O_2 and 79% CO_2. Compare the value to the value for propane in air. (Answer: 3.4%; this is higher than in air, because CO_2 has a higher heat capacity than N_2.)

3.1.2 ADDITION OF GASES

The reasoning in Section 3.1.1 also allows understanding the effect of the addition of gases on the flammability limits. Consider Eq. (3.2), and suppose inert gas is added. Then, this gas appears unchanged at the left hand side and the right hand side of the reaction equation. In Eq. (3.3), the gas appears in the denominator, illustrating that the temperature becomes lower, all other quantities being fixed. In other words, it becomes more difficult to reach the threshold temperature and thus the flammability limits become narrower. The effect of the inert gas is larger as its specific heat value is larger. As mentioned in Section 3.1.1, this makes CO_2 an interesting inerting gas, given its high specific heat.

If the added gas also acts on the chemistry mechanism, the effect can also be interpreted through Figure 3.1. The addition of a chemical inhibitor can then be interpreted as an increase in activation energy. Figure 3.1 illustrates the substantial effect.

3.1.3 Flammability of Liquid Fuels

It was explained in Section 3.1.1, through Eq. (3.3), that the concentration of combustible gases in the oxidizer (oxygen or air) must be sufficiently close to stoichiometry in order to have a flammable mixture.

In the case of liquid fuels, the concentration of gaseous vapour in equilibrium conditions can be calculated from the Clausius–Clapeyron equation, Eq. (2.138). As such, the 'flammability' of a liquid fuel can be considered. In the case of mixtures, Eq. (3.7) can be used again.

3.2 PREMIXED FLAMES

Premixed combustion refers to conditions where fuel and oxidizer are mixed before reaction takes place. An example in the context of fire dynamics concerns the filling of an enclosure with gaseous fuel from, e.g., a leak. If the fuel-air mixture becomes flammable and there is an ignition source (e.g., a hot spot, or an electric spark from a light that is switched on), premixed combustion can take place. Another example in the context of fire dynamics concerns combustion inside a hot smoke layer underneath a ceiling (see also Section 4.5). Indeed, if combustion in the fire plume (Section 3.5) is incomplete, combustible gases such as CO can enter the smoke layer. If the smoke-air mixture becomes flammable and there is an ignition source, premixed combustion can take place inside the smoke layer. Premixed combustion is typically very hazardous, as it is very fast because the relatively slow mixing process has already been completed prior to combustion, in contrast to non-premixed combustion (Section 3.3).

3.2.1 Laminar Premixed Flame Structure

The structure of a laminar premixed flame is discussed in every textbook on combustion. Hence, only a few aspects are discussed here.

Figure 3.2 provides the sketch of a laminar premixed flame. As mentioned, the term 'premixed' refers to the fact that fuel and oxidizer are mixed before reaction takes place. The reaction takes place in the reaction zone (with thickness δ_r) in the flame. As explained in Section 2.2.2.1, the chemical enthalpy is transformed into sensible enthalpy. Due to thermal diffusion, this heat causes a temperature rise in the region of yet-to-burn fuel/oxidizer mixture near the flame. This zone is called the 'pre-heat zone'. Once the temperature has been raised beyond an 'ignition' temperature, the fuel–oxidizer mixture reacts and the cycle continues.

The process is initiated by ignition. Figure 3.2 shows idealized conditions of a perfectly planar flame, without any heat losses (i.e., adiabatic conditions) and without friction at the walls. In such conditions, the flame runs towards the unburnt mixture at a speed that is called the 'laminar burning velocity'. This is an important property of the fuel–oxidizer mixture, given the conditions (e.g., atmospheric conditions).

Obviously, the situation as sketched in Figure 3.2 can only occur if the mixture is flammable (see Section 3.1).

3.2.2 Laminar Burning Velocity

An important feature of a premixed flame is that it has its own dynamics. Indeed, as mentioned in Section 3.2.1, in a steady regime as sketched in Figure 3.2, the flame runs towards the unburnt

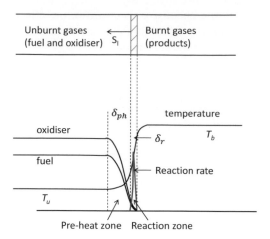

FIGURE 3.2   Sketch of the structure of a laminar premixed flame.

mixture. It is instructive to examine Figure 3.2 in more detail to provide a qualitative discussion on the phenomenon, based on the approach of Refs. [6,7]. The notion 'ignition temperature' is introduced at the border between 'pre-heat' and 'reaction' zones. This is a fictitious temperature, not to be confused with an 'auto-ignition' temperature in homogeneous conditions, but it serves the analysis in that no chemical reactions are assumed as long as $T < T_{ig}$ here (i.e., in the pre-heat zone).

In the pre-heat zone, assuming no chemical reaction (because the reactions are too slow due to the low temperature), there is competition (balance) of diffusion and convection. In the reaction zone, however, there are steeper gradients (in temperature and species). Given that convection relates to first-order derivatives and diffusion to second-order derivatives, it is reasonable to assume that in the reaction zone, there is a balance of chemical reaction and diffusion. Around the ignition temperature, all three (convection, diffusion and reaction) are of the same order.

This leads to the following derivation. The pre-heat zone moves towards the unburnt mixture with velocity equal to the laminar burning velocity S_l. As such, per unit cross area, there is a net flow of heat per unit area and per unit time through the pre-heat zone equal to $\rho_u S_l c_p (T_{ig} - T_u)$. Indeed, the mass flow rate of unburnt mixture into the pre-heat zone, per unit cross-sectional area, equals $\rho_u S_l$. This unburnt mixture is heated from the unburnt temperature, T_u, up to the ignition temperature, T_{ig}, so that multiplying with the specific heat yields the energy associated with this rise in temperature.

The heat required for this heating up stems from thermal diffusion from the reaction zone into the pre-heat zone. This relates to conduction, Eq. (2.5). Approximating the temperature gradient as $\dfrac{T_b - T_{ig}}{\delta_r}$, the heat flux per unit time and per unit cross-sectional area reads $\lambda \dfrac{T_b - T_{ig}}{\delta_r}$. The energy balance over the pre-heat zone leads to the laminar burning velocity:

$$\rho_u S_l c_p (T_{ig} - T_u) = \lambda \frac{T_b - T_{ig}}{\delta_r} \rightarrow S_l = \frac{\lambda}{\rho_u c_p} \frac{1}{\delta_r} \frac{T_b - T_{ig}}{T_{ig} - T_u}. \tag{3.8}$$

Equation (3.8) reveals that the laminar burning velocity is:

- Proportional to the thermal diffusivity of the mixture ($\lambda/\rho_u c_p$).
- Inversely proportional to the reaction zone thickness. The thinner the reaction zone, the higher the reactivity of the mixture and the higher the laminar burning velocity.

- Proportional to the temperature ratio. This does not have a direct physical meaning per se, as T_{ig} is an artificial temperature, but it does provide an indication on the temperature gradient and thus on the reactivity of the mixture.

Now, using the notation $\dot{\omega}$ for the reaction rate (with unit kg/(m³s)), the following approximation can be made: $\rho_u S_l = \delta_r \dot{\omega}$. Eliminating δ_r from Eq. (3.8) yields:

$$S_l = \sqrt{\frac{\lambda}{\rho_u c_p} \frac{T_b - T_{ig}}{T_{ig} - T_u} \frac{\dot{\omega}}{\rho_u}} \propto \sqrt{\frac{\lambda}{\rho_u c_p} \frac{\dot{\omega}}{\rho_u}}. \tag{3.9}$$

Assuming the chemical reaction rate to be determined by an Arrhenius expression, Eq. (3.1), this illustrates that the laminar burning velocity is proportional to the square root of the chemical reactivity:

$$S_l \propto \sqrt{\frac{\lambda}{\rho_u c_p}} \exp\left(-\frac{E_a}{2RT}\right). \tag{3.10}$$

Several aspects can be understood directly from Eq. (3.10):

- The more reactive the fuel, i.e., the lower the activation energy E_a, the higher the laminar burning velocity. Note that this corresponds to a lower threshold temperature for flammability (Section 3.1). Vice versa, if inhibitors are added, this corresponds to an apparent increase in activation energy E_a, and therefore a decrease in laminar burning velocity.
- The higher the temperature, the higher the laminar burning velocity. This has a couple of consequences, related to aspects mentioned in Section 3.1:
  - A mixture that is close to stoichiometry will have a higher laminar burning velocity than a mixture that is close to its flammability limits. This is typically presented as a dependence of the laminar burning velocity on the equivalence ratio, Eq. (2.12). The laminar burning velocity is at its maximum for $\Phi = 1$ and decreases monotonically as Φ deviates from 1. The laminar burning velocity suddenly drops to zero at values for Φ corresponding to the lower and upper flammability limits (Section 3.1). If nothing is specified, it is implicitly assumed that the 'maximum' laminar burning velocity is reported.
  - A fuel will have a higher laminar burning velocity as the oxygen concentration in the oxidizer increases.
  - A fuel–oxidizer mixture will have a higher laminar burning velocity if the mixture is pre-heated.

It is important to note that the existence of the laminar burning velocity, with unit m/s, along with the thermal diffusivity (with unit m²/s), results in an inherent 'flame thickness':

$$l_f = \frac{\alpha}{S_l} = \frac{\lambda}{\rho_u c_p S_l} \tag{3.11}$$

In Eq. (3.11), $l_f = \delta_r + \delta_{ph}$ is the sum of the thickness of the reaction zone and the pre-heat zone. For methane–air combustion in atmospheric conditions, l_f is in the order of 1 mm. It is also typical that δ_r / l_f is approximately 1/10.

There is an important caveat in that the laminar burning velocity must not be confused with the 'effective' flame speed. As long as all ends are 'open', as sketched in Figure 3.2, the difference need not be too large (as long as there is no turbulence, see Section 3.2.3). However, if, e.g., a tube is considered with one end closed and ignition at the closed end, the flame will run much faster than by its own laminar burning velocity, because the expansion of the hot gases in between the flame front and the closed tube end will push the flame forwards towards the open end. This is not discussed further here.

3.2.3 THE EFFECT OF TURBULENCE

In the context of turbulent premixed combustion, use is made of the 'Borghi' diagram to characterize the regime (e.g., [8–15]). Figure 3.3 provides a simplified sketch. The flame thickness is the sum of the pre-heat zone and reaction zone thicknesses (Figure 3.2). The integral and Kolmogorov turbulent scales have been introduced in Section 2.8.3.2. It is recalled here that the integral scales are determined by the geometry and the mean flow. The smallest 'Kolmogorov' scales relate to the integral scales through the Reynolds number.

From the integral scales, the 'turbulent' Reynolds number can be defined:

$$\text{Re}_t = \frac{\rho v'' l_{\text{int}}}{\mu} \propto \frac{\rho \sqrt{k} l_{\text{int}}}{\mu}. \tag{3.12}$$

In Eq. (3.12), k is the turbulent kinetic energy (m^2/s^2).

With respect to chemistry, the Damköhler number is important:

$$Da = \frac{\tau_{\text{flow}}}{\tau_{\text{chem}}} = \frac{\tau_{\text{int}}}{\tau_{\text{chem}}}. \tag{3.13}$$

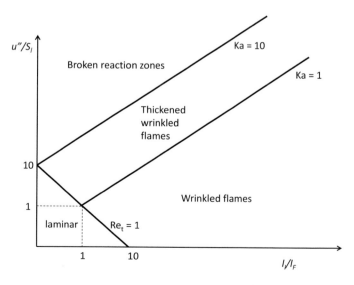

FIGURE 3.3 Simplified representation of the Borghi diagram for premixed combustion. Logarithmic scales on the axes.

The Damköhler number expresses how fast chemistry is, compared to characteristic flow time scales (also called 'residence times'). In Eq. (3.13), τ_{int} is the integral turbulent time scale, as mentioned in Section 2.8.3.2, and τ_{chem} is the characteristic chemical time scale. The chemical time scale relates to the reaction rate, Eq. (3.1):

$$\tau_{chem} \propto \exp\left(\frac{E_a}{RT}\right). \tag{3.14}$$

The faster the chemistry (i.e., the lower the activation energy and/or the higher the temperature), the smaller τ_{chem}. In other words, the Damköhler number becomes very large for fast chemistry, compared to the (turbulent) flow scales.

In a similar way, the Karlovitz number can be defined. However, the Karlovitz number is based on the smallest turbulent scales, the Kolmogorov scales (see Section 2.8.3.2):

$$Ka = \frac{\tau_{chem}}{\tau_\eta}. \tag{3.15}$$

In Eq. (3.15), τ_η is the Kolmogorov time scale, as mentioned in Section 2.8.3.2. If $Ka < 1$, this implies that the chemistry is faster than the fastest turbulent phenomena.

With these numbers as guidance, the 'Borghi' diagram is constructed. Many refinements have been made over the years [8–15], but a simplified version suffices for the context of the present discussion. In the bottom left corner, laminar premixed combustion takes place. There are hardly any fluctuations. For sufficiently large fluctuations, turbulence occurs. However, below the line $Ka = 1$, chemistry is faster than the fastest turbulent phenomena, as has just been explained. This means that the flame reacts very quickly to any changes in the local flow field, induced by turbulence. Consequently, the local flame structure resembles that of a planar laminar flame. However, the flame is no longer planer, but 'wrinkled'. This corresponds also to the sketch on the left in Figure 3.4, sketching possible situations of relative magnitude of the smallest turbulent scales and

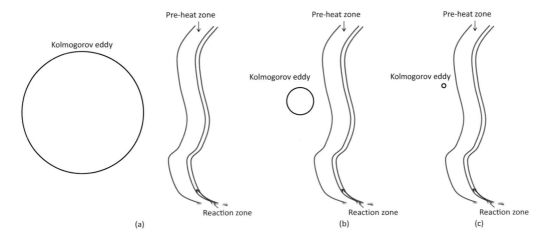

FIGURE 3.4 Sketch of different regimes, depending on the relative size of the smallest turbulent scales (Kolmogorov scales) and the flame thickness. (a) Wrinkled flames regime, (b) thickened wrinkled flames regime and (c) broken reaction zones regime.

flame thickness: even the smallest turbulent eddy is larger than the flame thickness. A fortiori, the integral turbulent scales are much larger than the flame thickness. The only effect of turbulence in this region is thus the wrinkling of the flame, thereby increasing the area available for reaction, as discussed below.

In the region $1 < Ka < 10$, the smallest turbulent scales are of the same order as the thickness of the pre-heat zone (Figure 3.2), but still larger than the thickness of the reaction zone. Indeed, as mentioned above, the ratio δ_r/l_f is approximately 1/10. In this region, the smallest turbulent eddies promote heat and mass transfers from the reaction zone into the unburnt mixture. This widens the pre-heat zone, hence the label 'thickened wrinkled flames'. They also stretch the reaction zone, and sometimes this region in the Borghi diagram is called the 'thin reaction zones' regime. Yet, none of the turbulent eddies are of such a small scale that they can enter into the reaction zone to effectively break it up. This regime corresponds to the middle sketch of Figure 3.4.

In the region $Ka > 10$, however, there is very strong interaction with the chemistry, directly at the level of the reaction zone. Indeed, the small turbulent eddies are of the same size as the reaction zone and direct interaction, and break-up is possible. This zone is therefore called the 'broken reaction zones' regime. However, another view point is that mixing is very intense, and thus, this regime is sometimes also called the 'well-stirred reactor' regime. This regime corresponds to the sketch on the right in Figure 3.4.

It is instructive to examine the position of fire-related hazards in this diagram. Flows are typically 'mildly' turbulent, i.e., turbulent velocity fluctuations v'' are of the same order of magnitude as the laminar burning velocities (which are in the order of 0.5 m/s for most fuels in air, with hydrogen (3.2 m/s) and acetylene (1.7 m/s) as exceptions). Also, the integral turbulent length scales (which are of the order of the geometry dimensions) are much larger than the flame thicknesses (which are of the order of 1 mm, as mentioned). Thus, fire-related hazards with premixed combustion are in the 'wrinkled flames' regime.

In such conditions, an increase in apparent burning velocity is observed as the level of turbulence increases (e.g., [3,15]). This can easily be understood: the flame is no longer planar, but wrinkled due to the turbulent motions. Consequently, for a given geometry, there is a larger area available for chemical reaction. Moreover, there is positive feedback in the loop: as the level of turbulence increases, the burning velocity increases and as such the turbulence becomes more violent, due to higher velocities. Thus, the situation can evolve towards the 'thickened wrinkled flames' regime, where turbulence also strongly enhances the efficiency of the transport processes of heat and chemical radicals from the flame zone into the pre-heat zone (Figure 3.2) due to eddy mixing. This reveals that a hazard typically becomes worse as it evolves: increasing the velocity of the running flame due to turbulence (flame wrinkling), the Reynolds number increases and the increase in transport due to eddy mixing adds to the wrinkling effect.

It is worth mentioning that in an enclosed space, but also in the open when a sufficient number of obstacles are present [3], the running premixed flame, thereby creating combustion products that are much hotter than the unburnt mixture (and that therefore want to expand), can lead to severe pressure build-up. This can lead to explosions (see, e.g., [3]).

To close this section, it is worth noting as a practical example that ill-designed flame arresters can have a strong adverse effect due to turbulence. Indeed, as mentioned in Section 3.1, the basic principle of flame arresters is to force the flow through very small holes. If the diameter of these holes is effectively less than the quenching diameter for the mixture and conditions at hand, the flame will be held at the arrester (and eventually quench). However, if the flame manages to pass through, then the primary effect of the holes is the creation of turbulence, so the flame will move even (much) faster than before meeting the arrester and a more severe hazard is the result.

3.3 DIFFUSION FLAMES

The structure of a laminar non-premixed, or 'diffusion', flame is discussed in every textbook on combustion. Hence, only a few aspects are discussed here.

3.3.1 LAMINAR DIFFUSION FLAME STRUCTURE

Figure 3.5 provides the sketch of a steady laminar planar non-premixed flame. Because diffusion is the fundamental process, as explained below, this type of flame is also called 'diffusion' flame. Note that diffusion is also important in premixed combustion, as explained in Section 3.2, though.

The flame 'thickness' again consists of a 'reaction zone', and a 'diffusion' zone. This is very similar to the reaction and pre-heat zones as shown in Figure 3.2 for premixed combustion. However, there are a few fundamental differences, compared to premixed combustion:

- Fuel and oxidizer are each at a different side of the reaction zone.
- The burning rate is controlled by the molecular diffusion of fuel and oxidizer towards the reaction zone. This mass diffusion is not necessary anymore in premixed combustion, as the mixing process has already been completed prior to the initiation of the combustion process. The reaction region extends from the LFL (at the side of the oxidizer) to the UFL (at the fuel side), but by far the reaction rate is the highest near stoichiometry, where the temperature is the highest (see Section 3.1). The thicknesses of the diffusion and reaction zones depend on the characteristic times for chemical reaction and diffusion.
- Diffusion flames do not have their own propagation dynamics. They are essentially mixing (or 'diffusion') controlled.
- The thickness of a diffusion flame is no inherent property of the flame itself. It depends on the local flow properties.

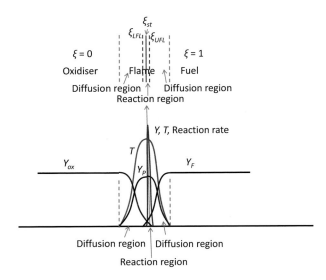

FIGURE 3.5 Sketch of a steady laminar planar non-premixed flame. The mixture fraction has been introduced in Section 2.3.5. ξ_{st} denotes the position of stoichiometry, ξ_{LFL} corresponds to the lower flammability limit (Section 3.1) and ξ_{UFL} corresponds to the UFL (Section 3.1). Y refers to mass fraction (Section 2.1.1.5) of oxidizer ('ox'), fuel ('F') and products ('P'). The sum of the mass fractions equals unity everywhere. Products mass fraction has a similar profile as temperature. Reaction rate is shown in red.

It is instructive to not only consider the structure of a diffusion flame in physical space, but also in mixture fraction space. Mixture fraction ξ evolves monotonically from zero to one (from left to right) in Figure 3.5. Introducing the scalar dissipation rate χ:

$$\chi = 2D \sum_{k=1}^{3} \left(\frac{\partial \xi}{\partial x_k} \right)^2, \tag{3.16}$$

it can be shown (e.g., [8–15]) that the transport equation for the mass fraction Y_i of species i reads:

$$\rho \frac{\partial Y_i}{\partial t} = \frac{1}{2} \rho \chi \frac{\partial^2 Y_i}{\partial \xi^2} + \dot{\omega}_i, i = 1,\dots,N. \tag{3.17}$$

Note that the dimension of χ is s^{-1}, i.e., the inverse of a time scale. A high value for χ corresponds to fast diffusion/mixing.

Similarly, with the assumption of unity Lewis number, the transport equation for temperature reads:

$$\rho \frac{\partial T}{\partial t} = \frac{1}{2} \rho \chi \frac{\partial^2 T}{\partial \xi^2} + \dot{\omega}_T. \tag{3.18}$$

In steady conditions, these equations boil down to:

$$\dot{\omega}_i = -\frac{1}{2} \rho \chi \frac{\partial^2 Y_i}{\partial \xi^2}, i = 1,\dots,N; \dot{\omega}_T = -\frac{1}{2} \rho \chi \frac{\partial^2 T}{\partial \xi^2}. \tag{3.19}$$

Clearly, the reaction rates for species and temperatures depend on mixture fraction and scalar dissipation rate only. Through the gradients in physical space (Eq. 3.16), the scalar dissipation rate contains information from the flow/mixing field. Chemical effects are incorporated through the flame structure in mixture fraction space.

The end result is that species mass fractions and temperature can be written as follows:

$$Y_i = Y_i(\xi, \chi), i = 1,\dots,N; T = T(\xi, \chi). \tag{3.20}$$

This is called the 'steady laminar flamelet' assumption. This will be discussed again in Section 8.4.

A very important consequence is that from a physical point of view as well as from a modelling point of view, the determination of the position, temperature and chemical composition of a diffusion flame consists of two sub-problems:

- Determination of the flow and mixing field. This concerns the solution of the conservation of mass, total momentum and energy, along with the transport equation for mixture fraction (Eq. 2.73).
- Determination of the flame structure (temperature and species mass fractions) by specifying the relations to mixture fraction (Eq. 3.20). Different models are mentioned in Section 8.4.

Having introduced the scalar dissipation rate, Eq. (3.16), a flame thickness can now be defined, using the diffusivity:

$$l_f = \sqrt{\frac{D}{\chi_{st}}}. \tag{3.21}$$

The subscript 'st' refers to the fact that the scalar dissipation rate is evaluated at the line of stoichiometric mixture fraction.

Also the Damköhler number, Eq. (3.13), can be defined, using the diffusion time scale as characteristic for the flow scale:

$$Da = \frac{\tau_{\text{flow}}}{\tau_{\text{chem}}} = \frac{\chi_{st}^{-1}}{\tau_{\text{chem}}} \propto \chi_{st}^{-1} \exp\left(-\frac{E_a}{RT}\right). \tag{3.22}$$

It is instructive to combine Eq. (3.19) with Eq. (3.22). Starting from a situation of fast chemistry (high Damköhler number), an increase in scalar dissipation rate (e.g., by introducing higher velocity gradients in the flow field) leads to an increase in reaction rate (Eq. 3.19), due to increased diffusion (steeper concentration gradients). However, there is a limit: for too high scalar dissipation rates, the chemistry cannot keep up anymore (lower Da value, Eq. 3.22), primarily due to too high heat losses from the flame. Indeed, also temperature gradients become steeper, so more heat is lost by diffusion and the temperature in the flame reduces. This point is addressed again in Section 3.4. The critical value of χ for which the flame 'quenches', is called χ_q.

3.3.2 THE EFFECT OF TURBULENCE

As mentioned in Section 3.3.1, a diffusion flame does not have its own propagation dynamics. In other words: it is not able to move 'against' a flow, in contrast to premixed flames. As a consequence, diffusion flames are much more sensitive to the effect of turbulence.

Turbulence strongly enhances transport in physical space, due to eddy mixing. On the one hand, this can increase the reaction rate of the diffusion flame, as the diffusion process is typically much slower than the chemistry. Thus, the limiting process becomes faster, increasing the burning rate. However, also heat is transported away faster from the reaction zone, thereby reducing the temperature. If the instantaneous local scalar dissipation rate exceeds the critical value χ_q (see Section 3.3.1), the flame will suffer from (local or global) extinction.

Incorporating the interaction between turbulence and chemistry in computer simulations is essential for accuracy reasons. This is addressed in Section 8.4.

3.3.3 JET FLAMES

Figure 3.6 provides a sketch of a canonical laminar jet flame. Fuel is supplied through a nozzle with diameter D (not to be confused with diffusivity D). The same lines are sketched as in Figure 3.5, but now the equivalence ratio Φ (Eq. 2.12) is added. In the central region, the local composition is too rich for reaction to take place ($\Phi > \Phi_{\text{UFL}}$). Sufficiently far away from the reaction zone, the local mixture is too lean for reaction to take place ($\Phi < \Phi_{\text{LFL}}$). The reaction zone is in between ($\Phi_{\text{LFL}} < \Phi < \Phi_{\text{UFL}}$), with the most intense reaction around stoichiometry ($\Phi = 1$). Again fuel and oxidizer are each at a different side of the reaction zone.

The sketch of Figure 3.6 corresponds to the situation where there is more than enough oxidizer available to burn all the fuel supplied (well-ventilated or over-ventilated conditions).

It is important to mention the formation and consumption of soot. Soot is formed at the rich side of stoichiometry, i.e., at the inner side of the flame in Figure 3.6. If the residence time in the reaction region is sufficiently long, the soot is consumed when it flows through the oxidizing region. However, depending on the fuel type and the residence time, not all soot may be consumed and soot can break through the flame. Once it reaches a zone of lower temperature, the soot does not oxidize anymore. Soot breaking through is called 'smoke'. This is addressed in more detail in Section 8.7.

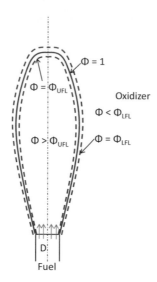

FIGURE 3.6   Sketch of a laminar jet flame.

Given the importance of diffusion for the burning rate of a diffusion flame (Section 3.3.1) and given the fact that a higher flow rate of fuel in the laminar diffusion flame of Figure 3.6 will cause higher velocities (stronger convection), it is not surprising that the flame height for the configuration of Figure 3.6 increases with the fuel flow rate and decreases with diffusivity of the fuel [3,16].

For sufficiently high fuel flow rate, the Reynolds number (Section 2.7.2) becomes high enough for the jet to become turbulent (Section 2.8). Turbulence strongly enhances mixing, and it is observed that the turbulent jet flame height becomes essentially independent of the jet velocity. Higher velocities indeed induce stronger mixing. In such conditions, the jet flame height becomes proportional to the nozzle diameter [3,15]. It also depends on the flame and fuel temperatures, as well as on the stoichiometric mixture fraction value. The flame height becomes independent of the molecular diffusivity. This is to be expected, because turbulence totally overwhelms the molecular diffusion process. Correlations for flame heights are presented in Section 3.5.1.6.

3.4   EXTINCTION OF FLAMES

3.4.1   PREMIXED FLAMES

In Section 3.2, the basic phenomenology of a premixed flame has been discussed. Essentially, assuming that the mixture is within the flammability limits, reactions take place in a narrow region (the reaction zone, Figure 3.2) and the heat, generated by the transformation of chemical enthalpy into sensible enthalpy, is transported by diffusion and convection into the pre-heat zone. In steady conditions, there is a balance.

Considering first laminar premixed flames only, the following situations can lead to flame extinction (which is often undesirable in combustion processes, but desirable in hazards):

- The atmosphere is leaner than the lower flammability limit (LFL) or richer than the upper flammability limit (UFL). In case of a hazard, the first option is the most likely. This concept is called 'inerting': the flame itself is not truly extinguished directly, but it cannot survive the ambient conditions.

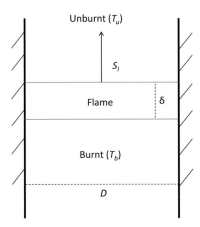

FIGURE 3.7 Sketch of premixed flame in tube with heat losses $\delta = \delta_r + \delta_{ph}$ (Figure 3.2).

- The chemical reactions are slowed down so much that the generation of heat cannot keep up with the heat loss through convection and diffusion:
    - Through introduction of a chemical inhibitor/suppressant, upstream of the moving flame front. The reaction mechanism is interrupted or slowed down chemically.
    - Through introduction of a physical suppressant, such as water mist. The evaporating water droplets consume much of the heat, generated at the flame front. Hence, this heat cannot be used to pre-heat the yet-to-burn mixture, and if the 'ignition' temperature (see Section 3.2) is not reached, the flame extinguishes. The water vapour, released by the evaporating water droplets, can also render the atmosphere more inert (and perhaps even below the LFL, see Section 3.1), which also assists in extinguishing the flame.
    - Through direct removal of heat by a solid structure. This is physical quenching and happens when the heat losses exceed the heat generated in the reaction zone.

The latter principle is used in flame arresters and is explained through Figure 3.7. This figure shows a laminar flame, which is assumed flat for simplicity (i.e., friction at the circular tube wall is ignored). The flame moves upwards with laminar flame speed S_l (Section 3.2). The flame thickness δ is the sum of the reaction zone (δ_r) and pre-heat zone (δ_{ph}) thicknesses. Now consider a heat balance, with the flame as control volume:

- The heat generated per unit time by the chemical reactions, reads:

$$\dot{Q} = \rho_u S_l c_p \left(T_b - T_u \right) \frac{\pi D^2}{4}, \tag{3.23}$$

- The heat lost to the tube wall per time unit is proportional to:

$$\dot{Q}_{loss} \propto \lambda \frac{T_b - T_u}{D} \pi D \delta. \tag{3.24}$$

In expressions (3.23) and (3.24), a few approximations have been made, namely:

- In Eq. (3.23), a uniform temperature rise is assumed from the unburnt temperature to a 'burnt' mixture temperature. The temperature T_b is lower than the adiabatic temperature, because heat losses are considered.
- The primary heat losses are essentially through convection from the flame to the tube wall. However, at the wall itself, in the end heat transfer is through conduction through a thin boundary layer (because the fluid stands still at the wall). This heat loss can then be written as the product of the fluid's conductivity and the radial temperature gradient at the tube wall. The latter is assumed to be inversely proportional to the tube diameter and proportional to $(T_b - T_u)$, assuming the wall temperature equal to T_u and assuming that the temperature rise on the symmetry axis in the flame is proportional to $(T_b - T_u)$.

While particularly the last assumption may be rough, it is acceptable of the present 'order of magnitude' analysis, since the principle of the 'quenching diameter' can be illustrated. Indeed, the flame is quenched if the heat loss (Eq. 3.24) exceeds the heat generated (Eq. 3.22),

$\lambda (T_b - T_u) \pi \delta > \rho_u S_l c_p (T_b - T_u) \dfrac{\pi D^2}{4}$, which leads to the 'critical' or 'quenching diameter':

$$D_q \propto \sqrt{\alpha} \sqrt{\dfrac{\delta}{S_l}}. \tag{3.25}$$

To close this discussion, note that the square root dependences in Eq. (3.25) also relate to the geometry: if Figure 3.7 had been a sketch of two flat plates, the dependence would be linear, because the area for the mass flow rate in Eq. (3.22) would then scale in a linear way with D (which would then denote the distance between the flat plates).

For turbulent premixed flames, all the above remains valid. As explained in Section 3.2.3, in the context of fire hazards, the primary effect of turbulence is an increase in flame surface area in a given geometry. The local flame structure remains essentially laminar (Figure 3.2). In the Borghi diagram (Figure 3.3), it is shown that interaction of small turbulent eddies with the reaction zone may cause the reaction zone to break up (which corresponds to local extinction). However, this is not a useful method to extinguish the flame in a hazard, because the (initial) Reynolds numbers are too low, so the smallest turbulent scales are too large to effectively extinguish the reaction zones.

3.4.2 Diffusion Flames

Clearly, one way to extinguish a non-premixed flame is through the addition of inert gases (like nitrogen or carbon dioxide) such that no flammable mixture can be formed anywhere. However, this is not always feasible and there are other ways to extinguish diffusion flames.

As mentioned in Section 3.3, diffusion flames do not have their own intrinsic dynamics. They exist at the virtue of fuel and oxidizer mixing. Therefore, an effective way of extinguishing a diffusion flame is by removing or blocking the fuel supply or the supply of oxidizer (e.g., by putting a cap over a candle flame).

Similar to premixed flames, though, is the fact that the heat generated in the flame region (Section 3.3) is removed from that region through diffusion and convection. An immediate consequence is that the methods of extinction, mentioned in Section 3.4.1 for premixed flames, remain valid for diffusion flames. Yet, the most instructive way to discuss extinction of diffusion flames is through the Damköhler number, Eq. (3.13):

$$Da = \frac{\tau_{\text{flow}}}{\tau_{\text{chem}}}. \tag{3.26}$$

The flame extinguishes if chemistry becomes too slow, compared to flow phenomena. Flow phenomena could be related to convection, diffusion/mixing and turbulence, as explained next. The point is that the flame extinguishes if the Damköhler number becomes too low, i.e., for:

$$Da < Da_q. \tag{3.27}$$

The subscript 'q' denotes 'quenching'.

From Eqs. (3.27) and (3.26), the following methods of flame extinction can be understood in the sense that a reduction in Da can lead to flame extinction:

- *Addition of inert gases*: this increases the effective thermal capacity of the atmosphere per unit fuel or per unit oxygen consumed. Consequently, the temperature in the flame region reduces and the chemical time scale (Eq. 3.14) increases. Thus, Da decreases. This is a reduction of Da through physics (i.e., through heat transfer). As mentioned in Section 3.1.1, the higher the specific heat of the inert gas, the more effective it is (explaining why CO_2 is attractive as inerting gas).
- *Addition of chemical inhibitors/suppressants*: the chemistry is slowed down through the chemistry mechanism. Thus, Da (Eq. 3.26) decreases due to the increase in the chemical time scale. This is a reduction of Da through chemistry.
- *Addition of water*: at the level of direct flame extinction, it is favourable to add the water as a mist. The small water droplets, with large 'area to volume' ratio, evaporate. As such they consume much energy (around 2.3 kJ/g latent heat of vaporization at 25°C), reducing the flame temperature and thus reducing Da. Moreover, the water vapour formed occupies much more volume than the liquid water droplets. Due to this expansion, the supply of oxidizer to the flame region is blocked (or at least substantially reduced), which further assists in extinguishing the flame. This point is addressed in more detail in Section 7.6.
- *Blowing out*: this concerns a reduction of Da through physics, although care must be taken concerning chemistry, as explained next.

  Consider first the situation where there is complete combustion in the flame region. In such circumstances, the addition of a forced flow of cold gases will lead to a reduction in temperature due to the removal of heat by the flow of cold gases. This leads to an increase in the chemical time scale and a reduction of Da.

  The forced flow of cold gas also acts at the level of diffusion/mixing, in that it distorts the flame region, such that the thickness of the flame region reduces. The scalar dissipation rate, Eq. (3.16), increases due to steeper gradients, which leads to a lower Da (Eq. 3.22). Heat is also taken away from the flame region by thermal diffusion, so the reduction in temperature in Eq. (3.22) also makes Da decrease. The flame extinguishes if $\chi > \chi_q$, as mentioned in Section 3.3.1.

  Moreover, the forced flow of cold gases can reduce the flow time scale related to convection. In this case, the flow time scale should be interpreted as a 'residence time', i.e., the time the fuel and oxidizer have available for reaction in the region of the flame. Reducing this time to below the chemical time scale means that the reaction becomes incomplete. Thus, this assists in extinguishing the flame.

  The forced flow can also relate to more intense turbulence. On the one hand, this intensifies mixing of fuel and oxidizer (which explains why turbulence is a desired flow feature in combustion devices, because the flames become more compact as the combustion of fuel is completed more quickly), but, on the other hand, it also intensifies the heat transfer away from the flame region due to turbulent thermal diffusion. This reduces the flame temperature and thus Da. Note that this was not mentioned in Section 3.4.1 because, as explained in Section 3.2, in the context of fire-related hazards with pre-mixed flames, turbulence is typically not such that flame extinction would occur. Rather, it would intensify the hazard due to increased reaction area (and perhaps intensified heat transfer, leading to higher turbulent flame speeds).

However, as mentioned, care must be taken concerning chemistry, particularly if the original hazardous situation is such that there is incomplete combustion in the flame region due to lack of oxidizer. In that case, if the forced flow of cold gases contains oxidizer, the primary effect may not be a reduction in temperature for the reasons just given, but an increase in temperature due to more complete combustion. If this is the case, the flame will be strengthened, rather than extinguished (unless the amount of cold gases is so large that the cooling effect dominates the fact that the combustion is more complete, so that the reasoning just given for complete combustion, again holds).

- *Direct removal of heat by a solid structure.* This is physical quenching and happens when the heat losses exceed the heat generated in the reaction zone. The reasoning is similar to what has been described in Section 3.4.1 (but the configuration of Figure 3.7 obviously does not hold for diffusion flames).
- For completeness, also the possibility of extinction through 'radiation weakening' is mentioned [17]. This corresponds to situations of low scalar dissipation rate, but in this case, as a consequence of excessive levels of heat radiation from the flame, the temperature becomes too low for sustained combustion.

3.5 FIRE PLUMES

3.5.1 FREE FIRE PLUMES

The canonical configuration for a free fire plume is a pool fire. Figure 3.8 shows an extremely simplified sketch, on which the fundamental quantities are depicted. For extensive discussions on the dynamics of pool fires, the reader is referred to, e.g., Refs. [2,3]. The focus here is on fluid mechanics aspects. For the sake of the fluid mechanics, it suffices to accept that a certain mass

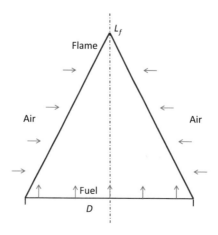

FIGURE 3.8 Simplified representation of a fire plume (circular geometry).

flow rate (i.e., the 'fuel mass loss rate') of fuel leaves the surface of the liquid pool (or solid) fire source, due to evaporation (or pyrolysis) of the liquid (or solid) fuel. The heat required for this evaporation (or pyrolysis) process stems from the flames and hot gases inside the flame envelope, mainly through convection and radiation.

The resulting expression for the total heat release rate, expressed in W or kW or MW, is:

$$\dot{Q} = \chi \dot{m}'' A \Delta H_c. \tag{3.28}$$

In Eq. (3.28), χ $(0 \leq \chi \leq 1)$ denotes the efficiency or completeness of the combustion. $\chi = 1$ corresponds to the theoretical situation of complete combustion, whereas $\chi = 0$ corresponds to no combustion at all. $\dot{m}''$ (in kg/(m²s)) refers to the fuel mass release rate per unit area. This is the amount of fuel released per unit time and per unit area as combustible gas. This is sometimes also called the 'burning rate', but this is not entirely correct, as this suggests burning of the fuel released, which need not necessarily be the case (i.e., fuel can be released without being burnt afterwards in the gas phase). A (in m²) is the fuel area involved, while ΔH_c (in J/kg or kJ/kg or MJ/kg) is the theoretical heat of combustion, i.e., the amount of energy released per kg fuel upon complete combustion (Eq. 2.16). Note that the symbol Δh_c has been replaced by ΔH_c, which is more often encountered in the literature. Values for heat of combustion of various fuels are found in the literature. This is often referred to as either 'lower' or 'upper' heating value. The difference between these two values is the latent heat of condensation of the water vapour, generated by the combustion of the fuel. Thus, in the context of fires, the lower heating value (LHV) should be used, as water vapour is not supposed to condensate during the fire.

As mentioned, the fuel mass release rate per unit area depends on the incident heat flux per unit area $\dot{Q}''_{in} = \dot{q}_{in}$, through a possible combination of conduction, convection and radiation, conduction being typically negligible. Whether convection or radiation is dominant primarily depends on the amount of soot formed in the flames. If more soot is formed, radiation dominates. For example, in alcohols, which are typically non-sooty, convection typically dominates. In any case, this incident heat flux per unit area $\dot{q}_{in}$ is used for the endothermic process of evaporation (of a liquid fuel) or pyrolysis (of a solid fuel), essentially at the fuel surface (see Sections 2.12 and 2.13, resp.). However, $\dot{q}_{in}$ will not be used entirely for this evaporation, resp. pyrolysis process. Indeed, part of $\dot{q}_{in}$ will be re-radiated from the surface, part is lost by in-depth conduction into the solid fuel or by convection into the liquid fuel from the surface, and part is lost by lateral conduction or convection. Therefore, the 'net' incident heat flux per unit area $\dot{q}_{in,net}$ is used for the endothermic process of evaporation, resp. pyrolysis, at the fuel surface. Noting the required energy to evaporate 1 kg of fuel as L_v, the fuel mass release rate per unit area reads:

$$\dot{m}'' = \frac{\dot{q}_{in,net}}{L_v}. \tag{3.29}$$

Note that the description is restricted from now on to liquid fuels. For solid fuels, L_v needs to be replaced by ΔQ_{pyr}, the heat of pyrolysis (see Section 2.14), keeping in mind that this reflects a simplified representation of a much more complex reality for solid fuels.

Equation (3.29) reveals that the larger the value of L_v, the lower the fuel mass release rate per unit area. In an enclosure (see Chapter 5), $\dot{q}_{in,net}$ stems not only from the flame or fire plume, but also from the ceiling, hot walls, hot smoke layer, other objects inside the enclosure, …

Combining Eqs. (3.28) and (3.29) yields:

$$\dot{Q} = \chi \dot{m}'' A \Delta H_c = \chi \dot{q}_{in,net} A \frac{\Delta H_c}{L_v}. \tag{3.30}$$

This expression reveals the most important factors affecting the total heat release rate:

- *The completeness of combustion* χ: this will strongly depend on the turbulent mixing of combustible gases and oxygen. In enclosures (see Chapter 5), reduced ventilation can reduce χ significantly.
- *The net incident heat flux per unit area* $\dot{q}_{in,net}$: in enclosures (see Chapter 5), $\dot{q}_{in,net}$ not only stems from the flame or fire plume, but also from the ceiling, hot walls, hot smoke layer, other objects inside the enclosure, ...
- *The fire area* A: in enclosures (see Chapter 5), the fire area is a very important parameter, determining the fire development.
- The ratio $\Delta H_c / L_v$ called the 'combustibility ratio' in Ref. [3]. This ratio expresses how much heat is released per unit fuel combusted, compared to the amount of heat required to turn 1 kg of fuel into the combustible gaseous phase. Fuels with high values of $\Delta H_c / L_v$ lead to more severe fire development, particularly in enclosures, than fuels with low $\Delta H_c / L_v$ values.

It is noted that for a gaseous fuel, expression (3.28) can also be used, but then $\dot{m}''$ is simply determined by the supply of gaseous fuel.

The flame configuration is clearly of the diffusion type: inside the flame envelope, the conditions are fuel-rich, while outside the conditions are lean. The gaseous fuel burns in the flame region, as sketched in Figure 3.6, which is positioned where the mixture fraction is close to stoichiometric conditions. A stable situation, with steady conditions, can be obtained.

Figure 3.8 is extremely simplified in the sense that in reality, the flow will be turbulent. This can be seen through the Reynolds number, Eq. (2.48). Indeed, for velocities in the order of 1 m/s and a characteristic length scale L of 1 m, the Reynolds number becomes Re = 10,000, using $\nu = 10^{-4}$ m²/s as the order of magnitude for the kinematic viscosity of air around 900°C. Another simplification is that straight lines are drawn, so the flame shape resembles a cone. Yet, these approximations are satisfactory for the first global discussion.

It is important to appreciate that the main flow features are not directly a consequence of the combustion phenomenon in the flames itself. Rather, it is the resulting density difference that governs the main flow phenomena. Therefore, the (strong) thermal plume is used as canonical configuration in the next sections, so that heat transfer is effectively avoided in the discussion. Only buoyancy-dominated situations are considered. The possible effect of wind is neglected first.

3.5.1.1 Average Flame Height

Consider the situation of Figure 3.8. The density inside the 'flame envelope' is assumed to be much lower than ambient density (which is equivalent to the assumption of much higher temperature inside the flame envelope than ambient temperature). Recall that only buoyancy-dominated situations are considered. This means that the Froude number, Eq. (2.93) or Eq. (2.94), is low.

Consider now the Navier–Stokes equations, Eq. (2.38), in steady-state conditions, with z as vertical direction and assuming viscous effects can be ignored (which is equivalent to the assumption of a sufficiently high Reynolds number):

$$\frac{\partial}{\partial t}(\rho v_1) + \rho v_1 \frac{\partial v_1}{\partial x_1} + \rho v_2 \frac{\partial v_1}{\partial x_2} + \rho v_3 \frac{\partial v_1}{\partial x_3} = -\frac{\partial p}{\partial x_1},$$

$$\frac{\partial}{\partial t}(\rho v_2) + \rho v_1 \frac{\partial v_2}{\partial x_1} + \rho v_2 \frac{\partial v_2}{\partial x_2} + \rho v_3 \frac{\partial v_2}{\partial x_3} = -\frac{\partial p}{\partial x_2},$$

$$\frac{\partial}{\partial t}(\rho v_3) + \rho v_1 \frac{\partial v_3}{\partial x_1} + \rho v_2 \frac{\partial v_3}{\partial x_2} + \rho v_3 \frac{\partial v_3}{\partial x_3} = -\frac{\partial p}{\partial x_3} - \rho g. \tag{3.31}$$

First, an order of magnitude analysis is performed. For gradients in the horizontal direction, D can be used as characteristic length scale. For gradients in the vertical direction, the (yet to be defined) average flame height L_f can be used. The main flow in the configuration of Figure 3.8 is upwards, due to buoyancy (see also Section 2.6): the low-density gas wants to rise into the higher-density ambient air. This implies that $v_3 \gg v_1 \approx v_2$, i.e., the vertical velocity component is much larger than the horizontal velocity components. The horizontal velocity refers to entrainment of ambient air into the low-density envelope.

Given that the horizontal velocity components are so small, the first two equations of Eq. (3.31) illustrate that pressure gradients in the horizontal direction are small. This implies that to a good approximation, pressure can be assumed constant in horizontal planes. This implies that throughout the flow field, the law of hydrostatics, Eq. (2.81), can be used (because this law is valid sufficiently far away from the envelope):

$$\frac{dp}{dz} \left(= \frac{dp}{dx_3} \right) = -\rho_{amb} g. \tag{3.32}$$

Using Eq. (3.32) and D as characteristic length scale in the third equation of Eq. (3.31), leads to:

$$\rho \frac{v_3^2}{D} \propto (\rho_{amb} - \rho) g. \tag{3.33}$$

This is not surprising, as this results from the key assumption that the flow is buoyancy dominated. The important result is that:

$$v_3 = v_z = \sqrt{\frac{\rho_{amb} - \rho}{\rho} g D} \propto \sqrt{D}. \tag{3.34}$$

For a circular pool, the fuel mass loss rate can now be reformulated as the product of density, velocity and surface area: $\dot{m}_F \propto \rho \sqrt{D} D^2$. Note that, through the heat of combustion, the fuel mass flow rate is proportional to the total heat release rate, Eq. (3.34), so that: $\dot{Q} = \chi \dot{m}_F \Delta H_c \propto \rho \sqrt{D} D^2$. Also note that, for the sake of simplification, a constant density ρ, or, equivalently, uniform flame temperature, is assumed inside the cone. In reality, this will obviously not exactly be the case, although the approximation is not too bad to first order. This point is discussed again in Section 3.5.1.2. Typical values for average fire plume temperatures would be in the order of 1,000–1,200 K.

Assume now that the flame height is determined as the position where stoichiometric mixture fraction (see Section 2.3.5) is obtained. The mixing happens through entrainment (see Figure 3.8). The area available for entrainment corresponds to the side area of the cone, which equals $\pi \dfrac{D}{2} \sqrt{L_f^2 + \dfrac{D^2}{4}}$. Thus, assuming that the horizontal velocity components are proportional to the vertical velocity component (which is proportional to the square root of the fire diameter, Eq. 3.34), the mass flow rate of entrained air can be approximated as: $\dot{m}_{entr} \propto \rho_{amb} \sqrt{D} D \sqrt{L_f^2 + \dfrac{D^2}{4}}$. Stated in another manner, the area $\pi \dfrac{D}{2} \sqrt{L_f^2 + \dfrac{D^2}{4}}$ is required to get to stoichiometric mixture fraction, given a fuel mass flow rate $\dot{m}_F \propto \rho \sqrt{D} D^2$. Clearly, there should be a proportionality $\dot{m}_F \propto \dot{m}_{entr}$ (or, equivalently, $Q \propto \dot{m}_{entr}$). This is fulfilled if $L_f \propto D$, which yields (with $\dot{m}_F \propto \rho \sqrt{L_f D^2}$):

$$L_f \propto \dot{m}_F^{2/5} \propto \dot{Q}^{2/5}. \tag{3.35}$$

TABLE 3.2

Average Flame Height, Calculated from Correlation (3.36), at 1,01,325 Pa and 15°C

D(m)/HRR(kw)	50	100	250	500	1,000	2,000	5,000	10,000
0.25	0.8	1.2	1.8	2.4	3.3	4.4	6.5	8.7
0.5	0.6	0.9	1.5	2.2	3.0	4.2	6.1	8.4
1	0.1	0.4	1.0	1.7	2.5	3.7	5.7	7.9
1.5	−0.5	−0.1	0.5	1.2	2.0	3.2	5.2	7.4
2	−1.0	−0.6	0.0	0.7	1.5	2.7	4.7	6.9
4	−3.0	−2.7	−2.0	−1.4	−0.5	0.6	2.7	4.8
6	−5.0	−4.7	−4.1	−3.4	−2.6	−1.4	0.6	2.8
10	−9.1	−8.8	−8.2	−7.5	−6.6	−5.5	−3.4	−1.3

Values marked in grey correspond to unlikely conditions (heat release rate per unit area less than 50kW/m² or more than 1 MW/m²).

Note that, in reality, much more air is entrained into the flame region than what is theoretically required to burn all the fuel. However, the assumption of proportionality is reasonable. Indeed, whereas this is an order of magnitude analysis, this exponent is recovered in 'Heskestad's correlation' [18]:

$$\frac{L_f}{D} = -1.02 + 3.7 \dot{Q}^{*2/5}. \tag{3.36}$$

In Eq. (3.36), $\dot{Q}^*$ is the non-dimensional total heat release rate (HRR):

$$\dot{Q}^* = \frac{\dot{Q}}{\rho_{amb} c_p T_{amb} \sqrt{gD} D^2}. \tag{3.37}$$

The total HRR has been defined in Eq. (3.28), with $\chi = 1$ when used in Eq. (3.37) [18]. Table 3.2 provides some values at 1,01,325 Pa and 15°C.

Note that the exponent '2/5' in Eq. (3.35) is indeed recovered in fire plumes (Eqs. 3.36 and 3.38). Expression (3.36) covers the entire range of $\dot{Q}^*$ values, except for the momentum-driven regime [18]. Note that Eq. (3.36), combined with Eq. (3.37), reads.

$$L_f = -1.02D + 3.7 \left(\frac{1}{\rho_{amb} c_p T_{amb} \sqrt{g}} \right)^{2/5} \dot{Q}^{2/5}. \tag{3.38}$$

Equation (3.38) reveals that as long as $\dot{Q} \propto D^{5/2}$ (see above), the flame height is proportional to the fire diameter.

The same reasoning can now be built for a rectangular configuration, as sketched in Figure 3.9. The diameter D is now replaced by the 'hydraulic diameter' D_h, which is defined as follows:

$$D_h = \frac{4LW}{2(L+W)}. \tag{3.39}$$

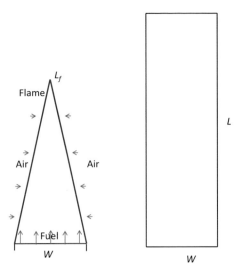

FIGURE 3.9 Simplified representation of a fire plume (rectangular geometry, with $W \ll L$). W and L are horizontal dimensions, and L_f is a vertical dimension.

Consider now the situation where $W \ll L$. This corresponds to a line plume of length L. Then the hydraulic diameter becomes $D_h = \dfrac{4LW}{2(L+W)} \approx 2W$. With Eq. (3.34) still in place, the fuel mass loss rate now becomes: $\dot{m}_F \propto \rho\sqrt{D_h}\,LW$. Entrainment now essentially occurs from both sides in Figure 3.9 (rather than circumferentially, as in Figure 3.8), so the entrainment mass flow rate becomes: $\dot{m}_{\text{entr}} \propto \rho_{\text{amb}}\sqrt{D_h}\,2L\sqrt{L_f^2 + \dfrac{D_h^2}{16}} \approx 2\rho_{\text{amb}}\sqrt{D_h}\,LL_f$. Again the proportionality $\dot{m}_F \propto \dot{m}_{\text{entr}}$ is fulfilled if $L_f \propto W \propto D_h$, and in this case, this results in (through $\dfrac{\dot{m}_F}{L} \propto \dfrac{\dot{m}_{\text{entr}}}{L} \propto \sqrt{D_h}\,L_f \propto L_f^{3/2}$):

$$L_f \propto \left(\frac{\dot{m}_F}{L}\right)^{2/3}. \tag{3.40}$$

This result is in line with Ref. [19], revealing the exponent '2/3' in the correlations for flame height with HRR per m (for a line source):

$$\frac{L_f}{W} = 3.64\dot{Q}^{*2/3}. \tag{3.41}$$

In Eq. (3.41), $\dot{Q}^*$ is defined as follows:

$$\dot{Q}^* = \frac{\dot{Q}}{\rho_{\text{amb}}c_p T_{\text{amb}}\sqrt{gW}\,WL}. \tag{3.42}$$

Note that nowhere in the discussion above, use was made of the concept 'combustion'. The only assumption was that the density inside the 'flame' envelope is lower than ambient density and that the flow is buoyancy dominated. This illustrates that physics is dominant, not chemistry, in the discussion of flame heights in fire plumes. Yet, the HRR is an important variable, in that it serves as input variable, determining the dynamics of the fire plume (and the flame height), as explained above.

Exercise:

Consider a circular $20\,\mathrm{m^2}$ pool fire. The fuel, hexane, is burning at a regression rate of 7 mm/min (i.e., the fuel surface goes down due to evaporation at a rate of 7 mm/min). Determine the average flame height of a free fire plume in the absence of wind, for ambient temperature equal to 15°C, if the pool is circular. (Answer: 14.2 m)

Repeat the exercise for ambient temperature equal to 35°C. (Answer: 14.2 m)

Repeat the exercise if the pool is rectangular, with one side equal to 20 m and one side equal to 1 m, for ambient temperature equal to 15°C, and assuming the same regression rate. (Answer: 7.8 m)

Compare the result to the flame height for circular pool. Explain the difference. (Answer: The flame height is lower for the rectangle. The reason is the difference in surface area for entrainment; the perimeter of the circle is 15.8 m, while the perimeter of the rectangle with the same area is 42 m)

3.5.1.2 Temperature Evolution

In order to provide an estimate for the temperature in the fire plume, the derivation as presented in Ref. [2] is adopted. The following simplifying assumptions are required here:

- The fire is idealized as a point source, with HRR $\dot{Q}$ and radiative fraction (i.e., the fraction of the total HRR that is radiated to the surroundings, and thus not given as convective driving force to the plume) χ_r;
- The ambient is at uniform temperature T_{amb};
- Properties are considered uniform at any height z; note that such 'top hat' profiles are less close to reality than Gaussian profiles;
- All material/fluid properties are constant;
- Combustion takes place uniformly throughout the plume, as long as there is fuel.

The last assumption cannot reflect reality, as can be understood from flame structures as presented in Sections 3.2 and 3.3. Yet, for the sake of simplicity, a uniform volumetric heat source is assumed here.

Consider now an infinitesimal element dz at height z and assume that the plume radius at that height z equals b. Using $w\,(=v_3=v_z)\,w$ as notation for the upward velocity in the plume at height z (recall that w is supposed to be uniform) and assuming that the radially inward velocity of entrained air into the plume is proportional to the upward velocity (i.e., $v_{entr}=\alpha w$, where α is the entrainment constant), the mass flow rate of air entrained into the element through the side surface of the cylinder (with radius b and height dz) reads:

$$d\dot{m}_{entr}=\rho_{amb}\alpha w 2\pi b dz. \tag{3.43}$$

If S denotes the stoichiometric mass ratio of air to fuel and assuming that all the entrained air reacts with fuel, the following HRR can be calculated in the element considered, assuming complete combustion (i.e., $\chi=1$ in Eq. 3.28):

$$d\dot{Q}=\rho_{amb}\alpha w 2\pi b dz \frac{\Delta H_c}{S}. \tag{3.44}$$

with ΔH_c the heat of combustion of the fuel (Section 2.2.2.1).

Knowing that the upward mass flow rate in the plume is given by $\dot{m} = \rho\pi b^2 w$, expressing that only the convective part $(1-\chi_r)$ of the heat released by combustion (Eq. 3.44) is given to the plume as a temperature rise, compared to ambient, leads to:

$$\frac{d}{dz}\left(\rho\pi b^2 w c_p\left(T - T_{amb}\right)\right) = \rho_{amb}\alpha w 2\pi b\frac{\left(1-\chi_r\right)\Delta H_c}{S}. \tag{3.45}$$

Furthermore, conservation of mass yields:

$$\frac{d}{dz}\left(\rho\pi b^2 w\right) = \rho_{amb}\alpha w 2\pi b. \tag{3.46}$$

Using this in Eq. (3.45) yields: $T - T_{amb} = \dfrac{\left(1-\chi_r\right)\Delta H_c}{c_p S}$. However, one assumption made is too far off reality for this expression to be a reasonable estimate, namely that all the entrained air would react completely. Due to the slowness of turbulent mixing, there is a high level of 'unmixedness', i.e., air that is entrained into the plume, but does not mix to react [2]. In Ref. [2], it is suggested that the following expression provides a 'spatial average flame temperature':

$$T - T_{amb} = \frac{\left(1-\chi_r\right)\Delta H_c}{n c_p S}. \tag{3.47}$$

In Eq. (3.47), n is the ratio of the amount of air entrained by the plume to the amount of air that actually reacts with the fuel. The value for n is mentioned to be around 10 in Ref. [2]. However, note that this leads to relatively low temperatures. Indeed, for many fuels, the ratio $\Delta H_c/S \approx 3$ MJ/kg. Thus, even in the absence of radiation, i.e., $\chi_r = 0$, and with $c_p = 1$ kJ/(kg K) (although the actual value will probably be higher inside the plume, depending on the mixture), Eq. (3.47) yields: $T - T_{amb} = 3000/n$ K. Thus, the value $n = 10$ leads to a 'spatial average flame temperature' equal to about 600 K, which is low. The reason is that the additional entrained air does not mix completely with the combustion products inside the fire plume, so that the 'spatial average flame temperature' does not have a strong physical meaning.

An interesting result of Eq. (3.47) is that the fire plume temperature is approximately constant in the fire plume. In other words, the temperature does not vary substantially with height, as long as fuel reacts. The actual value of the temperature is also shown to primarily depend on the radiative fraction of the HRR. Expression (3.47) indeed shows that the higher the radiative fraction, i.e., the sootier the flame (Section 8.7), the lower the temperature in the fire plume.

Exercise:

Consider a fire plume of propane in air, with the ambient temperature equal to 25°C. Calculate the plume temperature for unmixedness levels 'n' equal to 2, 4, 6 and 10, assuming that 10% of the HRR is emitted as radiation. (Answer: 1,375°C; 700°C; 475°C; 295°C)

3.5.1.3 Kelvin–Helmholtz Instability

In Section 3.5.1.1, only a global analysis has been presented. In real fire plumes (as well as in strong thermal or variable density plumes), often turbulence is involved, as explained above. Turbulence is triggered by instabilities. Typical for fire plumes is the fact that these instabilities are generated by density differences and by the resulting flow field.

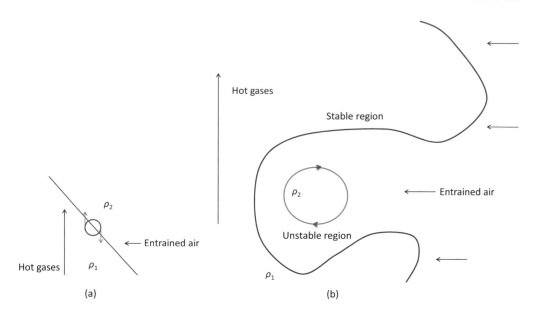

FIGURE 3.10 Sketch of: zoom in on the flame region for a situation as sketched in Figure 3.8 (a), indicating the differences in velocity and density; sketch of vortex roll-up with unstable and stable regions (b). Density $\rho_1 < \rho_2$.

Initially, Rayleigh–Taylor instabilities can be formed. Such finger-like instabilities occur if a heavier fluid is on top of a lighter fluid. In such a situation, due to buoyancy the lighter fluid wants to rise and the heavier fluid flows downwards.

However, in steady situation, which is typically of interest when fire plumes are considered, the Kelvin–Helmholtz instabilities are more important. Kelvin–Helmholtz instabilities occur when there are local differences in velocity in the flow. The reader is referred to dedicated textbooks on turbulence for advanced stability analyses. However, the main phenomena can be explained on the basis of Figure 3.10, as the density difference will provide a substantial contribution in the turbulent flow dynamics of fire plumes.

The global motion is an upward motion of the low-density gas (labelled 'hot gases' in Figure 3.10, as this is the typical case for fire plumes) and an inward (and upward) motion of entrained air, as explained in Section 3.5.1.1. Thus, there is clear velocity shear near the fire envelope. Moreover, due to turbulence, there are fluctuations in velocities. These fluctuations can grow to form larger-scale structures (Kelvin–Helmholtz instability). This process is strengthened by the density differences, as sketched in Figure 3.10. Starting from the sketch on the left, which is a zoom of the situation sketched in Figure 3.8, it is clear that locally higher-density fluid is on top of lower-density fluid. Thus, superposed on the global upward and inward motion, local buoyancy effects are such that the higher-density fluid wants to make a relative downward motion, whereas the lower-density fluid wants to make an additional upward motion. This causes a vortex (clockwise in Figure 3.10) that deforms the flame region. This is sketched at the right hand side in Figure 3.10. These large-scale vertical structures are typical for fire plumes. There are 'unstable' regions, where higher-density fluid is on top of lower-density fluid, and there are 'stable' regions where the opposite is true.

Obviously, the situation is not 'steady': there is a global upward motion, during which a number of vortex roll-ups take place before the flame tip is reached. In Ref. [20], the puffing mechanism is explained as follows. A toroidal vortex is formed at the flame envelope, as explained above, with strong enough circulation to affect the entrainment velocity in the contracting flame region (with an increase in entrainment velocity as the vorticity increases). This squeezes the inner upward flowing hot gases. In its upward convection motion, the vortex quickly loses strength and the flame region opens outwards. The hot flame vortex moves further upwards, affected by the global upward flow motion and the buoyancy force in the cold ambient air (Figure 3.10). This process repeats itself continuously. Clearly this phenomenon strongly affects the entrainment process. Note that the only argument used so far is the presence of buoyancy, so there is no direct impact of combustion.

The puffing phenomenon creates a so-called 'puffing' frequency. Again, this behaviour is not related to combustion directly. It is merely the density difference that is the main driving force, i.e., physics is governing the flow and mixing field, not chemistry. This is reported in, e.g., Ref. [20], concluding from experiments in isothermal helium plumes that 'the puffing phenomenon is associated with the instability of the buoyant plume'. Consequently, correlations for the puffing frequency are also found outside the fire literature, in the context of (strong) thermal plumes or, e.g., helium plumes. The non-dimensional number, related to these frequencies, is the so-called *Strouhal* number:

$$St = \frac{fD}{\sqrt{gD}}. \tag{3.48}$$

In Eq. (3.48), the frequency f is expressed in Hz (s^{-1}).

It turns out that for fires, the Strouhal number is around $St \approx 0.5$ (see, e.g., [2,3]). In other words, the puffing frequency is inversely proportional to the square root of the fire diameter.

It must be noted that there is strong amplification of puffing in terms of amplitude in a fire plume, compared to a non-reacting buoyant plume [20]. Indeed, the buoyancy force is diminished very rapidly due to the intense turbulent mixing in a non-reacting plume, whereas the local HRR, due to combustion, maintains the buoyancy force in a fire plume. At this point, combustion does play an important role.

This puffing phenomenon also causes the flame height not to be constant in time, but fluctuating between minimum ($L_{f,\min}$) and maximum ($L_{f,\max}$) values. In fact, the average flame height is relatively loosely defined as the height on the fire plume axis where half of the time a flame is 'seen'. The reader is referred to, e.g., Refs. [2,3] for more details.

Figure 3.11 provides an illustration of a fire plume, with rectangular fuel pool. The plume is slightly tilted due to forced mechanical horizontal ventilation, but this is not an essential feature of the figure (see also Section 3.5.1.4). The fuel is hexane. The Kelvin–Helmholtz induced vortex structures, as sketched in Figure 3.10, are readily recognized (some are indicated through white arrows).

Upon closer examination, it can be seen in the bottom left corner of the fire plume that there are regions where no flame is observed. This is even more so towards the centre of the plume, close to the fuel surface. The reason is that the mixture there is too rich, i.e., above the UFL. Indeed, Figure 3.8 illustrates that pure gaseous fuel (mixture fraction equal to 1) evaporates from the surface and mixing with entrained air is required for a flammable mixture to be formed. The existence of a relatively cool fuel-rich region on top of the fuel surface, may effectively block heat transfer (particularly by radiation) from the flames, if that fuel-rich vapour absorbs part of the radiation. Consequently, this can affect the HRR. The reader is referred to, e.g., Ref. [3] for a more detailed discussion. For the sake of the present section, focusing on fluid mechanics aspects, the HRR is merely considered an input variable for the fire plume flow dynamics, as discussed above.

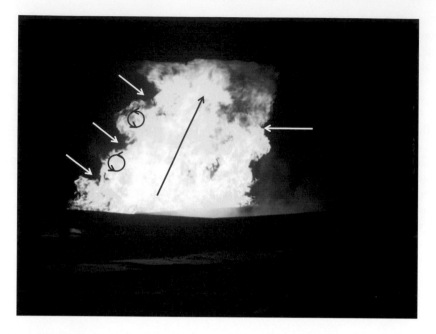

FIGURE 3.11 Picture of a pool fire (rectangular pool, hexane fuel). The white arrows indicate 'unstable' situations as sketched in Figure 3.10. The red arrows indicate vertical motion and the global upward motion. The fire plume is slightly tilted due to forced ventilation.

Exercise:

Consider the situation as for the exercise in Section 3.5.1.1. Determine the puffing frequency. (Answer: 0.7 Hz; rectangle (using the hydraulic diameter): 1.1 Hz)

3.5.1.4 The Effect of Wind

The effect of wind is complex and multifold. Not only is the axisymmetry of Figure 3.8 disturbed, the effect of wind also affects the turbulent entrainment process as sketched in Figure 3.12. From an average point of view, cross sections of the fire plume in horizontal planes will no longer be circular, either, which affects the area available for entrainment as well. In general, wind can be expected to reduce the flame length, due to enhanced entrainment of fresh air, and to tilt the plume axis over an angle θ from the vertical axis. Obviously, the effect of wind will depend on the wind velocity, compared to a characteristic velocity in the fire plume. Actually, to be more precise, it depends on the sideward momentum of the wind flow through the fire plume, compared to the upward momentum of the fire plume itself. Figure 3.12 provides a sketch.

There are, thus, two effects: on the one hand, a reduction in L_f, compared to quiescent conditions, and on the other hand a tilting of the fire plume, resulting in an additional reduction in vertical height (through multiplication with the cosine of the angle of deviation from the vertical axis).

The topic is briefly touched upon in Refs. [3, 21], though. The data presented in those references reveal much scatter, indicating the complexity of the problem and the difficulties of obtaining accurate and repeatable experimental data. The following correlation, proposed by the American Gas Association, provides an expression for the cosine of the angle θ:

$$v^* \leq 1 : \cos\theta = 1; \; v^* > 1 : \cos\theta = v^{*-1/2}. \tag{3.49}$$

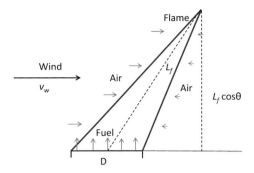

FIGURE 3.12 Simplified representation of a fire plume (circular geometry) with the effect of wind.

In Eq. (3.49), the non-dimensional quantity v^* is the ratio of the wind velocity (measured at a height of 1.6 m [22]) to a 'characteristic plume velocity', defined as follows:

$$v^* = \frac{v_w}{v_c} = \frac{v_w}{\left(g \dot{m}_F'' D \rho_{\text{amb}}^{-1} \right)^{1/3}}.$$ (3.50)

In Eq. (3.50), $\dot{m}_F''$ is the fuel mass loss rate per unit area (kg/m²s).

The definition of v_c seems somewhat arbitrary and solely based on dimensional analysis. Indeed, from a physics point of view, the buoyancy-induced velocity, which is related to the convective part of the total HRR, seems a more natural quantity to compare to the wind velocity. Particularly at a height of 1.6 m, where the wind velocity is measured, the upward plume velocity can be ten times higher than the velocity at the fuel surface (which relates to v_c). This may also explain to some extent the scatter observed in the data, because for the same value of $\dot{m}_F''$, strongly different upward plume velocities can be obtained, depending on the fuel, through ΔH_c and χ_r. Both factors will affect the driving force for the upward plume velocity, as explained in Section 3.5.1.2. Moreover, from a fluid mechanics point of view, the tilting of the plume is a consequence of balances in momentum, not due to a ratio of velocities. More recent research on this topic indeed takes this into account, leading to updated correlations, also for the variation in L_f due to wind (e.g., [23]).

3.5.1.5 Transition from Buoyancy-Driven to Momentum-Driven Jets

So far, in Section 3.5.1, the implicit assumption was that the fire plume is a buoyancy-driven flow. This is true for most fire situations. However, this will depend on the Froude number (Eq. 2.93):

$$Fr = \frac{\rho v^2}{\Delta \rho g L}$$ (3.51)

In Eq. (3.51), v is a characteristic velocity (e.g., the upward vertical velocity near the fire source), ρ is the characteristic hot gas mass density in the fire plume and $\Delta \rho$ is the density difference between the characteristic hot gas mass density in the fire plume and the mass density of ambient air. Only for sufficiently small values of Fr, the flow is buoyancy-driven. In the fire literature, typically Eq. (2.94) is used for the Froude number, using the (hydraulic) diameter of the fire source as characteristic length scale:

$$Fr = \frac{v^2}{g D_h}.$$ (3.52)

An order of magnitude analysis reveals that Froude numbers for typical fire configurations are indeed low. Velocities are in the order of a few metres per second, and diameters are in the order of 1 m, so with g about $10\,\mathrm{m/s^2}$, the Froude number is in the order of unity (or less). In jet flames, on the other hand, velocities can go up to $100\,\mathrm{m/s}$ (or more) and nozzles can be as small as $5\,\mathrm{mm}$ (or less), so that Froude numbers can be 1,00,000 times higher.

An important consequence is that the flame length of a turbulent momentum-driven jet flame no longer increases with fuel flow rate (in contrast to Eq. 3.35). It is merely determined by the nozzle diameter and the fuel [24].

This can be understood, repeating the reasoning of Section 3.5.1.1, but now with the following assumptions: the fuel velocity v_F is high and independent of the nozzle diameter D, and the flame length L_f is much larger than D. The flame height is again determined by the position where stoichiometric mixture fraction is obtained (at the tip of the cone, sketched in Figure 3.8). The fuel mass flow rate $\dot{m}_F \propto \rho_F v_F D^2$. The entrainment velocity is assumed to be proportional to the vertical velocity (which is determined now by the momentum jet velocity v_F, not by buoyancy). The area for entrainment of air into the fire cone (Figure 3.8), $\pi \dfrac{D}{2}\sqrt{L_f^2 + \dfrac{D^2}{4}}$, can be approximated as $\pi L_f \dfrac{D}{2}$ (using the fact that $L_f \gg D$). Thus, the entrainment mass flow rate is proportional to $\dot{m}_{\mathrm{entr}} \propto \rho_{\mathrm{amb}} v_F D L_f$. Expressing now that $\dot{m}_F$ must be proportional to $\dot{m}_{\mathrm{entr}}$ to determine the position of stoichiometric mixture fraction, leads to:

$$L_f \propto D. \tag{3.53}$$

Indeed, the flame height is directly proportional to the nozzle diameter. The proportionality factor is determined by the fuel (through its mass density and its chemical composition, i.e., the stoichiometric fuel/oxygen ratio). A correlation is provided in Section 3.5.1.6.

3.5.1.6   Correlations

In Ref. [25], correlations are provided for flame lengths, considering buoyancy-driven and momentum-driven flames, where turbulence is primarily generated outside the fuel nozzle, by buoyancy or shear. This is typical for fire plume configurations.

Using the notation of [16] for the stoichiometric mixture fraction f_{st}, defined as:

$$f_{st} = \frac{1}{S+1}, \tag{3.54}$$

with S the air to fuel mass ratio at stoichiometric conditions, the following 'flame Froude number' is defined:

$$\mathrm{Fr}_f = \frac{v_F}{\sqrt{gD}}\, f_{st}^{3/2} \left(\frac{\rho_{\mathrm{amb}}}{\rho_F}\right)^{1/4} \left(\frac{\Delta H_c\left(\chi - \chi_r\right)}{c_p\left(1+S\right)T_{\mathrm{amb}}}\right)^{-1/2}. \tag{3.55}$$

In Eq. (3.55), v_F is the fuel exit velocity at the nozzle (with diameter D), ρ_F is the fuel density at the nozzle exit, ΔH_c is the heat of combustion of the fuel, χ is the combustion efficiency/completeness, and χ_r is the radiative fraction of the HRR. In Ref. [16], the final factor is rewritten, using the 'characteristic temperature rise resulting from combustion' ΔT_f:

$$\mathrm{Fr}_f = \frac{v_F}{\sqrt{gD}}\, f_{st}^{3/2} \left(\frac{\rho_{\mathrm{amb}}}{\rho_F}\right)^{1/4} \left(\frac{\Delta T_f}{T_{\mathrm{amb}}}\right)^{-1/2}. \tag{3.56}$$

Using the flame Froude number, the following correlation for flame length is provided in Ref. [25]:

$$L_f = 13.5 \left(1+S\right) D \sqrt{\frac{\rho_F}{\rho_{\text{amb}}}} \frac{\text{Fr}_f^{2/5}}{\left(1+0.07\text{Fr}_f^2\right)^{1/5}}. \tag{3.57}$$

It is instructive to plot the final factor of Eq. (3.57). It is observed then that this factor becomes constant for $\text{Fr}_f > 20$ and the deviation from the constant value becomes $<10\%$ for $\text{Fr}_f > 5$. This explains why in Ref. [16] Eq. (3.57) is reformulated as:

$$L_f = 13.5 \left(1+S\right) D \sqrt{\frac{\rho_F}{\rho_{\text{amb}}}} \frac{\text{Fr}_f^{2/5}}{\left(1+0.07\text{Fr}_f^2\right)^{1/5}}, \text{Fr}_f < 5; \; L_f = 23 \left(1+S\right) D \sqrt{\frac{\rho_F}{\rho_{\text{amb}}}}, \text{Fr}_f \geq 5. \tag{3.58}$$

Equation (3.58) clearly indicates the distinction between configurations where buoyancy is negligible ($\text{Fr}_f > 5$) and configurations where this is not the case.

Also note that in Eq. (3.58), the exponent 2/5 is recovered for small values of Fr_f (i.e., for buoyancy-dominated plumes, Section 3.5.1.1), while the proportionality to the fire/nozzle diameter D (Eq. 3.53) is recovered for momentum-driven flows (large-enough values of Fr_f).

3.5.2 Interaction with Non-Combustible Walls

As repeatedly explained above, air entrainment is the crucial process, determining the flame height for a given fire size (in terms of HRR and geometrical dimensions). For obvious reasons, the presence of a wall near the fire can disturb the entrainment process.

For circular fires, however, the effect turns out to be small. There are, indeed, counteracting effects. On the one hand, there is blockage of entrainment near the wall. This effect is assumed to cause an increase in flame height. However, due to the reduced entrainment at one side, the fire plume is no longer circular in horizontal planes. Indeed, the plume is tilted towards the wall (but typically the flames do not attach to the wall) and stretched into an ellipse-like shape in horizontal planes, thereby increasing the area for entrainment into the fire plume. Both effects essentially cancel out each other.

For a line fire against a non-combustible wall, correlation Eq. (3.41) is confirmed, but with almost doubled coefficient [26,27]:

$$\frac{L_f}{W} = 6\dot{Q}^{*2/3}. \tag{3.59}$$

This is not surprising, as the reasoning held in Section 3.5.1.1, based on Figure 3.9, still holds, but entrainment is only possible from one side (namely the side that is not blocked by the wall) as the fire plume attaches to the wall.

For a square burner in a non-combustible corner, an exponent ½ is reported in Ref. [28]:

$$\frac{L_f}{D} = C\dot{Q}^{*1/2}. \tag{3.60}$$

Interestingly, this exponent is in between 2/5 (Eq. 3.36) and 2/3 (Eq. 3.59). This need not be surprising, as indeed the configuration can be interpreted as something in between the configurations of Figures 3.8 and 3.9. The exponent ½ is also reported in Ref. [28] for an L-shaped line burner, set flush with the corner. Yet, the importance of the coefficient C in Eq. (3.60) must not be underestimated. It is set to $C = 3.9$ for $L_{f,\min}$ (i.e., the top of the 'continuous flame region') for the square burner ($C = 4.3$ for the L-shaped line burner) and to $C = 5.9$ for $L_{f,\max}$ (i.e., the 'flame tip') for both the square burner and the L-shaped line burner. Indeed, the data in Figure 27 of Ref. [28] reveals that earlier developed correlations (refs. 4 and 14 in Ref. [28]), containing the exponent 2/3 (i.e., like Eq. 3.59), would also envelope the data by simply increasing the coefficient.

3.5.3 Interaction with Non-Combustible Ceiling

Interaction of flames with a ceiling is important, as the flow and flames underneath the ceiling will strongly affect the activation of detection and/or sprinkler systems and will affect possible ignition and flame spread underneath the ceiling. In Ref. [29], an overview is presented of available data, focusing on heat fluxes and flame lengths. It is instructive, though, to first discuss fluid mechanics aspects involved.

The global flow configuration concerns a vertically upward flow, impinging onto a ceiling and then turning into a horizontal flow underneath the ceiling, as described in Section 2.12. The driving force is the convective part of the fire HRR.

There are two important aspects. First of all, as the assumption here is that the fire plume is interacting with the ceiling (and not the buoyant non-reacting smoke plume, see Section 4.5 for that), the ceiling is supposed to be relatively close to the fire source. In other words, L_f is supposed to be higher than (or at least similar to) the distance H between the fire and the ceiling. Consequently, the upward flow impinges relatively strongly onto the ceiling, and no fully developed buoyant plume has developed by the time the (fire) plume impinges onto the ceiling. In such circumstance, there is strong turbulence generation in the impingement and turning region (i.e., the region where the flow turns from a vertically upward motion into a horizontal motion underneath the ceiling), as discussed for momentum-driven flows in, e.g., Ref. [30]. This strong increase in turbulence causes intense entrainment of air.

The second important aspect concerns the fact that entrainment of air into the horizontal flow of hot gases underneath the ceiling is far less effective than the entrainment of air into the vertically rising fire plume. Indeed, buoyancy keeps the hot gases up, floating in a stable manner on the more dense ambient air. This is quantified through the Richardson number:

$$Ri = \frac{\left(\rho_{\mathrm{amb}} - \rho_g\right)gh}{\rho_g v_g^2}.$$ (3.61)

The Richardson number is the ratio of the buoyancy force to the kinetic energy of the flow of hot gases. In Eq. (3.61), ρ_{amb} refers to the density of ambient air, ρ_g refers to the density of the hot gases underneath the ceiling, h is the thickness of the layer of hot gases and v_g is the characteristic velocity of the hot gases underneath the ceiling. In case of high values of the Richardson number, turbulent mixing is suppressed.

These two aspects form the basis for the explanation of the strong difference between situations where the flame height L_f in the absence of a ceiling is not much higher than the distance H between the fire and the ceiling on the one hand, and situations where $L_f \gg H$. In the former case, most of the fuel has already been consumed by the time the ceiling is reached, so the conditions in the impingement region are relatively lean. The remainder of the fuel is mainly burnt in the turning region, where air entrainment is intense due to increased turbulence generation. In other words, the first aspect mentioned is dominant and the reduced entrainment in the horizontal flow underneath the ceiling (for high Richardson number values) is of secondary importance. In such circumstances, the total flame length (defined as the sum of the vertical part and the horizontal extension from the fire plume axis) is comparable to L_f.

In situations where $L_f \gg H$, however, conditions are relatively rich in the impingement region and a large portion of the fuel remains to be burnt. In that case, the reduced intensity in air entrainment into the horizontal flow underneath the ceiling in case of high Richardson number values, leads to strong flame extension, compared to the flame height for the same fire in the absence of a ceiling. Still, while the coefficients in the correlations for flame height will vary, the exponents remain unchanged, in particular with respect to the fire HRRs. Indeed, in Ref. [29], the following dependencies are reported (for total flame tip lengths):

$$\text{Unbounded ceilings:} \quad L_f \propto \dot{Q}^{2/5}, \tag{3.62}$$

$$\text{Corner fires:} \quad L_f \propto \dot{Q}^{1/2}, \tag{3.63}$$

$$\text{Corridor fires:} \quad L_f \propto \dot{Q}^{2/3}. \tag{3.64}$$

Note the similarity with expressions (3.35), (3.60) and (3.41), respectively.

3.5.4 THE EFFECT OF VENTILATION

In the previous sections, it was implicitly assumed there is always enough oxygen available for the consumption of the fuel. The air entrained was assumed to be at ambient temperature and atmospheric pressure, and was assumed to contain ~21% oxygen by volume. In case of reduced ventilation conditions, however, this need not be the case. This will affect the dynamics of the fire (and smoke) in enclosures (see Chapter 5). In the present section, a few aspects are discussed, related to the fire plume itself.

3.5.4.1 Reduced Oxygen at Ambient Temperature

One possible situation is a reduction in oxygen at ambient temperature. This can be the case in, e.g., regions at high altitude. Indeed, assuming constant ambient temperature T_{amb}, integration of the law of hydrostatics (2.81) from sea level to high altitude ('h.a.') reveals an exponential decrease in atmospheric pressure with height:

$$p_{h.a.} = p_{sea} \exp\left(-\frac{g}{RT_{amb}}(z_{h.a.} - z_{sea})\right). \tag{3.65}$$

Assuming a constant oxygen volume fraction, this leads to a reduction in oxygen mass supply for a given entrainment velocity: $\dot{m}_{ox,entr} \propto \rho_{amb} v_{entr}$. As was done in Section 3.5.1.1, the entrainment velocity is assumed to be proportional to the upward velocity. The reasoning of Section 3.5.1.1 can be repeated now, i.e., the flame height is determined as the position of the tip of the cone where stoichiometric mixture fraction is found. This reasoning, leading to Eq. (3.35), implies that for a given HRR (or fuel mass flow rate), the flame height increases with decreasing pressure with exponent $-2/5$, because ambient density is proportional to ambient atmospheric pressure (for constant ambient temperature). This is also reported in, e.g., Ref. [31]. In Ref. [32], the empirically determined exponent is $-1/2$. This may be related that the buoyancy force is also slightly weaker at high altitude (due to the lower-density difference between the hot gases and the ambient density). In any case, the flame height increases with decreasing pressure, at least for pressures above 80 kPa [31,32].

It is important to appreciate that a reduction in oxygen concentration also leads to a reduction in flame temperature, due to stronger dilution of the combustion products, as relatively more 'air' needs to be entrained. Consequently, Eqs. (3.14) and (3.13) reveal that the Damköhler number decreases. This makes finite rate chemistry effects more important and can lead to incomplete combustion and eventually flame extinction for sufficiently low oxygen concentration, as explained in Section 3.4.2. Indeed, if the oxygen concentration drops below a threshold value, the value of which depends on the fuel, the fuel–oxidizer mixture is no longer within the flammability limits (Section 3.1), because the mixture becomes too rich. Conditions in a fire plume are harsher than premixed conditions (which is the case in the apparatus in which the flammability limits are determined): mixing needs to happen before combustion can take place, and the flow and turbulence take away heat from the reaction zone by convection and diffusion. Thus, flame extinction will occur more easily than what would be calculated from flammability limits only. Typically extinction starts by blow-off of the flame from the base (where the conditions are the richest and there is relatively strong velocity shear, causing high levels of turbulence), after which very rapidly the flames extinguish due to the absence of heat supply.

A forced reduction of oxygen concentration, e.g., by diluting the atmosphere with nitrogen or carbon dioxide, can be a fire suppression measure. Indeed, the temperature decreases by the dilution and a relative increase of thermal capacity (so that the same HRR results in a lower temperature increase) and the Damköhler number can be reduced sufficiently to extinguish the flame (see Section 3.4.2). Addition of suppression agents, affecting the formation of radicals in the chemistry, can strongly assist this thermal process in extinguishing the flame.

Finally, it is noted that a reduction in oxygen concentration leads to a flame position further away from the fire source. In the case of solid or liquid fuel, the resulting reduction in heat transfer by convection and radiation from the flame to the fuel leads to a reduction in pyrolysis or evaporation, respectively. Consequently, the fuel mass flow rate decreases and the flame length can also decrease.

3.5.4.2 Oxygen-Enriched Fire Plumes

For a given HRR, flame heights are reduced in oxygen-enriched conditions. Indeed, as the oxygen fraction in the air increases, stoichiometric mixture fraction is obtained closer to the fire source, for a given entrainment mass flow rate and for a given HRR (or fuel mass flow rate). Also the buoyancy force slightly increases, because flame temperatures increase. The increased temperatures also lead to increased Damköhler numbers (Eqs. 3.13 and 3.14), but that effect is not very strong: as long as the Damköhler number is large enough, its exact value is not important.

It is noteworthy, though, that in the case of combustion of solids or liquids, often the HRR also increases as the oxygen concentration in the air increases. Indeed, the flames become hotter and lie closer to the surface, causing more intense heat transfer by convection and radiation to the fuel. This causes more intense pyrolysis or evaporation, leading to a higher mass flow rate of fuel and thus a higher heat release rate. This can cause an increase in flame length again.

3.5.4.3 Viated Conditions

The situation described in Section 3.5.4.1 is not the most typical situation of reduced oxygen supply in fire situations. A more typical situation for reduced oxygen is the situation where the air supply (per unit time) into a compartment is not sufficient for complete combustion of all the fuel released (per unit time) in that compartment.

This leads to a more complex situation than described in Section 3.5.4.1, in that not only the oxygen concentration of the 'air' entrained into the fire plume decreases, but also the composition and temperature of the entrained 'air' change. Indeed, the entrained gas becomes a mixture of complex (incomplete) combustion products, nitrogen and oxygen, potentially at elevated temperatures. This implies a double effect: on the one hand, the reduction in oxygen concentration reduces the combustion efficiency and leads to higher flame lengths (as explained in Section 3.5.4.1), while, on the other hand, the increased temperature (and the possible presence of radicals) stimulates chemical reaction kinetics, as explained in Section 2.2.3. At the same time, flame temperatures reduce due to the dilution of combustion products, compared to combustion in clean air. Such circumstances are called 'vitiated' conditions.

In controlled combustion processes, dilution of flames with combustion products is used to reduce the maximum flame temperatures, as a technique to reduce NO_x emissions. Completeness of combustion is ensured by increasing the residence time, in order to overcome the reduction in Damköhler number by the reduced temperature (Eqs. 3.13 and 3.14). In fires, on the other hand, residence times are not controlled. The reduced Damköhler number leads to incomplete combustion, so that unburnt fuel and intermediate combustion products (e.g., CO) are airborne in the compartment (and typically collected in the smoke layer underneath the ceiling, as explained in Section 5.2.1).

Different situations are possible, all affecting the fire plume. One possible situation is that there is 'normal' air supply to the bottom part of the fire plume, but the upper part is inside smoke (where smoke is a hot mixture of air, combustion products and particles). In such a case, the analysis as described in Section 3.5.1 essentially holds for the bottom part of the fire plume. Indeed, there is little effect of 'downstream' phenomena on what happens 'upstream' (where 'upstream' and 'downstream' are defined from the fire point of view). The conditions around the upper part of the fire plume will determine whether the flame becomes longer or shorter, compared to the well-ventilated fire plume.

Another possible situation is that there is lack of oxygen even at the flame base for complete combustion to occur. In such cases, the entire fire plume is affected. If the fuel supply sustains, conditions can become very rich, and eventually too rich for combustion to take place. Depending on the way air is supplied to the compartment, a pulsating phase is typical before possible extinction of the fire. This point is addressed in Section 5.5. Note also that the absence of (visible) flames does not imply an absence of fuel supply. This can lead to dangerous situations in case of sudden fresh supply of oxygen, such as backdraft. This is discussed in Section 5.6.

3.5.5 Fire Whirls

Fire whirls are an important phenomenon in wildland and urban fires. They occur in regions where flow circulation interacts with a fire plume. Fire whirls are dangerous, in that they involve a substantial increase in burning rate (fuel mass flow rate), flame length, upward velocities and radiation, compared to 'normal' free fire plumes as discussed in Section 3.5.1. The increase in flame length and radiant fluxes are dangerous for fast fire spread, e.g., by distant additional spot fires, hence the interest in fire whirls.

In Ref. [33], an interesting analytical discussion on the fluid mechanics is presented. The full derivation is beyond the scope of this book, but an interesting observation in Ref. [33] is that the existence of a viscous core is taken into account, leading to lower upward velocities at the periphery of the fire plume and increased values near the centreline, explaining the flame stretch.

We restrict ourselves here to a few observations and results, explained in Refs. [34] and [35]. First of all, an increased burning rate, compared to free fire plumes without rotation, is explained by an increased flame area near the fire surface (note that liquid fuel is used in Refs. [34,35]). The increased heat transfer by convection and radiation to the fuel explains the increased burning rate, which by itself would already lead to an increase in flame length, because the position of stoichiometric mixture fraction will be higher for a fixed entrainment rate of air into the plume.

However, also the entrainment of air into the fire plume is reduced. Indeed, it is shown in Ref. [35] that the flow circulation significantly reduces the fire plume radius, thereby leaving less area for entrainment to take place. It is suggested in Ref. [35] that the fire plume can be seen as a combination of a cylinder and a cone, rather than a cone. Additionally, it is illustrated in Ref. [35] that the flow rotation causes laminarization in the lower region of the fire whirl flame, which additionally reduces the entrainment, because laminar mixing is far less effective than turbulent mixing. At higher regions, this effect disappears and the entrainment becomes turbulent again. The laminarization phenomenon is claimed to be essential for the increase in flame height [35]. Correlations in Ref. [35] suggest that the effect of rotation can be taken into account through the dimensionless 'circulation', defined as:

$$\Gamma^* = \frac{\pi D^2 \Omega}{\sqrt{gD}D},$$
(3.66)

where Ω is the vorticity. The expression for flame length with whirl then takes the form:

$$L_{f,\text{whirl}} = L_f f\left(\Gamma^*\right).$$
(3.67)

The exact function $f\left(\Gamma^*\right)$ will depend on the choice made for the correlation for L_f, but the results in Ref. [35] suggest that the effect of rotation can be isolated.

Finally, it is mentioned that also the temperatures become higher than in free fire plumes without whirl (Section 3.5.1.2), due to the reduction in entrainment rate of air into the whirling plume.

3.6   FLAME SPREAD

The phenomenon of flame spread over a liquid or solid fuel surface is typically discussed by taking the heat transfer and fuel evaporation (liquid fuel) or pyrolysis (solid fuel) as dominant features (e.g., [3,36–38]). This is justified, as indeed the radiative and convective heat transfer from the flames determines the rate of evaporation or pyrolysis of the fuel and thereby the rate of release of combustible volatiles. Flame spread can be interpreted as an 'advancing ignition front'. Here, we focus on fluid dynamics aspects involved, but in Section 3.6.1, we first derive the basic expressions for flame spread velocity under various circumstances, through expression of heat balances.

Before doing so, it is interesting to mention that in a pool of liquid fuel, the flame spread mechanism observed strongly depends on whether or not the liquid fuel temperature is below its flashpoint. If the liquid fuel temperature is higher than the flashpoint, a flammable fuel–oxidizer mixture exists near the fuel surface. In such circumstances, premixed combustion (Section 3.2) will occur upon ignition and the flame spread is determined by phenomena in the gas phase, which are relatively fast. After this process, steady non-premixed combustion can establish, as described in Section 3.5: the heat transfer from the flames sustains the evaporation of the liquid fuel. Combustion takes place as the fuel vapour mixes with the oxidizer. If the liquid fuel is initially below its flashpoint temperature, the liquid fuel needs to be heated up to the flashpoint temperature by the spreading flames for a flammable fuel–oxidizer mixture to appear near the fuel surface and the flame spread is much slower. Complex interactions can take place, involving surface tension-driven flows within the liquid fuel [3]. These aspects are not discussed here.

Focus is put now on flame spread over solid surfaces, which is the most common situation in case of fire in an enclosure.

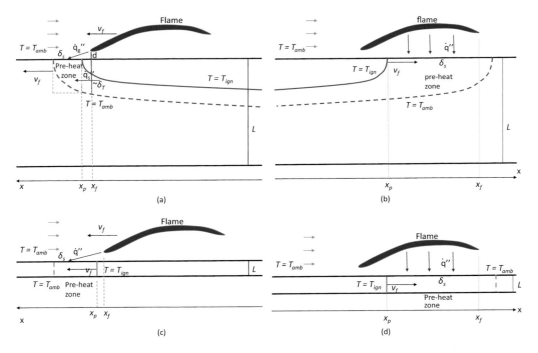

FIGURE 3.13 Sketch of four representative situations for flame spread over solid fuels: opposed flow (a and c) and concurrent flow (b and d) flame spread over thermally thick (a and b) and thermally thin (c and d) fuels. Solid blue line corresponds to the 'ignition temperature', T_{ig}, of the fuel. Dashed blue line corresponds to the thermal penetration depth, Eq. (3.69). Blue arrows denote flow field velocity vectors. The green dashed lines in the top left figure illustrate the control volume for the heat balance as described in Section 3.6.1.1.

3.6.1   FLAME SPREAD VELOCITY – A HEAT BALANCE

As the focus of this book is on the gas-phase phenomena, only some basic expressions are derived here. The reader is referred to the literature for more in-depth discussion and correlations for flame spread velocities for thermally thin and thermally thick fuels. The theoretical derivation by de Ris [39] is particularly worth mentioning.

Figure 3.13 illustrates four representative situations for flame spread: opposed flow (Figure 3.13a and b) and concurrent flow (Figure 3.13c and d) flame spread for thermally thick (Figure 3.13a and c) and thermally thin (Figure 3.13b and d) fuels. Expressions for the flame spread velocity for these configurations are discussed in the present section, based on a heat balance.

The concept of the 'advancing ignition front' [3,36–38] is adopted. Labelling the position where the solid fuel attains the 'ignition temperature' T_{ign} as x_p (where the subscript 'p' refers to 'pyrolysis'), the flame spread velocity v_f can be defined as follows:

$$v_f = \frac{dx_p}{dt} \tag{3.68}$$

Note thus that 'flame' spread effectively refers to the motion of the pyrolysis front from now on in this section. Note that in Section 5.4, a distinction will be made between flame spread in the gas phase (noted as v_s) and the velocity at which the 'burnout front' moves (noted as v_{BO} in Section 5.4). The latter corresponds to Eq. (3.68). Hence, a technically more correct term would be 'pyrolysis front spread velocity', rather than 'flame spread velocity'. However, the term 'flame spread velocity' is commonly used.

The concept of the advancing ignition front boils down to expressing that as the pyrolysis front advances with velocity v_f, a certain amount of mass of unburnt fuel is heated up from its original temperature T_{amb} to its ignition temperature T_{ign}. This amount of mass depends on the length of the 'pre-heat' zone, δ_s, as well as on the in-depth thickness of the pre-heat zone. Note that the concept of the pre-heat zone is very similar to what was described in Section 3.2.2 in the discussion of the laminar burning velocity, but now it refers to the solid phase.

3.6.1.1   Opposed Flow Flame Spread over a Thermally Thick Fuel

Consider first the opposed flow flame spread over a thermally thick fuel (Figure 3.13a). The notion 'thermally thick' refers to the situation where the thermal penetration depth, δ_T, is smaller than the thickness, L, of the fuel (see Section 2.3.3.2). The thermal penetration depth, Eq. (2.54), refers to unsteady conduction inside a solid and can be defined as follows:

$$\delta_T = C\sqrt{\alpha_F t} = C\sqrt{\frac{k_F}{\rho_F c_F}t}. \tag{3.69}$$

As before, the subscript 'F' refers to 'fuel' (which is solid here). In Eq. (3.69), the value of C ranges from 1 to 4, depending on the source consulted [2]. Physically it refers to the depth over which the solid 'feels' inside that there is a heat flux at its surface. In Figure 3.13, this is sketched as the dashed blue line. The depth increases with the square root of time and is proportional to the square root of the thermal diffusivity, Eq. (2.6), of the solid material.

The 'thickness' of the pre-heat zone is proportional to the thermal penetration depth, Eq. (3.69).

Consider now the heat balance over the control volume, sketched in green dashed lines, in the pre-heat zone.

At the left and bottom boundary, the temperature inside the solid can be assumed uniform, equal to ambient temperature T_{amb}. At the top and right boundary of the control volume, the temperature is not uniform. For the sake of simplicity of the analysis, the average temperature will be used at these boundaries:

$$T_{s,av} = \frac{1}{2}\left(T_{ign} + T_{amb}\right). \tag{3.70}$$

As the pre-heat zone moves forwards with velocity v_f, Eq. (3.68), the amount of mass (per unit length in the transverse direction) heated up per unit time in the control volume is proportional to: $\rho_F \delta_T v_f$, where δ_T is the thermal penetration depth at the flame tip position (see Figure 3.13a). Thus, using the specific heat c_F of the solid material, the amount of energy consumed per unit time for heating up this mass from T_{amb} to the average temperature, Eq. (3.70), reads:

$$\dot{Q}' \propto \frac{1}{2}\rho_F \delta_T v_f c_F\left(T_{ign} - T_{amb}\right). \tag{3.71}$$

The unit of $\dot{Q}'$ is W/m. This heat, required for the heating process, stems from the flame and from conduction inside the solid.

The heat transfer from the flame is primarily due to conduction in the gas phase, as the view factor for radiation to the unburnt fuel is low and the flame is not in contact with the unburnt fuel (ruling out convection). This conductive heat transfer will depend on the flame temperature, T_f, and the distance between the flame tip and the pre-heat zone. The latter will depend on the flame 'stand-off distance', d (see Figure 3.13a), i.e., the distance between the flame and the surface. Note that this stand-off distance is never equal to zero, because the mixture of combustible pyrolysis gases and oxidizer is too rich for combustion to take place (see Section 3.1). Ignoring the fact that the flame is positioned downstream of x_p (see Figure 3.13a), the distance between the flame tip and the middle of the pre-heat zone on the surface can be determined as follows:

$$d_f = \sqrt{d^2 + \frac{\delta_s^2}{4}}. \tag{3.72}$$

Approximating the temperature gradient inside the gas phase as $T_f - T_{s,av}/d_f$, the conductive heat transfer through the gas phase, Eq. (2.5), can be approximated as follows:

$$\dot{q}_g \approx \frac{k_g \left(T_f - T_{s,av}\right)}{d_f}. \tag{3.73}$$

This heat flux per unit area needs to be multiplied with δ_s to obtain the heat flux per metre, entering the control volume by conduction through the gas phase:

$$\dot{Q}'_g = \dot{q}_g \delta_s. \tag{3.74}$$

The conduction inside the solid can be approximated as follows:

$$\dot{q}_s \approx \frac{k_F \left(T_{s,av} - T_{amb}\right)}{\delta_s} = \frac{1}{2} \frac{k_F \left(T_{ign} - T_{amb}\right)}{\delta_s}. \tag{3.75}$$

In Eq. (3.75), the temperature gradient at the right boundary of the control volume in Figure 3.13a has been approximated by the average temperature gradient over distance δ_s: $\left(T_{s,av} - T_{amb}\right)/\delta_s$. The heat flux per unit area, Eq. (3.75), needs to be multiplied with δ_T to obtain the heat flux per metre, entering the control volume by conduction through the solid phase:

$$\dot{Q}'_s = \dot{q}_s \delta_T. \tag{3.76}$$

Combining Eqs. (3.71), (3.73) and (3.76), the heat balance reads:

$$\frac{1}{2} \rho_F \delta_T v_f c_F \left(T_{ign} - T_{amb}\right) = \frac{k_g \left(T_f - T_{s,av}\right)}{d_f} \delta_s + \frac{1}{2} \frac{k_F \left(T_{ign} - T_{amb}\right)}{\delta_s} \delta_T. \tag{3.77}$$

The right hand side of Eq. (3.77) consists of the sum of conduction in the gas phase and conduction in the solid phase. Assuming that the length of the pre-heat zone is of the same order of magnitude as the thermal penetration depth at the flame tip position ($\delta_s \approx \delta_T$), the following aspects determine which term will dominate:

- The flame stand-off distance, d_f, compared to the size of the pre-heat zone, δ_s. Typically these distances are of the same order of magnitude.
- The flame temperature, T_f, compared to the ignition temperature of the solid fuel, T_{ign}. Typically the flame temperature is much higher than the ignition temperature (which is in the order of 700 K for many materials).

- The conduction coefficient in the gas phase (k_g, which is in the order of 0.02 W/(m K) for air), compared to the conduction coefficient of the solid, k_F.

In any case, the incident heat flux can be written as follows:

$$\dot{Q}' = \dot{q}\delta_s. \tag{3.78}$$

In Eq. (3.78), the heat flux per unit area reads:

$$\dot{q} = \frac{k_g\left(T_f - T_{s,av}\right)}{d_f} + \frac{1}{2}\frac{k_F\left(T_{ign} - T_{amb}\right)}{\delta_s}\frac{\delta_T}{\delta_s}. \tag{3.79}$$

In many analyses, the latter term of Eq. (3.79) is ignored, assuming that the conduction in the gas phase is dominant.

Combining Eqs. (3.71) and (3.78), using Eq. (3.69), yields:

$$\frac{1}{2}\rho_F v_f C\sqrt{\frac{k_F}{\rho_F c_F}}t c_F\left(T_{ign} - T_{amb}\right) = \dot{q}\delta_s. \tag{3.80}$$

The characteristic time t in Eq. (3.80) is determined from the flame spread velocity and the length of the pre-heat zone:

$$t = \frac{\delta_s}{v_f}. \tag{3.81}$$

Using this in Eq. (3.80) reveals the following expression for the flame spread velocity for opposed flow flame spread over a thermally thick fuel:

$$v_f \propto \delta_s \frac{1}{\rho_F k_F c_F}\left(\frac{\dot{q}}{T_{ign} - T_{amb}}\right)^2. \tag{3.82}$$

Equation (3.82) reveals that the flame spread velocity increases as the thermal inertia, $\rho_F k_F c_F$, of the solid material decreases, as the incident heat flux, $\dot{q}$, increases and as the difference between the ignition temperature and ambient temperature, $T_{ign} - T_{amb}$, decreases.

Note that Eq. (3.82) can be rewritten as follows:

$$v_f \propto \frac{\delta_s}{\tau}. \tag{3.83}$$

In Eq. (3.83), τ is a characteristic time to ignition:

$$\tau = \rho_F k_F c_F\left(\frac{T_{ign} - T_{amb}}{\dot{q}}\right)^2. \tag{3.84}$$

In opposed flow flame spread, steady conditions (i.e., constant values for δ_s and v_f) can be obtained.

3.6.1.2 Opposed Flow Flame Spread over a Thermally Thin Fuel

Consider now the configuration of opposed flow flame spread over a thermally thin fuel (Figure 3.13c). It is noted that for thermally thin fuels, it is typical for flames to appear at both sides of the fuel (not sketched in Figure 3.13c). For obvious reasons, this increases the incident heat flux, and thus the flame spread velocity, by a factor of approximately 2.

The reasoning of Section 3.6.1.1 can be repeated, but now the thermal penetration depth, Eq. (3.69), needs to be replaced by the thickness L of the fuel. Also, the temperature at the right boundary of the control volume can now be approximated to be uniform at T_{ign}.

Thus, Eq. (3.71) becomes:

$$\dot{Q}' \propto \rho_F L v_f c_F \left(T_{ign} - T_{amb}\right).$$
(3.85)

While Eq. (3.73) remains in place, Eq. (3.75) now becomes:

$$\dot{q}_s \approx \frac{k_F \left(T_{ign} - T_{amb}\right)}{\delta_s}.$$
(3.86)

This flux needs to be multiplied by L, which is assumed to be (much) smaller than δ_s (which is still of the same order as δ_T). Thus, under these circumstances, typically the conduction in the gas phase is the dominant term (see Eq. 3.79).

Using Eqs. (3.73), (3.74), (3.85) and (3.86) yields:

$$v_f \propto \delta_s \dot{q}_g \frac{1}{\rho_F L c_F \left(T_{ign} - T_{amb}\right)} \propto \delta_s \frac{2k_g \left(T_f - T_{s,av}\right)}{d_f} \frac{1}{\rho_F L c_F \left(T_{ign} - T_{amb}\right)}.$$
(3.87)

The factor 2 in the last term in Eq. (3.87) refers to the appearance of flames at both sides of the solid fuel, as mentioned above.

Equation (3.87) reveals that the flame spread velocity increases as the mass density, specific heat and geometrical thickness of the solid material $(\rho_F L c_F)$ decrease, as the incident heat flux $(\dot{q}_g)$ increases and as difference between the ignition temperature and ambient temperature $\left(T_{ign} - T_{amb}\right)$ decreases. The latter two now appear with exponent equal to 1 in Eq. (3.87), in contrast to the quadratic dependence in Eq. (3.82). This is an important observation.

The characteristic time to be used in Eq. (3.83) now reads:

$$\tau = \rho_F L c_F \frac{T_{ign} - T_{amb}}{\dot{q}_g}.$$
(3.88)

As for thermally thick fuels, in opposed flow flame spread, steady conditions (i.e., constant values for δ_s and v_f) can be obtained.

3.6.1.3 Concurrent Flow Flame Spread over a Thermally Thick Fuel

Consider now the configuration of concurrent flow flame spread over a thermally thick fuel (Figure 3.13b).

The reasoning held in Section 3.6.1.1 still holds, but a key difference is that the size δ_s is now primarily determined by the flame size. Indeed, there is now direct heat transfer from the flame to the unburnt fuel by radiation, as well as by convection from the hot combustion products. This heat transfer from the gas phase dominates the conduction inside the solid material.

As a consequence, labelling the location of the flame tip (see Section 3.5.1.1 for a discussion on how this position can be determined) as x_f, the size δ_s is proportional to:

$$\delta_s \propto \left(x_f - x_p\right).$$
(3.89)

Thus, Eq. (3.82) becomes:

$$v_f \propto \left(x_f - x_p\right) \frac{1}{\rho_F k_F c_F} \left(\frac{\dot{q}}{T_{\text{ign}} - T_{\text{amb}}}\right)^2 . \tag{3.90}$$

Alternatively, Eq. (3.83) becomes:

$$v_f \propto \frac{\left(x_f - x_p\right)}{\tau} . \tag{3.91}$$

Whereas the expression for the characteristic time is still given by Eq. (3.84), the key difference with opposed flow flame spread is that there is now typically accelerating flame spread. Indeed, there is a positive feedback loop. The incident heat flux per unit length, Eq. (3.78), increases as $\delta_s \propto \left(x_f - x_p\right)$ increases. As a consequence, the pyrolysis process becomes more intense, thereby releasing a higher fuel mass flow rate. As a consequence, the flame length increases (see Section 3.5), so that $\delta_s \propto \left(x_f - x_p\right)$ increases even more, which illustrates the positive feedback loop. Also note that as the incident heat flux increases, the characteristic time, Eq. (3.84), decreases, which leads to an additional increase in flame spread velocity (Eq. 3.83 or 3.91).

3.6.1.4　Concurrent Flow Flame Spread over a Thermally Thin Fuel

The reasoning held in Section 3.6.1.3 can be repeated for a thermally thin fuel, but now using the discussion as presented in Section 3.6.1.2 as the starting point.

3.6.2　Gas-Phase Phenomena

Obviously, combustion will be a fundamental process in the gas phase. Indeed, the heat required to heat up and pyrolyse the solid fuel stems from the flames, as explained above. Regardless of whether the combustion is of premixed or non-premixed type, the Damköhler number, Eq. (3.13), plays a central role:

$$Da = \frac{\tau_{\text{flow}}}{\tau_{\text{chem}}} . \tag{3.92}$$

The time scale related to the flow can be interpreted in terms of residence time: the shorter the residence time, the lower the Damköhler number. The chemical time scale can be related to temperature, Eq. (3.14): the higher the temperature, the higher the Damköhler number. As an example: in an oxygen-enriched atmosphere, the Damköhler number will increase, due to higher flame temperatures (because less-inert nitrogen is heated up by the combustion reactions). Moreover, the flame will lie closer to the surface, because stoichiometric conditions will be met closer to the surface, due to higher stoichiometric mixture fraction values. As a consequence, the heat transfer will intensify strongly and flame spread will be faster.

For non-premixed combustion, it must not be forgotten that the turbulent mixing of fuel and oxidizer is typically the rate-limiting process, not the chemical time scale of the chemical reaction. An increase in turbulence can therefore assist the combustion process and accelerate flame spread. On the other hand, the residence time decreases with increased flow velocity as well, as discussed below. Moreover, heat is transported away faster from the reaction zone in case of increased turbulence, thereby reducing the temperature. The latter two effects lead to an effective reduction in the Damköhler number.

It is important to appreciate that the combustible gases, released from the pyrolysing fuel, are at relatively low temperature, in relation to the activation energy. Indeed, pyrolysis temperatures are below 450°C, so Figure 3.1 reveals that heating up is required before combustion in a flammable mixture can take place. This implies that there will be a fundamental difference between the situation where the volatiles are released in a region of cool fresh air (with which they mix to form a flammable mixture) and in a region with a mixture of hot combustion products and air (where it may be more challenging to find oxygen).

Finally, it is recalled that the temperature increase due to combustion will lead to a vertically upward buoyant flow of the combustion products.

3.6.3 Horizontal Surface

3.6.3.1 Natural Convection

Figure 3.14 shows a sketch of flame spread over a horizontal surface, facing upwards, in a quiescent atmosphere. Even though the atmosphere is assumed quiescent, in the flame region near the solid surface a counter-current flow situation is observed due to the air entrainment. Ahead of the flame, volatile fuel mixes with the atmosphere and premixed combustion can occur in a flammable mixture near the surface. This is not a steady process: once consumed, a new flammable air-fuel mixture must form again. Behind the flame (or within the fire plume), a non-premixed situation exists. The volatiles mix with combustion products and need to find oxygen. They rise in the buoyancy-dominated flow and burn at the edge of the fire plume in diffusion combustion regime, as discussed in Section 3.5. Non-premixed combustion is the dominant type of combustion in the configuration at hand. Heat is transferred by radiation and conduction in the gas phase to the unburnt fuel ('virgin material'), by radiation and convection to the pyrolysing fuel and by conduction inside the solid material.

Figure 3.15 shows a sketch for a horizontal surface, facing downwards, in a quiescent atmosphere. A fundamental difference from Figure 3.14 is that buoyancy tries to push the combustion products back to the ceiling, rather than away from the reaction zone. As a consequence, flame spread underneath a combustible ceiling typically needs to be supported by a flow, e.g., created by a ceiling jet, which relates to the configuration of Section 3.6.2.2. Otherwise, the hot products stagnate underneath the ceiling and the flame loses so much heat that it quenches at the surface [3]. However, if the flame spreads, it is closer to the fuel surface than in Figure 3.13, so that the heat transfer is more intense and flame spread is faster than in similar conditions for a horizontal

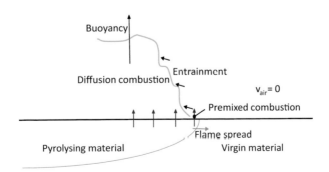

FIGURE 3.14 Sketch of flame spread over a horizontal surface, facing upwards, in a quiescent atmosphere. Orange solid line: edge of the fire plume (flame). Red arrows: volatile fuel entering the gas phase. Black arrows: flow velocities.

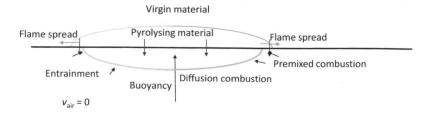

FIGURE 3.15   Sketch of flame spread over a horizontal surface, facing downwards, in a quiescent atmosphere. Orange solid line: edge of the fire plume (flame). Red arrows: volatile fuel entering the gas phase. Black arrows: flow velocities. Note: this type of flame spread typically only occurs in forced flow conditions.

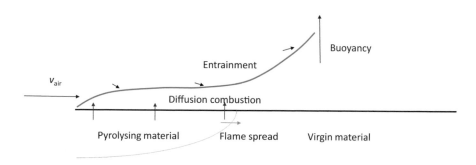

FIGURE 3.16   Sketch of flame spread over a horizontal surface, facing upwards, with concurrent airflow. Orange solid line: edge of the fire plume (flame). Red arrows: volatile fuel entering the gas phase. Black arrows: flow velocities.

surface facing upwards. Wall quenching of the flames can still occur, though, along with incomplete combustion with CO formation [40].

3.6.3.2   Concurrent Airflow

Figure 3.16 provides a sketch for flame spread over a horizontal surface, facing upwards, with a forced airflow in the direction of the flame spread.

   As opposed to the situation in Figure 3.14, the volatiles are now released into a forced flow. In Figure 3.16, the material is pyrolysing at the leading edge, but it is also possible that a boundary layer, as described in Section 2.9, develops ahead of the pyrolysing zone. Depending on the oncoming air flow, the volatiles are released into a laminar or a turbulent flow. As explained in Section 2.9, the boundary layer becomes turbulent as soon as it is sufficiently thick (or sufficiently far away from the leading edge). Yet, the volatiles first enter a laminar sub-layer, close to the solid surface, where all turbulence is damped by the viscous forces. Thus, the pyrolysis gases enter (and disturb) the laminar viscous sub-layer, to mix and react with the oncoming air at the upstream flame edge. Downstream of the flame tip, the pyrolysis gases enter a flow consisting of hot combustion products, typically diluted with air. The combustion is clearly of the non-premixed type. The volatiles are convected downstream as they mix with the oxidizer to form the flame. Meanwhile, they also start moving upwards due to buoyancy. As explained in Section 3.5, buoyancy-generated turbulence can appear through the entrainment of cold air.

   The flame shape is important. Due to the momentum of the oncoming horizontal airflow and the improved mixing with fuel, compared to the mixing in a quiescent atmosphere, the flame is closer to the surface than in quiescent atmosphere. Indeed, the natural vertically upward flow, induced by buoyancy, competes with the horizontal airflow (cfr. fire plume tilting, discussed in Section 3.5.1.4.). As a consequence, the flame does not rise as quickly as without forced airflow. The heat transfer is dominated by radiation, primarily due to an increase in view factor from the flame towards the surface. There is also heat transfer by convection to the solid fuel: the hot combustion products are blown over the surface, whereas they rise with the buoyant flow in quiescent conditions. Clearly, flame spread with concurrent airflow will be much faster than without concurrent airflow. Note, though, that flame blow-out is possible for too large air velocities, due to too much cooling of the flame and/or too high scalar dissipation rates. This can particularly happen at early stages, when the flame is not strong yet.

As the fire area grows, the buoyant force becomes stronger, turning the flames more vertical, in spite of the forced horizontal flow. This phenomenon is substantially different if entrainment from the sides is blocked (i.e., in a tray or channel configuration, rather than an open floor). The horizontal momentum then becomes channelled (and therefore stronger, relative to the buoyancy), the flames become longer stretched horizontally, the heat transfer is intensified (both through radiation and convection) and the flame spreads faster.

Flame spread with concurrent airflow underneath a ceiling does not differ too much from the situation sketched in Figure 3.16, but buoyancy now pushes the flame still closer to the surface. This not only intensifies heat transfer, but can also weaken the flames due to increased heat losses.

3.6.3.3 Counter-Current Airflow

Figure 3.17 sketches the situation for counter-current airflow (or 'opposed flow' flame spread). At first sight, the situation may not look too different from Figure 3.16. However, a key difference is that the flame now does not lie above virgin material, but above already pyrolysing material. Consequently, the view factor for radiation heat transfer from the flame to the virgin material is much lower now. This makes flame spread with counter-current airflow much slower. Moreover, also the convection heat from the hot combustion products towards the solid fuel is now transferred towards the already pyrolysing fuel, not to the virgin material. On the contrary, the virgin material is effectively cooled by the forced convection with cold air. Also this causes the flame spread in counter-current air flow to be slower than in concurrent airflow.

Note that the oncoming (laminar or turbulent) boundary layer flow of air, as described in Section 2.9, is not disturbed by the pyrolysis gases.

As explained in Section 3.5.1.3, buoyancy-generated turbulence can appear due to the entrainment of cold air into the flame.

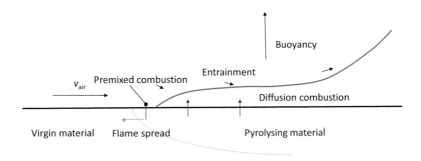

FIGURE 3.17 Sketch of flame spread over a horizontal surface, facing upwards, with counter-current airflow. Orange solid line: edge of the fire plume (flame). Red arrows: volatile fuel entering the gas phase. Black arrows: flow velocities.

As explained in Section 3.6.3.1, there can be a region of premixed combustion, preceding the actual diffusion flame.

The effect of the oncoming airflow is threefold. On the one hand, the flame is kept closer to the surface, compared to the situation in quiescent atmosphere (Figure 3.14). However, as stated above, the main part of the heat transfer is given to the already pyrolysing part of the fuel, from where it needs to be transferred towards the virgin material by conduction inside the solid. Even more so, the view factor for radiation towards the virgin material effectively becomes smaller, compared to quiescent atmosphere conditions. As such, there is no a priori clear increase in heat transfer towards the virgin material.

A second effect of the oncoming air flow is the supply of oxygen to the flames, effectively increasing the mixing of volatile fuel and oxidizer. This would lead to an increase in flame spread, provided that a lack of oxygen is the limiting factor in the process.

A third effect, however, is a cooling effect on the flame and a reduction of the residence time of volatile fuel in the flame region. Both effects lead to a reduction in the Damköhler number, Eq. (3.92). Indeed, if the counter-current air flow is too strong (i.e., the velocity or momentum is too high), the flame can effectively be quenched if too much heat is taken away from the flame (see Section 3.4).

It has been observed in Ref. [41] that an increase in flame spread in counter-current flow conditions is only observed in oxygen-enriched conditions. Clearly, the third effect mentioned is important. Regardless of the oxygen concentration in the atmosphere, the flame is quenched if the counter-current flow velocity is too high. The limiting velocity becomes higher for higher oxygen concentrations. This can be understood from the fact that the flame temperature is higher for higher oxygen concentrations, so that τ_{chem} becomes smaller in Eq. (3.92) and it takes longer for the Damköhler number to drop below its critical value Da_q for flame quenching.

3.6.4 VERTICAL SURFACE

Only quiescent atmosphere conditions are considered, because they are most relevant for fire dynamics. A major distinction is to be made between vertically upward and vertically downward flame spread.

Figure 3.18 provides a sketch of vertically downward flame spread over a vertical surface in a quiescent atmosphere. Near the flame base, the configuration is very much like the one sketched in Figure 3.14, namely naturally induced counter-current flame spread. A region of premixed combustion can precede the actual non-premixed flame.

Most of the heat is transferred by radiation and convection to already pyrolysing material, and hence, flame spread is slow. The classical example for this type of flame spread is a match, held vertically upwards with the flame at the tip. It is well known that such a flame travels downwards very slowly and often even self-extinguishes. The mechanisms and reasons have been explained in Sections 3.6.3.1 and 3.6.3.3. The main difference from the configuration of a horizontal surface, facing upwards, is that buoyancy now does not try to move the flame away from the surface. On the contrary, it keeps the flame well aligned with the surface, leading to increased convective heat transfer, compared to Figure 3.14. On the other hand, the view factor for radiation from the flames to the virgin material is lower than in Figure 3.14. Buoyancy-generated turbulence can appear by entrainment of cool air into the flame, very similar to what has been explained for free fire plumes in Section 3.5.1.3.

Vertically downward flame spread is observed to be steady, with flame spread velocities in the order of 1 mm/s or lower, depending on the fuel type and the atmosphere.

Figure 3.19 provides a sketch of vertically upward flame spread over a vertical surface in a quiescent atmosphere. This configuration resembles the one sketched in Figure 3.16, namely

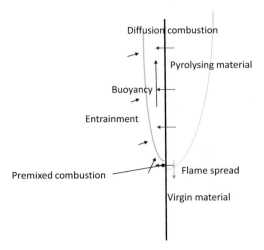

FIGURE 3.18 Sketch of vertically downward flame spread over a vertical surface in a quiescent atmosphere. Orange solid line: flame. Red arrows: volatile fuel entering the gas phase. Black arrows: flow velocities.

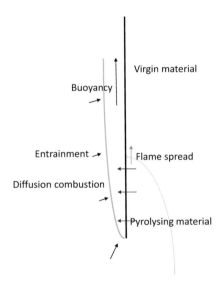

FIGURE 3.19 Sketch of vertically upward flame spread over a vertical surface in a quiescent atmosphere. Orange solid line: flame. Red arrows: volatile fuel entering the gas phase. Black arrows: flow velocities.

concurrent flame spread (which is now naturally induced by the buoyancy). Much heat is transferred by radiation to virgin material, due to a high view factor. This view factor can even be higher than for horizontal flame spread (depending on the velocity of the forced air flow there), because buoyancy keeps the flame aligned with, and close to, the virgin solid fuel. Also the heat transfer by convection towards the virgin fuel is effective. The hot combustion products are indeed in direct contact with virgin material. The convection coefficient is not very high, though (typically in the order of 10–20 W/(m²K)). In any case, the effective heat transfer makes vertically upward flame spread very fast, as can be seen holding the match vertically upside down. Vertically upward flame spread is not steady, due to a positive feedback loop, as mentioned in Section 3.6.1. Indeed, as heat transfer increases with larger flame lengths, more volatile fuel is released per unit time due to increased pyrolysis rates. This larger amount of fuel results in yet larger flames: the volatiles are convected upwards due to buoyancy, while mixing with the oxygen from the atmosphere, so an increase in volatile fuel mass flow rate leads to longer flames for a given mixing rate.

The pyrolysis gases disturb the boundary layer flow as they leave the solid material (thereby making the mixing more effective) and buoyancy-generated turbulence can appear by entrainment of cool air into the flame, again similar to what has been explained for free fire plumes in Section 3.5.1.3.

As closing note, it is mentioned that flow conditions for vertical flame spread in quiescent ambient conditions typically correspond to low values of Reynolds (Eq. 2.92) or Grashof (Eq. 2.97)/ Rayleigh numbers (Eq. 2.96) at the flame base. Indeed, the high temperatures lead to a reduction in mass density and an increase in viscosity/thermal diffusivity. Consequently, flow conditions are typically laminar at the flame base, and it takes some distance for turbulence to develop. This poses challenges to modelling in numerical simulations, as reported in, e.g., Ref. [42]. The reader is referred to specialized literature for more details.

3.6.5 INCLINED SURFACE

Flame spread over inclined surfaces poses an interesting problem. Consider first a surface, facing upwards, in quiescent atmosphere. As explained in Section 3.6.3.1, horizontal flame spread corresponds to essentially counter-current conditions, because the entrainment velocity opposes the flame spread velocity (Figure 3.14). If the surface is in vertical position, the flame spread can be either in counter-current conditions (vertically downward flame spread, Figure 3.18) or in concurrent conditions (vertically upward flame spread, Figure 3.19).

Not surprisingly, only a small decrease in flame spread rate is observed when changing the configuration from horizontal to vertical, with downward flame spread (e.g., [43]). Indeed, the flow and heat transfer configuration being similar, the main difference is a reduction in view factor for the radiation towards the virgin material in the vertically downward flame spread configuration. Indeed, buoyancy tries to move the flame upwards, so it aligns with the vertical surface, while it rises from the horizontal surface.

A much stronger variation in flame spread rate is observed when changing the configuration from horizontal to vertically upward flame spread. In Ref. [3], it is reported that the switch from counter-current to concurrent flame spread takes place at an angle of about 15°–20°. As explained in Sections 3.6.1 and 3.6.3, this results in a fundamental change in heat transfer and consequently in much faster flame spread. A more recent work [44] illustrates this very well. In Ref. [44], it is further illustrated that maximum flame spread rate values are observed for near-vertical surfaces, with even slightly higher flame spread rates observed for an angle of about 30° from the vertical direction, with the flames at the downward side of the surface (i.e., for a downward facing surface). The reason is presumably intensified heat transfer due to buoyancy, pushing the flame closer to the virgin material. Flame spread rates decrease again for further inclination (i.e., further modification of the configuration towards a horizontal surface, facing downwards, Section 3.6.3.1). It is reported in Ref. [44] that the heat flux reduces. As mentioned in Section 3.6.3.1, flame spread over a horizontal surface, facing downwards, needs to be supported by an external flow [3].

It is important to mention the 'trench' effect in the context of upward flame spread over inclined surfaces [3]. The term refers to suppressed entrainment of air from the sides into the flame, e.g., due to the presence of side walls. In such situations, significant flame extension can occur [45], as the pyrolysis gases need to mix with air to react. An obvious consequence is a strong increase in heat transfer, particularly when the flow configuration becomes concurrent [46]. An increase in flame spread rate, due to the presence of side walls, by a factor of 4 has been reported in Ref. [46] for an inclination angle of 20° (with reference to the horizontal plane).

A very instructive video is found on *http://player.vimeo.com/video/111735846*, illustrating many of the aforementioned effects.

3.6.6 PARALLEL VERTICAL PLATES CONFIGURATION

The parallel vertical plates configuration has been used to study vertically upward flame spread (e.g., [47,48]). In Ref. [47], the configuration consists of two plates (60 cm wide, 2.4 m high), 30 cm apart from each other with a 60 kW propane sandstone burner at the bottom. This corresponds to the set-up of Ref. [49]. In Ref. [48], the plates are 40 cm wide and 2.5 m high and two distances (10.5 and 30.5 cm) between the plates have been examined. The burner at the bottom has a power of 30 kW in Ref. [48].

The parallel panel set-up clearly affects the flame spread rate in different ways. One aspect is an increase in radiative heat transfer towards the virgin material. Indeed, part of the radiation from the flame (Figures 3.18 or 3.19) now is not lost to the surroundings, but incides onto the parallel panel, which in its turn reflects part of the radiation. The parallel panel also heats up by convection and by absorption of radiation, so that it also emits radiation. Thus, the virgin material heats up more rapidly, thereby increasing the flame spread rate. This is much more pronounced for vertically upward flame spread than for vertically downward flame spread (where the view factor for radiation is smaller). This effect also becomes smaller if the distance between the plates increases, due to a decrease in view factor for the radiation (among the plates, as well as from the flames to the plates).

Secondly, also the convective heat transfer increases. Indeed, the hot combustion products are 'trapped' in between the plates, so their upward velocity increases. There is also less entrainment of cold air into the flames. This effect disappears when the plates are too far away from each other and can also reduce if the plates are too close to each other. Indeed, in Ref. [48], lower temperatures are reported for the inter-plate distance of 10.5 cm, compared to 30.5 cm, presumably due to incomplete combustion in between the plates due to lack of oxygen. This corresponds to under-ventilated conditions (see Section 5.2.1.2) in between the plates and combustion of 'excess fuel' (i.e., combustible gases that do not find oxygen in between the plates) outside the gap between the plates. The heat released by the combustion of the excess fuel does not assist the flame spread.

The forced upward flow, trapped in between the plates, in combination with the reduced air entrainment, compared to a single vertical plate, due to the blocking of flow by the presence of the parallel panel, also leads to an increase in flame length. This in its turn leads to a strong increase in radiative and convective heat transfer from the flames to the virgin material and thus to a higher flame spread rate. This effect is stronger for smaller inter-plate distances [48].

Although not reported in the literature, it is likely that for given size of the plates, there is a certain inter-plate distance for which the flame spread rate reaches a maximum value. Indeed, putting the plates closer together increases the radiation heat transfer (for a given flame temperature) and convection heat transfer, up to a point where the combustion becomes too incomplete due to lack of oxygen supply and temperatures drop so much that the heat transfer decreases. Indeed, the mere presence of the plates blocks air entrainment (sketched in Figure 3.19 in the absence of a vertical plate). Under such circumstances, the flames emerge outside the channel formed by the plates [48], so their heat is lost to the surroundings, i.e., that heat does not contribute to the flame spread process. Also turbulence can be damped if the plates are too close to each other, but presumably this is only a second-order effect.

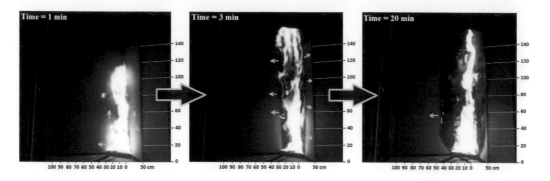

FIGURE 3.20 Pictures of flame spread in a corner configuration in an SBI set-up [51]. The material is MDF. Arrows indicate flame spread direction.

3.6.7 CORNER CONFIGURATION

The same aspects prevail as mentioned in Section 3.6.6. On the on hand, there is increased radiation heat transfer, due to a reduction in radiation losses to the surroundings and re-radiation among the plates. On the other hand, there is a reduction in air entrainment, as also explained in Section 3.5.2. In contrast to what was explained in Section 3.6.6, though, there is no risk of under-ventilated conditions (except perhaps for sharp wedges). The main effect of the reduced air entrainment, compared to flame spread over flat surfaces, is that temperatures increase and flames become longer, in line with what was described in Section 3.5.2 [50]. In particular for upward flame spread, this results in a strong increase in flame spread rate [51]. The relative contribution of the impact of reduced air entrainment, on the one hand, and re-radiation from the panels, on the other hand, has been discussed in Ref. [52], emphasizing the importance of the geometry of a corner set-up.

Figure 3.20 shows a few pictures of a test performed on MDF (medium-density fibreboard) in an SBI set-up [53]. Two vertical boards of 1.5 m high (one of 1.0 m wide and one 0.5 m wide) are perpendicular to each other to form a corner. A triangular propane sandstone burner of 30 kW puts a thermal attack at the bottom corner. The instantaneous pictures reveal fast vertical upward flame spreading in the early stages. Lateral flame spreading also occurs, but at a much slower rate [51]. The lateral (horizontal) flame spread rate over the vertical surface is partly counter-current flame spread (horizontal air entrainment from the direction opposite to the flame spread direction), although the flow field is essentially vertically upwards due to buoyancy. This makes the flow field different from what has been described in Sections 3.6.3.1 and 3.6.4. As a consequence, the horizontal lateral flame spread is slightly faster at higher heights, due to the addition of convective heat transfer to the radiation heat transfer from the flames [51].

REFERENCES

1. C.L. Beyler (2002) "Flammability limits of premixed and diffusion flames", in M.J. Hurley et al. (Eds.), *The SFPE Handbook of Fire Protection Engineering*, Chapters 2-7. Springer: New York, pp. 2-121 – 2-188.
2. J.G. Quintiere (2006) *Fundamentals of Fire Phenomena*. John Wiley & Sons, Ltd: Hoboken, NJ.
3. D. Drysdale (2011) *An Introduction to Fire Dynamics,* 3rd Ed. John Wiley & Sons, Ltd: Hoboken, NJ.
4. Zabetakis (1965) US Bureau of Mines – Bulletin 627.
5. H.F, Coward and G.W. Jones (1952) US Bureau of Mines – Bulletin 503.

6. E. Mallard and H.L. Le Chatelier, *"Combustion des melanges gaseux explosifs"* (1883) Annales des Mines 4, 379–568.

7. G. Jomaas (2016) *The SFPE Handbook of Fire Protection Engineering*, 5th Ed. Springer: New York.

8. P.A. Libby and F.A. Williams (1994) *Turbulent Reacting Flows*. Academic Press: Cambridge, MA.

9. K.K. Kuo (2005) *Principles of Combustion*, 2nd Ed. John Wiley & Sons, Inc: Hoboken, NJ.

10. C.K. Law (2006) *Combustion Physics*. Cambridge University Press: Cambridge.

11. F.A. Williams (1985) *Combustion Theory*, 2nd Ed. Addison-Wesley Publishing Company: Boston, MA.

12. T. Poinsot and D. Veynante (2005) *Theoretical and Numerical Combustion*. R T Edwards, Inc.: Dallas, TX.

13. N. Peters (2000) *Turbulent Combustion*. Cambridge University Press: Cambridge.

14. J. Warnatz, U. Maas and R.W. Dibble (1996) *Combustion*. Springer Verlag: Berlin/Heidelberg, Germany.

15. D. Veynante and L. Vervisch (2002) "Turbulent combustion modeling", *Progress in Energy and Combustion Science*, Vol. 28, pp. 193–266.

16. S.R. Turns (2000) *An Introduction to Combustion – Concepts and Applications*, 2nd Ed. McGraw – Hill: New York.

17. P. Narayanan and A. Trouvé (2009) "Radiation-driven flame weakening effects in sooting turbulent diffusion flames," *Proceedings of the Combustion Institute,* Vol. 32, pp. 1481–1489.

18. G. Heskestad (2002) "Fire plumes, flame height, and air entrainment", in M.J. Hurley et al. (Eds.), *The SFPE Handbook of Fire Protection Engineering*, Chapter 2-1. Springer: New York, pp. 2-1–2-17.

19. L.M. Yuan and G. Cox (1996) "An experimental study of some line fires", *Fire Safety Journal*, Vol. 27, pp. 123–139.

20. B.M. Cetegen and T.A. Ahmed (1993) "Experiments on the periodic instability of buoyant plumes and pool fires", *Combustion and Flame,* Vol. 93, pp. 157–184.

21. C.L. Beyler (2002) "Fire hazard calculations for large, open hydrocarbon fires", in M.J. Hurley et al. (Eds.), *The SFPE Handbook of Fire Protection Engineering*, Chapter 3-11. Springer: New York, pp. 3-268–3-314.

22. American Gas Association (1974) "LNG Safety Research Program" Report IS 3-1.

23. Y. Lin, M.A. Delichatsios, X. Zhang and L. Hu (2019) "Experimental study and physical analysis of flame geometry in pool fires under relatively strong cross flows" *Combustion and Flame*, Vol. 205, pp. 422–433. doi: 10.1016/j.combustflame.2019.04.025.

24. W.R. Hawthorne, D.S. Weddell and H.C. Hottel (1949) Mixing and combustion in turbulent gas jets. In: *Third Symposium on Combustion, Flame, and Explosion Phenomena*, Baltimore: Williams and Wilkins, pp. 266–288.

25. M.A. Delichatsios (1993) "Transition from momentum to buoyancy-controlled turbulent jet diffusion flames and flame height relationships", *Combustion and Flame,* Vol. 92, pp. 349–364.

26. Y. Hasemi (1984) "Experimental Wall Flame Heat Transfer Correlations for the. Analysis of Upward Flame Spread", *Fire Science and Technology*, Vol. 4, pp. 75–90.

27. K.-C. Tsai and D.D. Drysdale (2002) " Flame height correlation and upward flame spread modelling", *Fire and Materials*, Vol. 26, pp. 279–287.

28. B.Y. Lattimer and U. Sorathia (2003) "Thermal characteristics of fires in a non-combustible corner", *Fire Safety Journal*, Vol. 38, pp. 709–745.

29. B.Y. Lattimer (2013) "Heat fluxes and flame lengths from fires under ceilings", *Fire Technology*, Vol. 49, pp. 269–291.

30. B. Merci and E. Dick (2003) "Heat transfer predictions with a cubic K-epsilon model for axisymmetric turbulent jets impinging onto a flat plate", *International Journal of Heat and Mass Transfer*, Vol. 46(3), pp. 469–480.

31. L. Hu, Q. Wang, M. Delichatsios, F. Tang, X. Zhang and S. Lu (2013) " Flame height and lift-off of turbulent buoyant jet diffusion flames in a reduced pressure atmosphere", *Fuel*, Vol. 109, pp. 234–240.

32. J.M. Most, P. Mandin, J. Chen, P. Joulain, D. Durix and C. Fernandez-Pello (1996) *Proceedings of the Combustion Institute*, Vol. 26, pp. 1311–1317.

33. A.Y. Klimenko and F.A. Williams (2013) " On the flame length in firewhirls with strong vorticity", *Combustion and Flame*, Vol. 160, pp. 335–339.

34. J. Lei, N. Liu, L. Zhang, Z. Deng, N.K. Akafuah, T. Li, K. Saito and K. Satoh (2012) " Burning rates of liquid fuels in fire whirls", *Combustion and Flame*, Vol. 159, pp. 2101–2114.

35. K. Zhou, N. Liu, J.S. Lozano, Y. Shan, B. Yao and K. Satoh (2013) "Effect of flow circulation on combustion dynamics of fire whirl", *Proceedings of the Combustion Institute*, Vol. 34, pp. 2617–2624.

36. J.G. Quintiere (2002) Surface flame spread, in M.J. Hurley et al. (Eds.), *The SFPE Handbook of Fire Protection Engineering*, 3rd Ed. Springer: New York, pp. 2-246–2-257.

37. F.A. Williams (1976) "Mechanisms of fire spread", *Proceedings of the Combustion Institute*, Vol. 16, pp. 1281–1294.

38. A.C. Fernandez-Pello (1995) The solid phase, in G. Cox (Ed.), *Combustion Fundamentals of Fire*, pp. 31–100. Academic Press: London.

39. J.N. de Ris (1969) "Spread of a laminar diffusion flame" *Proceedings of the Combustion Institute*, Vol. 12, pp. 241–252.

40. Y. Hasemi, D. Nam and M. Yoshida (2001) *Proceedings of the 5th AOSFST*, pp. 379–390.

41. A.C. Fernandez-Pello, S.R. Ray and I. Glassman (1981) "Flame spread in an opposed forced flow: The effect of ambient oxygen concentration", In: *Eighteenth Symposium (International) on Combustion*, The Combustion Institute, pp. 579–589.

42. N. Ren, Y. Wang and A. Trouvé (2013) "Large eddy simulation of vertical turbulent wall fires", *Procedia Engineering*, Vol. 62, pp. 443–452.

43. A.C. Fernandez-Pello and F.A. Williams (1974) "Laminar flame spread over PMMA surfaces", *Proceedings of the Combustion Institute*, Vol. 15, pp. 217–231.

44. M.J. Gollner, X. Huang, J. Cobian, A.S. Rangwala and F.A. Williams (2013) "Experimental study of upward flame spread of an inclined fuel surface", *Proceedings of the Combustion Institute*, Vol. 34, pp. 2531–2538.

45. G.H. Markstein and J.N. de Ris (1972) "Upwardfire spreadover textiles", *Proceedings of the Combustion Institute*, Vol. 14, pp. 1085–1097.

46. D.D. Drysdale and A.J.R. Macmillan (1992) "Flame spread on inclined surfaces", *Fire Safety Journal*, Vol. 18, pp. 245–254.

47. J.L. de Ris and L. Orloff L. (2005) *Proceedings of the 8th IAFSS Symposium*, pp. 999–1010.

48. S. Wasan, P. van Hees and B. Merci (2011) "Study of pyrolysis and upward flame spread on charring materials: Part I: Experimental study," *Fire and Materials*, Vol. 35(4), pp. 209–229.

49. "FM Approvals, Cleanroom Materials, Flammability Test Protocol," Class Number 4910, FM Approvals, 1151 Boston-Providence Turnpike, Norwood, Massachusetts, U.S.A., September 1997.

50. D. Zeinali, S. Verstockt, T. Beji, G. Maragkos, J. Degroote and B. Merci (2018) "Experimental study of corner fires: Part I: Inert panel tests", *Combustion and Flame*, Vol. 189, pp. 472–490. doi: 10.1016/j.combustflame.2017.09.034.

51. D. Zeinali, S. Verstockt, T. Beji, G. Maragkos, J. Degroote and B. Merci (2018) "Experimental study of corner fires: Part II: Flame spread over MDF panels." *Combustion and Flame*, Vol. 189, pp. 491–505. doi: 10.1016/j.combustflame.2017.10.023.

52. D. Zeinali, E. Vandemoortele, S. Verstockt, T. Beji, G. Maragkos and B. Merci (2021) "Experimental study of corner fires: Part III: Flame spread over MDF panels." *Fire Safety Journal*, Vol. 121, p. 103265. doi: 10.1016/j.firesaf.2020.103265.

4 Smoke Plumes

4.1 INTRODUCTION

A smoke plume results from the release of hot combustion products by a fire. The temperature difference with the surrounding ambient air induces a buoyant force that drives the combustion products upwards. A smoke plume is therefore also called a buoyant plume. The study of buoyant plumes is of pinnacle importance in fire dynamics and fire safety because the release of toxic gases (such as carbon monoxide, the 'chief killer' in fires) is the first cause of fatalities, well before the exposure to radiant heat from flames. Furthermore, a good understanding of smoke plume dynamics is fundamental in the design of efficient smoke management systems (Chapter 6). The fire source (thus the smoke plume source) can be of several idealized shapes, e.g., circular, square and rectangular. If the source is circular, the plume is referred to as an axisymmetric plume. If the source is rectangular with an aspect ratio great enough (i.e., long and narrow source), the plume is called a line plume. In addition to the geometrical shape of the fire source, the plume dynamics depend on the interaction with nearby or bounding walls (see Section 3.5). When a line or axisymmetric plume is remote from any walls, ambient air is entrained from all sides. In the presence of a wall, air entrainment is inhibited from its side, yielding a different mixing process (see Chapter 3). This effect can be exemplified by spill plumes, which are a typical example of line plumes generally encountered in atria. The smoke layer moves first horizontally beneath the ceiling of a room or a balcony before it is released to the open space at the spill edge and rises upwards. If the air entrainment occurs at both sides, the plume is called free or double-sided plume. If the plume is 'attached' on one side to a wall, it is called an adhered plume. The plume may also interact with a cross-flow. Typical examples would be a plume in open atmosphere conditions and subjected to a wind or when a smoke control system with longitudinal ventilation is activated in a tunnel or a car park.

The objective of this chapter is to gain an understanding of the main phenomenological aspects of non-reacting plume dynamics. This is achieved by first addressing a thorough mathematical description (along with the underlying assumptions such as the Boussinesq approximation) of a simple canonical flow – the axisymmetric buoyant plume. The cases of interaction of a plume with walls, ceiling and a cross-flow are addressed later. Furthermore, a Computational Fluid Dynamics (CFD) simulation of a non-reacting turbulent buoyant plume is discussed in Chapter 8.

4.2 AXISYMMETRIC PLUME

In order to understand the basic principles in the development of a smoke plume, we start by addressing a canonical flow that has been extensively examined (analytically, experimentally and numerically) by many researchers, namely the axisymmetric plume configuration (Figures 4.1–4.3).

DOI: 10.1201/9781003204374-4

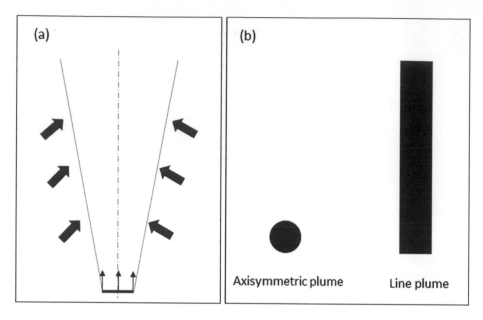

FIGURE 4.1 A 2-D schematic of free axisymmetric and line plumes. (a) Front view (the blue arrows denote the entrainment of fresh air). (b) Top view of the fire source.

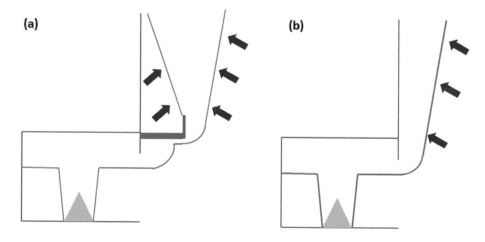

FIGURE 4.2 A 2-D schematic of (a) a free double-sided plume (b) an adhered plume. The blue arrows denote the entrainment of fresh air.

4.2.1 THEORY AND MATHEMATICAL MODELLING

4.2.1.1 Description of the Configuration

Figure 4.4 shows a round buoyant plume characterized by an initial radius, R_0, a vertical velocity, w_0, and a density difference, $\Delta\rho_0 = \rho_{amb} - \rho_0$, at the inlet (where ρ_{amb} and ρ_0 are, respectively, the ambient density and the density at the inlet). At the source, both momentum and buoyancy govern the flow.

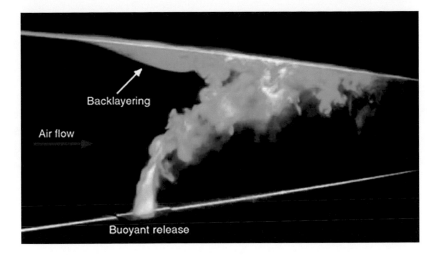

FIGURE 4.3 The influence of a longitudinal smoke control system (red arrow) on a smoke plume in a tunnel [1].

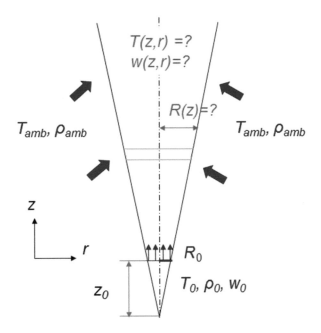

FIGURE 4.4 Sketch of a round buoyant plume using the point source model.

The kinematic (or specific) momentum and buoyancy rates at the source are, respectively, expressed as follows:

$$M_0 = 2\pi \int_0^{R_0} w_0^2(r)\,r\,dr \qquad (4.1)$$

and

$$B_0 = 2\pi \int_0^{R_0} w_0(r) g \frac{\Delta\rho_0(r)}{\rho_{\text{amb}}} r \, dr, \tag{4.2}$$

where r is the radial direction, R_0 is the radius of the source, and g is the gravitational acceleration.

Top hat vertical velocity and density profiles at the source (i.e., $w_0(r) = w_0 = $ constant and $\rho_0(r) = \rho_0 = $ constant) lead to the following expressions of, respectively, the momentum (in m^4/s^2) and buoyancy (in m^4/s^3) rates:

$$M_0 = \pi R_0^2 w_0^2 \tag{4.3}$$

and

$$B_0 = \pi R_0^2 w_0 \frac{\Delta\rho_0}{\rho_{\text{amb}}} g. \tag{4.4}$$

The source could also be characterized by a convective heat release rate, $\dot{Q}_{\text{conv}}$, which is expressed as follows:

$$\dot{Q}_{\text{conv}} = \pi R_0^2 \rho_0 w_0 c_p \cdot (T_0 - T_{\text{amb}}). \tag{4.5}$$

where c_p is the specific heat (that could be taken as constant, $c_p = 1$ kJ/(kg K) and T_0 and T_{amb} are, respectively, the inlet) and ambient temperatures.

By virtue of the ideal gas law (i.e., $\rho_{\text{amb}} T_{\text{amb}} = \rho_0 T_0$), one can write:

$$\frac{\Delta\rho_0}{\rho_{\text{amb}}} = 1 - \frac{\rho_0}{\rho_{\text{amb}}} = 1 - \frac{T_{\text{amb}}}{T_0} = \frac{T_0 - T_{\text{amb}}}{T_0}. \tag{4.6}$$

Inserting Eq. (4.6) into Eq. (4.4) and multiplying both the numerator and denominator by $\rho_0 c_p$ gives:

$$B_0 = \pi R_0^2 w_0 \frac{(T_0 - T_{\text{amb}})}{T_0} g = \pi R_0^2 \rho_0 w_0 c_p \frac{(T_0 - T_{\text{amb}})}{\rho_0 T_0 c_p} g. \tag{4.7}$$

Inserting Eq. (4.5) into Eq. (4.7) and using the ideal gas law, one obtains:

$$B_0 = \frac{g \dot{Q}_{\text{conv}}}{\rho_{\text{amb}} T_{\text{amb}} \cdot c_p}. \tag{4.8}$$

As the flow evolves, buoyancy overwhelms inlet momentum. After a certain distance from the inlet (i.e., further downstream), the momentum of the flow becomes solely driven by buoyancy, not by the momentum at the inlet. The distance over which this evolution occurs can be characterized by the so-called Morton length scale [2]:

$$L_M = \frac{M_0^{3/4}}{B_0^{1/2}}. \tag{4.9}$$

Studies have shown that the plume-like behaviour (flow governed solely by buoyancy) is achieved for at least $z / L_M > 5$ (z being the height above the source).

The objective of the next section is to address the development of an analytical model that provides the steady-state density (and thus temperature) and velocity profiles in the axisymmetric buoyant plume. The source is considered first as a point. This assumption is called the 'point source model'.

4.2.1.2 Conservation Equations of Mass, Momentum and Energy

In this section, the governing equations for mass, momentum and energy conservation are derived based on average quantities within the plume. Consider to that purpose the control volume in Figure 4.4 (grey zone), which represents a plume segment. Simplifying assumptions are used in order to derive analytical correlations for mean vertical velocity and buoyancy acceleration, as well as other parameters, such as the entrainment rate. The set of equations for a CFD approach will be specified later in Chapter 8.

4.2.1.2.1 Conservation of Mass

Figure 4.5 shows a visual description of the mass balance for the considered control volume where u is the horizontal velocity of entrained air and $\bar{\rho}$ and $\bar{w}$ are, respectively, the average density and vertical velocity across the plume segment. The latter is characterized by a radius, R, and a very small height Δz. The bottom section of the control volume is at height z.

- The mass flow rate (in kg/s) entering through the bottom boundary of the control volume is $\left(\pi R^2 \bar{\rho} \bar{w}\right)_z$.
- The mass flow rate entrained through the side boundary of the control volume (for a small Δz and thus an almost constant R, i.e., cylindrical shape) is $2\pi R \, \Delta z \, \rho_{\text{amb}} \, u$. The average value of the radius is in fact $R + \Delta R / 2$, but $\Delta R \cdot \Delta z \ll R \cdot \Delta z$ (because $\Delta R \ll R$) for small enough Δz. The same applies to all other related equations.
- The mass flow rate exiting from the top boundary of the control volume is $\left(\pi R^2 \bar{\rho} \bar{w}\right)_{z+\Delta z}$.

Hence, the mass balance for the control volume in steady-state conditions yields:

$$\left(\pi R^2 \bar{\rho} \bar{w}\right)_{z+\Delta z} = \left(\pi R^2 \bar{\rho} \bar{w}\right)_z + 2\pi R \, \Delta z \, \rho_{\text{amb}} \, u. \tag{4.10}$$

In differential form, the mass conservation equation reads:

$$\frac{d\left(R^2 \bar{\rho} \bar{w}\right)}{dz} = 2R\rho_{\text{amb}} u. \tag{4.11}$$

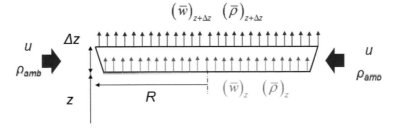

FIGURE 4.5 A schematic description of mass and momentum transfer for a plume segment.

It can be seen that the variation of mass across the control volume is solely induced by the air entrained laterally with (entrainment) velocity u.

4.2.1.2.2   Conservation of Momentum

In Figure 4.5, the mass of the displaced fluid (i.e., ambient air at density ρ_{amb}), is expressed as follows:

$$m_{df} = \pi R^2 \rho_{\text{amb}} \Delta z. \tag{4.12}$$

The actual mass of the plume segment (i.e., the hot volume of air at density $\bar{\rho} = \rho_{\text{amb}} - \Delta\bar{\rho}$) is:

$$m_{ps} = \pi R^2 \left(\rho_{\text{amb}} - \Delta\bar{\rho}\right)\Delta z. \tag{4.13}$$

By virtue of Archimedes' principle, the upward buoyancy force is expressed as follows:

$$F_B = \left(m_{df} - m_{ps}\right)g = \pi R^2 \Delta\bar{\rho}\, g\, \Delta z = \pi R^2 \rho_{\text{amb}}\bar{g}'\, \Delta z, \tag{4.14}$$

where

$$\bar{g}' = \frac{\Delta\bar{\rho}}{\rho_{\text{amb}}} g \tag{4.15}$$

is referred to as the 'reduced gravity'.

Furthermore, the momentum entering through the bottom boundary and the momentum exiting from the top boundary can be written as $\left(\pi R^2 \bar{\rho}\bar{w}^2\right)_z$ and $\left(\pi R^2 \bar{\rho}\bar{w}^2\right)_{z+\Delta z}$.

Since there is no vertical momentum due to lateral entrainment (i.e., the ambient fluid is at rest), the momentum balance for the control volume reads:

$$\left(\pi R^2 \bar{\rho}\bar{w}^2\right)_{z+\Delta z} = \left(\pi R^2 \bar{\rho}\bar{w}^2\right)_z + \pi R^2 \rho_{\text{amb}}\bar{g}'\, \Delta z. \tag{4.16}$$

In differential form, the momentum conservation equation reads:

$$\frac{d\left(R^2 \bar{\rho}\bar{w}^2\right)}{dz} = R^2 \rho_{\text{amb}}\bar{g}'. \tag{4.17}$$

It is interesting to recall here the more general expression for conservation of momentum, which has been derived in Chapter 2 (see Eq. 2.37). If we neglect all the shear stresses and pressure differences in the horizontal directions (i.e., quiescent environment), the remaining force per unit volume (which is in the vertical direction) becomes:

$$F'''_{\text{tot},z} = -\frac{\partial p}{\partial z} - \bar{\rho}g. \tag{4.18}$$

Combining the forces due to pressure and gravity as discussed in Section 2.6 and expressed in Eq. (2.83) gives:

$$F'''_{\text{tot},z} = \left(\rho_{\text{amb}} - \bar{\rho}\right)g. \tag{4.19}$$

By considering the volume, V, of the plume segment considered above and expressed as $V = \pi R^2 \Delta z$, the buoyancy force in Eq. (4.14) is recovered through:

$$F_B = F'''_{\text{tot},z} V. \tag{4.20}$$

It is important to note that the plume does not rise indefinitely; it extends vertically up to a certain height where the buoyancy force becomes too weak to overcome the viscous drag which has been ignored in the present analysis. This maximum height is expressed as follows:

$$z_{\text{max}} = 15.5 \, \dot{Q}_{\text{conv}}^{2/5} \, \Delta T_{\text{amb}}^{-3/5}, \tag{4.21}$$

where ΔT_{amb} is the increase in ambient temperature between the level of the fire source and the ceiling.

4.2.1.2.3   Conservation of Energy

If we take T_{amb} as a 'reference temperature', the entrained air does not bring energy to the plume. Furthermore, under the following considerations:

- no outflow through the side boundaries;
- no radiation; and
- no work against pressure, gravity and viscosity,

the energy conservation becomes equivalent to the conservation of the buoyancy flux, B_0 (see Eq. 4.8). Therefore, the energy conservation reads:

$$dB/dz = 0, B_0 = B = \pi R^2 \overline{w} \, \overline{g'}. \tag{4.22}$$

The system of three conservation equations is then:

- **Mass**

$$\frac{d\left(R^2 \overline{\rho} \overline{w}\right)}{dz} = 2R\rho_{\text{amb}} u. \tag{4.23}$$

- **Momentum**

$$\frac{d\left(R^2 \overline{\rho} \overline{w}^2\right)}{dz} = R^2 \rho_{\text{amb}} \overline{g'}. \tag{4.24}$$

- **Energy (buoyancy)**

$$B_0 = B = \pi R^2 \overline{w} \, \overline{g'}. \tag{4.25}$$

4.2.1.3   Model Development under the Boussinesq Approximation

4.2.1.3.1   The Boussinesq Approximation

The Boussinesq approximation states that if the density difference between the plume and the ambient is relatively small, it could be neglected. This implies that for the conservation equations developed above, the plume density can be replaced by the ambient density, except for the term where $\Delta \bar{\rho} / \rho_{amb}$ is multiplied by the gravitational force, g, to give the reduced gravity, $\bar{g}'$.

4.2.1.3.2   Conservation Equations

In the development of the mass and momentum conservation equations for a free axisymmetric plume without considering the Boussinesq approximation, the density of the control volume was taken as $\bar{\rho}$, as shown on the left-hand side of Eqs. (4.23) and (4.24). Under the Boussinesq approximation, the density of the control volume is approximated by the density of the surrounding ambient air, i.e., $\bar{\rho} = \rho_{amb}$. Since the density ρ_{amb} appears on both sides of Eqs. (4.23) and (4.24), a simplification is made and the new system of conservation equations becomes:

- **Mass**

$$\frac{d\left(R^2 \bar{w}\right)}{dz} = 2Ru. \tag{4.26}$$

- **Momentum**

$$\frac{d\left(R^2 \bar{w}^2\right)}{dz} = R^2 \bar{g}'. \tag{4.27}$$

- **Energy**

$$B = \pi R^2 \bar{w}\, \bar{g}'. \tag{4.28}$$

The set of equations obtained above results from an additional assumption according to which the ambient density does not vary with height, i.e., $d\rho_{amb}/dz = 0$. The surrounding ambient fluid is then called *not stratified* or *neutral*. In the case of a *stratified* environment (i.e., $d\rho_{amb}/dz \neq 0$) there is an additional local increase or decrease in buoyancy.

In order to close the obtained system of conservation equations, we assume:

$$u = a\,\bar{w}, \tag{4.29}$$

where a is a dimensionless coefficient called the entrainment constant. The unknowns can then be expressed as a function of the remaining parameters of the problem, i.e., z and B.

4.2.1.3.3   Model Solution and Comparison to Other Correlations in the Literature

From a dimensional analysis, one can write:

$$R = b\,z, \tag{4.30}$$

$$\overline{w} = c\,\frac{B^{1/3}}{z^{1/3}}, \tag{4.31}$$

$$u = a\,\overline{w}, \tag{4.32}$$

$$\overline{g'} = d\,\frac{B^{2/3}}{z^{5/3}}. \tag{4.33}$$

Inserting Eqs. (4.30), (4.31) and (4.33) in Eq. (4.28) gives:

$$\pi b^2 c d = 1. \tag{4.34}$$

Inserting Eqs. (4.30) to (4.33) in Eq. (4.26) gives:

$$\frac{5}{3}b = 2a. \tag{4.35}$$

Inserting Eqs. (4.30), (4.31) and (4.33) in Eq. (4.27) gives:

$$\frac{4}{3}c^2 = d. \tag{4.36}$$

The set of Eqs. (4.34) to (4.36) gives:

$$b = \frac{6}{5}a, \tag{4.37}$$

$$c = \left(\frac{25}{48\pi a^2}\right)^{1/3}, \tag{4.38}$$

$$d = \frac{4}{3}\left(\frac{25}{48\pi a^2}\right)^{2/3}. \tag{4.39}$$

The solution for the system of conservation equations for mass, momentum and energy becomes therefore a function of solely the height, z, the buoyancy, B, at the source and the entrainment coefficient, a:

$$R = \frac{6}{5}az, \tag{4.40}$$

$$\overline{w} = \left(\frac{25}{48\pi a^2}\right)^{1/3}\frac{B^{1/3}}{z^{1/3}}, \tag{4.41}$$

$$u = a \left(\frac{25}{48\pi a^2} \right)^{1/3} \frac{B^{1/3}}{z^{1/3}}, \tag{4.42}$$

$$\overline{g'} = \frac{4}{3} \left(\frac{25}{48\pi a^2} \right)^{2/3} \frac{B^{2/3}}{z^{5/3}}. \tag{4.43}$$

Using Eqs. (4.41) and (4.8), the average vertical velocity can be expressed as follows:

$$\overline{w} = \left(\frac{25g}{48\pi a^2 \rho_{amb} T_{amb} c_p} \right)^{1/3} \dot{Q}_{conv}^{1/3} z^{-1/3} \tag{4.44}$$

or

$$\overline{w} = \left(\frac{25}{48\pi a^2} \right)^{1/3} \sqrt{gz} \left(\dot{Q}_{conv}^* \right)^{1/3}, \tag{4.45}$$

where the dimensionless heat release rate is expressed as follows:

$$\dot{Q}_{conv}^* = \frac{\dot{Q}_{conv}}{\rho_{amb} T_{amb} c_p z^2 \sqrt{gz}}. \tag{4.46}$$

Using the ideal gas law and the Boussinesq approximation, the average temperature difference across the plume width can be expressed as follows:

$$\frac{\Delta \overline{T}}{T_{amb}} = \frac{4}{3} \left(\frac{25}{48\pi a^2} \right)^{2/3} \left(\rho_{amb} T_{amb} c_p \sqrt{g} \right)^{-2/3} \dot{Q}_{conv}^{2/3} z^{-5/3} \tag{4.47}$$

or

$$\frac{\Delta \overline{T}}{T_{amb}} = \frac{4}{3} \left(\frac{25}{48\pi a^2} \right)^{2/3} \left(\dot{Q}_{conv}^* \right)^{2/3}. \tag{4.48}$$

The plume entrainment mass flow rate (in kg/s) is calculated as follows:

$$\dot{m}_{ent} = \rho_{amb} \overline{w} \pi R^2. \tag{4.49}$$

Inserting Eqs. (4.44) and (4.40) into the previous equation gives:

$$\dot{m}_{ent} = \frac{3}{4} \rho_{amb} \left(\frac{25}{48\pi a^2} \right)^{-2/3} \left(\frac{g}{\rho_{amb} T_{amb} c_p} \right)^{1/3} \dot{Q}_{conv}^{1/3} z^{5/3}. \tag{4.50}$$

By taking $\rho_{amb} = 1.2 \text{ kg/m}^3$, $T_{amb} = 293 \text{ K}$, $c_p = 1 \text{ kJ}/(\text{kg K})$ and $g = 9.81 \text{ m/s}^2$ and setting the entrainment coefficient to $a = 0.15$ (as found from experiments for a top hat velocity profile [3]), the following simple expressions are obtained for the average vertical velocity, temperature difference and entrainment rate across the plume width:

- **Average vertical velocity**

$$\bar{w} \simeq 0.59 \dot{Q}_{conv}^{1/3} z^{-1/3}. \tag{4.51}$$

- **Average temperature difference**

$$\Delta \bar{T} \simeq 14 \dot{Q}_{conv}^{2/3} \; z^{-5/3}. \tag{4.52}$$

- **Plume entrainment**

$$\dot{m}_p \simeq 0.072 \dot{Q}_{conv}^{1/3} \; z^{5/3}. \tag{4.53}$$

Equation (4.53) for plume entrainment is commonly referred to in the literature [3] as the correlation of Zukoski who carried out hood experiments for plume gases [4] and confirmed the ideal plume theory.

According to Eq. (4.40), the virtual origin can be calculated as follows:

$$z_0 = -\frac{5}{6a} R_0. \tag{4.54}$$

which implies that it is a function of only the fire source diameter, $D_0 (= 2R_0)$.

In terms of fuel bed area, A_F, Eq. (4.54) could be rewritten as follows [5]:

$$z_0 = -\frac{5}{6a\pi^{1/2}} A_F^{1/2} \simeq -1.5 A_F^{1/2} \left(\text{for } a = 0.15 \right). \tag{4.55}$$

If the reference level for height is not the virtual origin but the fire source level, the variable z in the previous expressions for velocity, temperature and plume entrainment should be replaced by $z - z_0$.

However, several studies have shown that the location of the virtual origin depends not only on the fire source diameter (as suggested in Eqs. 4.54 and 4.55) but also on the total [6] heat release rate of the fire, $\dot{Q}$ [7–9]. Heskestad [10] provides the following expression:

$$z_0 = -1.02 D_0 + 0.083 \dot{Q}^{2/5}. \tag{4.56}$$

which is based on pool fire experiments with diameters in the range of 0.16–2.4 m.

4.2.1.3.4 Self-similarity Profiles

So far, we have assumed a top hat distribution of vertical velocity and density (and thus reduced gravity) across the plume width. However, numerous experiments (e.g., [11–13]) have shown that these profiles match Gaussian functions. Therefore, the local velocity and reduced gravity can be expressed as follows:

$$w = w_{max}(z) \exp\left(-\frac{r^2}{2\sigma^2} \right), \tag{4.57}$$

$$g' = g'_{max}(z) \exp\left(-\frac{r^2}{2\sigma^2} \right), \tag{4.58}$$

where σ is taken as:

$$\sigma = \frac{R}{2}. \tag{4.59}$$

The peak values along the plume centreline can then be related to their respective averages by:

$$\bar{w} = \frac{1}{\pi R^2} \int_0^\infty w 2\pi r dr, \tag{4.60}$$

$$\overline{g'} = \frac{1}{\pi R^2} \int_0^\infty g' 2\pi r dr. \tag{4.61}$$

The coefficients for Eqs. (4.57) and (4.58) will be provided in the next section for a number of experimental studies.

4.2.1.4 List of Assumptions

It is very important to recall at this stage the list of assumptions and approximations that led to the solution presented above for steady-state axisymmetric buoyant plumes in order to bear in mind the conditions for its validity.

- **Unbounded buoyant plume:** Fresh air is entrained from all the sides of the plume without being altered by nearby obstacles (typically walls). If a fire develops in an enclosure with a large floor area away from walls, one can reasonably assume *unbounded* conditions.
- **Quiescent environment:** This means that the ambient fluid is at rest and thus there are no perturbations induced by a local or distributed cross-stream (e.g., wind).
- **Small density differences:** This is generally the case for the far-field region of a fire plume. This allowed using the Boussinesq approximation (thanks to which the plume density is replaced by the ambient density, except where the density difference is multiplied by gravity).
- **Neutral, non-stratified environment:** The ambient density, in this case, is constant; it does not vary with height. If this is not the case, there will be additional sink or source terms for buoyancy that will influence the flow.
- Absence of shear stress
- **No fluctuations:** Turbulence is accounted for in the entrainment coefficient.
- No chemical reactions
- No radiation
- **Constant entrainment coefficient:** Some studies, e.g., [14], have shown that the entrainment coefficient might vary with the local Richardson number though.

4.2.2 EXPERIMENTS

The model developed above has been confirmed through a number of experiments where buoyancy was generated from either a hot air source (e.g., [13]) or a gas burner.

4.2.2.1 Hot Air Plumes

In their experiments, Shabbir and George [13] used a plume generator that consists of a source of heated air issued from a nozzle with diameter $D = 0.0635$ m. The exit temperature was $T_0 = 292\,°C \pm 1\,°C$. The ambient temperature varied between $23\,°C$ and $25\,°C$, and it was monitored to ensure that the facility was not stratified, so that the experiment was conducted in a neutral environment. The exit velocity measured by a two-wire probe was $U_0 = 0.98$ m/s. The velocity exit profile was almost uniform (i.e., a top hat profile). Based on the measured data, the reported flow parameters at the inlet are $B_0 = 0.0127$ m^4/s^3, $M_0 = 0.003$ m^4/s^2 and $L_M = 0.114$ m (see, respectively, Eqs. 4.1, 4.2 and 4.9). A square screen enclosure was placed around the plume, which was 2 m × 2 m in cross section and 5 m in height. Thus, the enclosure was far enough from the flow in order not to interfere with it.

The mean flow results for, respectively, the buoyancy flux (reduced gravity) and vertical velocity were expressed as follows [13]:

$$g'(r,z) = g\frac{\Delta T}{T} = A_T \frac{B_0^{2/3}}{z^{5/3}} \exp\left(-B_T \frac{r^2}{z^2}\right), \tag{4.62}$$

$$w(r,z) = A_W \frac{B_0^{1/3}}{z^{1/3}} \exp\left(-B_W \frac{r^2}{z^2}\right), \tag{4.63}$$

where A_T, B_T, A_W and B_W are constants provided in Table 4.1. Note the similarity of these two equations with Eqs. (4.33) and (4.31) as well as Eqs. (4.58) and (4.57) where $g'_{max}(z) = A_T B_0^{2/3} z^{-5/3}$ and $w_{max}(z) = A_W B_0^{1/3} z^{-1/3}$.

Shabbir and George calculated also [13]:

- the non-dimensional half-width of the buoyancy profile, $\ell_{\Delta T/2}$, defined as the location where the normalized buoyancy is half its centreline value, and
- the non-dimensional half-width of the vertical velocity profile, $\ell_{W/2}$, defined as the location where the normalized velocity is half its centreline value.

The non-dimensional half-widths of the buoyancy and vertical velocity profiles are calculated (based on their definitions and according to Eqs. 4.62 and 4.63) as:

$$\ell_{\Delta T/2} = \left[\frac{-\ln(1/2)}{B_T}\right]^{1/2} \tag{4.64}$$

TABLE 4.1

Measurements of Mean Flow Parameters and Turbulence Intensities for Buoyant Plumes [13]

References	A_T	A_W	B_T	B_W	$\ell_{\Delta T/2}$	$\ell_{W/2}$	$\sqrt{\overline{T'^2}}/\Delta T_c$	$\sqrt{\overline{w'^2}}/W_c$
Rouse et al. [12]	11.0	4.7	71	96	0.095	0.084	–	–
George et al. [11]	9.1	3.4	65	55	0.104	0.112	0.38	0.28
Nakagome and Hirata [15]	11.5	3.89	48.1	63	0.105	0.12	0.33	0.25
Papanicolaou and List [16]	14.28	3.85	80	90	0.093	0.0877	0.42	0.25
Chen and Rodi [17]	9.35	3.5	65	55	0.10	0.112	–	–
Shabbir and George [13]	9.4	3.4	68	58	0.10	0.107	0.40	0.33

and

$$\ell_{W/2} = \left[\frac{-\ln(1/2)}{B_W} \right]^{1/2}.$$ (4.65)

Table 4.1 shows the mean flow results as well as the turbulence intensities measured by Shabbir and George [13] and other researchers. The differences in the mean flow results for buoyancy (A_T, B_T and $\ell_{\Delta T/2}$) are mainly attributed to either the calculation of the buoyancy, B_0, or the experimental technique.

4.2.2.2 Smoke Plumes from Fires

The analytical equations developed above and verified for hot air plumes have also been examined for fire plumes. Some of the results are provided here.

4.2.2.2.1 Heskestad's Correlations

The far-field (smoke) plume region has been defined by Heskestad [10] as the region *immediately above the flame height*, L_f (i.e., $z > L_f$) (see Eq. 3.30 for the calculation of L_f). In this region, the centreline plume velocity and temperature difference are respectively expressed as follows:

$$w_c = 3.4 \left(\frac{g}{c_p T_{amb} \rho_{amb}} \right)^{1/3} \dot{Q}_{conv}^{1/3} (z - z_0)^{-1/3} \simeq 1.03 \, \dot{Q}_{conv}^{1/3} (z - z_0)^{-1/3}$$ (4.66)

and

$$\Delta T_c = 9.1 \left(\frac{T_{amb}}{g c_p^2 \rho_{amb}^2} \right)^{1/3} \dot{Q}_{conv}^{2/3} (z - z_0)^{-5/3} = 25 \, \dot{Q}_{conv}^{2/3} (z - z_0)^{-5/3}.$$ (4.67)

4.2.2.2.2 McCaffrey's Correlations

The far-field (smoke) plume region has been defined by McCaffrey [18] as the region where $z / \dot{Q}^{2/5} > 0.2 \, \mathrm{m/kW}^{2/5}$. In this region, the centreline plume velocity and temperature difference are, respectively, expressed as follows:

$$w_c = 1.1 \dot{Q}^{1/3} z^{-1/3}$$ (4.68)

and

$$\Delta T_c = T_{amb} \left(\frac{1.1}{0.9\sqrt{2g}} \right)^2 \dot{Q}^{2/3} z^{-5/3} \simeq 22 \dot{Q}^{2/3} z^{-5/3}.$$ (4.69)

Note that the four equations above for centreline values have the same form as Eqs. (4.51) and (4.52) but higher coefficients because the latter expressions provide *average quantities* across the plume width.

4.3 LINE PLUME

4.3.1 DESCRIPTION OF THE CONFIGURATION

The source of a line plume is characterized by a width, $W = 2b_0$, significantly smaller than its length, L (i.e., $2b_0 \ll L$) (see Figure 4.6). Such a configuration is typical for balcony spill plumes

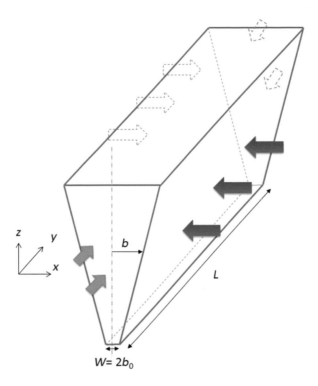

FIGURE 4.6   3-D sketch of a line plume.

(in atria for example), where the smoke volume is extended beneath the full length of a large balcony before rising into the large open space. Additionally, note that the plume is contained from
both ends in the y-direction so that entrainment occurs only in the x-direction.

Similarly to the axisymmetric plume, the source can be characterized by a kinematic momentum and buoyancy flux, in this case per unit length, that are, respectively, expressed as follows:

$$M_0' = 2b_0 \bar{w}_0^2 \qquad (4.70)$$

and

$$B_0' = 2b_0 \bar{w}_0 g \left(\frac{\rho_{\text{amb}} - \bar{\rho}_0}{\rho_{\text{amb}}} \right), \qquad (4.71)$$

where b_0, $\bar{w}_0$ and $\bar{\rho}_0$ are, respectively, the half-width, average vertical velocity and average density
at the source of the plume.

The Morton length scale is expressed as follows:

$$L_M = \frac{M_0'}{\left(B_0' \right)^{2/3}}. \qquad (4.72)$$

The exponent 2/3 is in line with the correlations derived in Section 3.5.1.1 for flames from a line
source.

The convective heat release rate per unit length is related to the buoyancy flux per unit length through:

$$B_0 = \frac{g\dot{Q}'_{conv}}{\rho_{amb}T_{amb}\,c_p L}.$$

(4.73)

4.3.2 CONSERVATION EQUATIONS

Similarly to the axisymmetric plume, the following conservation equations are obtained for a line plume under the Boussinesq approximation:

- **Mass**

$$\frac{d(b\bar{w})}{dz} = u.$$

(4.74)

- **Momentum**

$$\frac{d(b\bar{w}^2)}{dz} = b\bar{g}'.$$

(4.75)

- **Energy (buoyancy)**

$$B'_0 = B' = 2b\bar{w}\bar{g}'.$$

(4.76)

where b and $\bar{w}$ are, respectively, the half-width and average vertical velocity of the plume at a height z.

It can be seen that the mass and momentum conservation equations are analogous to the axisymmetric case (using the same entrainment model). However, in this case, the area of the plume segment is expressed as $L \times 2b$ as opposed to πR^2 for the axisymmetric plume. Also, the area of entrainment, in this case, is $2 \times L \times \Delta z$ as opposed to $2\pi R\Delta z$ for the axisymmetric plume. Therefore, the constant length L that appears on both sides of the equations gets cancelled.

The solution of the previous system of equations is:

- **Plume width**

$$b = Ez.$$

(4.77)

- **Average vertical velocity**

$$\bar{w} = 2E^{-1/3}\left(B'_0\right)^{1/3}.$$

(4.78)

- **Horizontal velocity**

$$u = E\bar{w}.$$

(4.79)

- **Average reduced gravity**

$$\bar{g}' = 2E^{-2/3}\left(B'_0\right)^{2/3} z^{-1},$$

(4.80)

where E is the entrainment coefficient.

By using Eqs. (4.77), (4.78) and (4.73), the entrainment mass flow rate is expressed as follows:

$$\dot{m}_{ent} = 2bL\rho_{amb}\bar{w} = \left(\frac{4E^2\rho_{amb}^2 g}{T_{amb}c_p}\right)^{1/3} \dot{Q}_{conv}^{\prime 1/3} L^{2/3} z. \tag{4.81}$$

Lee and Emmons [19] provide the following correlation for a line fire in the open:

$$\dot{m}_{ent} = 0.21\dot{Q}^{1/3}L^{2/3}(z-z_0). \tag{4.82}$$

which corresponds to Eq. (4.81) by taking $\rho_{amb} = 1.2\,\text{kg/m}^3$, $T_{amb} = 293\,\text{K}$, $c_p = 1\,\text{kJ/kg K}$ and $g = 9.81\,\text{m/s}^2$ and setting the entrainment coefficient to $E = 0.22$ (in addition to the use of a vertical origin, z_0).

4.3.3 Experimental Studies

In Rouse et al. [12], the vertical velocity and reduced gravity are expressed in the self-similarity region as follows:

$$w(B_0')^{-1/3} = 1.8\exp(-32x^2/z^2) \tag{4.83}$$

and

$$g\frac{\Delta T}{T}(B_0')^{-1/3}z = 2.6\exp(-41x^2/z^2). \tag{4.84}$$

The mixing structure of a self-preserving turbulent buoyant line plume has also been investigated experimentally by Sangras et al. [20] by simulating the plumes with helium/air sources in a still and unstratified environment. Measurements of the mean and fluctuating mixture fraction yielded, respectively, the following profiles:

$$F\left(\frac{x}{z-z_0}\right) = 2.1\exp\left[-60\left(\frac{x}{z-z_0}\right)^2\right] \tag{4.85}$$

and

$$F'\left(\frac{x}{z-z_0}\right) = \overline{f'}(B_0')^{2/3}(z-z_0)\left|1 - \frac{\rho_{amb}}{\rho_0}\right|, \tag{4.86}$$

where the maximum non-dimensional mixture fraction located at the centre is $F(0) = 2.1$ and the fluctuation intensity at the centreline is $\overline{f'} = 42\%$. These results will be compared later (Section 4.4.1) to an analogous experiment where the plume was placed against an adiabatic wall in order to quantify the influence of the latter on the flow field.

4.3.4 Transition from Line to Axisymmetric Plume

As the width of a line plume increases with height, it can reach a certain level where the cross section of the plume becomes almost an ellipse, leading to an axisymmetric plume-like behaviour. One must therefore bear in mind the change in plume entrainment rate for this type of situation.

4.4 WALL AND CORNER INTERACTION WITH PLUMES

Walls and corners can have a substantial effect on plume entrainment. This is illustrated for several cases in Figure 4.7. One can see, for instance, that if the square source is placed against a wall, air is entrained from only three sides (instead of four). Having a source in a corner results in an even lesser entrainment (namely from two sides only). In this context, the shape of the source is also important, as illustrated by the first two left figures. One can see that when a circular source is placed tangent to a wall, the entrainment remains 'similar' to axisymmetric conditions, which is not the case for the square source. Indeed, while there is some obstruction for entrainment from the back (where the wall is), there is at least still some room for entrainment into the initially circular plume. Moreover, due to the induced flow, the oncoming flow from the front can stretch the plume in the lateral direction, causing a more ellipse-like shape providing more area (perimeter) for entrainment, compared to a circular shape with the same cross-sectional area. This effect is less prominent in a corner.

4.4.1 DETAILED EXAMPLE: LINE PLUME BOUNDED BY AN ADIABATIC WALL

Wall plumes are encountered above fires along surfaces and, in general, by a source of buoyancy along the base of flat walls (e.g., see Figure 4.2b). Wall plumes have the characteristics of both boundary layers and free jets. Figure 4.8 shows a 2-D sketch of a line plume against a wall. The additional feature shown in Figure 4.8, in comparison to Figures 4.4–4.6 for axisymmetric and line plumes, is the shear stress, τ_w, exerted by the wall.

4.4.1.1 Conservation Equations

The conservation equations for a line plume bounded by an adiabatic wall are provided in the following.

4.4.1.1.1 Mass

$$\frac{d}{dz}\left(\bar{\rho}\,b\,\bar{w}\right) = \rho_{\mathrm{amb}}u = \rho_{\mathrm{amb}}\alpha\,\bar{w},\tag{4.87}$$

where b is the width of the plume (see Figure 4.6) and $\bar{w}$ its average vertical velocity.

It can be seen that the mass conservation equation is analogous to the axisymmetric case (where α is the entrainment coefficient). However, in this case, the surface of the plume segment is expressed as $L \times b$ (as opposed to πR^2 for the axisymmetric plume) where L is the assumed constant plume length that appears on both sides of the equation and thus gets cancelled.

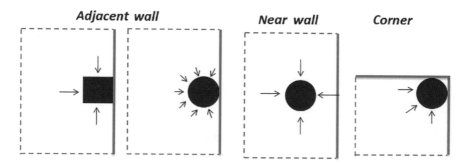

FIGURE 4.7 Influence of walls and corners on plume entrainment (the solid lines represent obstructions and the dashed lines represent open boundaries).

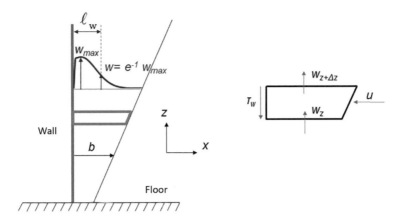

FIGURE 4.8   2-D sketch of a line plume bounded by a wall (the shear stress is ignored).

4.4.1.1.2   Momentum

$$\frac{d\left(b\,\overline{\rho}\,\overline{w}^{2}\right)}{dz} = \Delta\overline{\rho}\,g - \tau_{w},$$ (4.88)

where τ_{w} is the wall shear stress.

The same comment regarding the area of the plume segment applies to the momentum equation. Also, the entrainment area considered here is half of the area considered for the free-line plume. Furthermore, a second term is added on the right-hand side: $\tau_{w} = \mu_{w}\left.\dfrac{\partial\overline{w}}{\partial x}\right|_{x=0}$ where μ_{w} is the dynamic viscosity at the wall. This term expresses the wall friction effect of the solid surface [21], which reduces the momentum. The wall friction reduces the momentum. As opposed to free plumes, wall plumes exhibit 'less' large-scale disturbances, due to the stabilizing effect of the wall. As opposed to free-line plumes, which become (at sufficiently high location from the buoyancy source) self-preserving, adiabatic wall plumes do not reach formally this behaviour because the flow developments at the near-wall boundary layer and the outer plume-like region are not the same. Self-preserving behaviour of adiabatic wall plumes is an approximation [22–24].

4.4.1.1.3   Energy (Buoyancy)

If the wall is adiabatic (i.e., no heat losses by conduction), the energy equation reads:

$$\frac{dB'}{dz} = 0, \quad B' = \frac{g\dot{Q}'_{\text{conv}}}{\rho_{\text{amb}}T_{\text{amb}}c_{p}} = R\overline{w}\overline{g}'.$$ (4.89)

It is important to note here that the buoyancy flux, B', and the convective heat release rate, $\dot{Q}'$, are expressed per unit length since the configuration addressed here is essentially two-dimensional. The corresponding units are thus, respectively, m^{3}/s^{3} and kW/m.

Under the Boussinesq assumption, the mass and momentum equations become, respectively:

$\dfrac{d}{dz}\left(b\,\overline{w}\right) = u = \alpha\,\overline{w}$ and $\dfrac{d\left(b\overline{w}^{2}\right)}{dz} = \overline{g}'b - \dfrac{\tau_{w}}{\rho_{\text{amb}}}$, where $\overline{g}'$ is the average reduced gravity in the plume, Eq. (4.15).

Note that all the assumptions used for the axisymmetric plume in Section 4.2.1 are used here. Only the wall shear stress is added in the momentum equation and the area for entrainment develops in a different way with height.

4.4.1.2 Experiments

Several experimental investigations have been carried out to study buoyant turbulent wall plumes. For instance, Sangras et al. used helium/air mixtures in still air along a smooth vertical wall [23,24]. They carried out measurements of mixture fraction statistics [24] using laser-induced fluorescence (LIF) and measurements of velocity statistics [23] using laser velocimetry (LV). The source was 9.4 mm wide and 876 mm long. Two density ratios were examined at the source, 0.5 and 0.7. Lai and Faeth [25] also performed similar experiments using a mixture of carbon dioxide/air with density ratios of 1.02 and 1.04 and injection velocities of 0.31 and 0.43 m/s.

In order to understand the effect of the wall on the flow field, some of the results reported in Refs. [23,24] will be addressed in the following and compared to the free-line plume described earlier.

4.4.1.2.1 Mean and Fluctuating Velocity Profiles

The vertical velocity distribution, w, is written in Ref. [23] in the form:

$$f\left(\frac{x}{z-z_0}\right)=\frac{\bar{w}}{\left(B_0'\right)^{1/3}},\tag{4.90}$$

where the function f is referred to as an appropriately scaled cross-stream profile function, which approximates a universal function far from the source. It can be written explicitly here as follows:

$$\bar{w}\left(B_0'\right)^{-1/3}=2.84\exp\left[-270\left(\frac{x}{z-z_0}-0.02\right)^2\right].\tag{4.91}$$

The maximum non-dimensional velocity, $W_{max}=2.84$, occurs at a distance $x=0.02(z-z_0)$ from the wall. It must be noted that Eq. (4.91) is only valid at distances sufficiently far from the wall (i.e., $x\geq0.01(z-z_0)$) because at the boundary, $\bar{w}_{x=0}=0$ m/s.

The characteristic plume width from the vertical velocity distribution is defined as the length ℓ_W which gives:

$$f\left(\frac{\ell_W}{z-z_0}\right)=e^{-1}W_{max}.\tag{4.92}$$

For the experiment considered, the obtained value is $\ell_W/(z-z_0)=0.081$.

The velocity fluctuations, w', can be written in a form similar to the mean velocity:

$$f'\left(\frac{x}{z-z_0}\right)=\frac{\overline{w'}}{\left(B_0'\right)^{1/3}}.\tag{4.93}$$

The entrainment constant is calculated as follows:

$$E_0=-\frac{\bar{u}}{\bar{w}_{max}}=0.068.\tag{4.94}$$

4.4.1.2.2   Mean and Fluctuating Mixture Fraction Profiles

The mixture fraction distribution, F, is written in the form:

$$F\left(\frac{x}{z-z_0}\right) = \overline{f}\left(B_0'\right)^{2/3}(z-z_0)\left|1-\frac{\rho_{\text{amb}}}{\rho_0}\right|. \tag{4.95}$$

Similarly to the mean velocity profile, the function F can be written explicitly as follows:

$$F\left(\frac{x}{z-z_0}\right) = 5.9\exp\left[-130\left(\frac{x}{z-z_0}+0.015\right)^2\right], \tag{4.96}$$

where the maximum non-dimensional mixture fraction located at the boundary is $F(0)=5.73$.

The characteristic plume width from the mixture fraction distribution is defined as follows:

$$F\left(\frac{\ell_f}{z-z_0}\right) = e^{-1}F(0). \tag{4.97}$$

For the experiment considered, the obtained value is $\ell_f / (z-z_0) = 0.076$.

The fluctuating mixture fraction profile is written similarly to the mean in the form:

$$F'\left(\frac{x}{z-z_0}\right) = \overline{f'}\left(B_0'\right)^{2/3}(z-z_0)\left|1-\frac{\rho_{\text{amb}}}{\rho_0}\right|, \tag{4.98}$$

where the measured value of $\overline{f'}$ is $\overline{f'}=37\%$.

4.4.2   GENERAL CORRELATIONS FOR WALL AND CORNER CONFIGURATIONS

A general correlation for the entrainment in wall and corner configurations has been proposed in Ref. [26]. This correlation reads:

$$\dot{m}_p = k_m\left[\frac{\left(k_{LF}\dot{Q}\right)^{1/3}}{k_{LF}}\right](z-z_0)^{5/3}, \tag{4.99}$$

where $k_m = 0.076$ (a slightly different value than in Eq. (4.53), i.e., $k_m = 0.072$) and k_{LF} is a 'location' factor taken as the following:

- For free axisymmetric plumes, $k_{LF} = 1$.
- For wall plumes, $k_{LF} = 2$.
- For corner plumes, $k_{LF} = 4$.

Therefore, according to the above formula, fires along walls and in corners entrain 63, resp. 40% of the air they would entrain if they were 'free' in the open, with the same Heat Release Rate (HRR).

The concept of reflection states according to [26] that:

A fire located along a wall can be treated as a fire twice as large with entrainment around only one-half of its perimeter, and a fire located in a corner can be treated as a fire four times as large with entrainment around only one-fourth its perimeter. While this concept neglects the influence of friction along the wall on the entrainment, it has been found to yield reasonably accurate results for design purposes.

It is important to note that when the fire source is circular and the burner is placed with one edge tangent to a vertical wall (see the second sketch from the left in Figure 4.7), it has been found that there is very little influence on plume geometry and plume entrainment [3]. Further research is required on this topic.

4.5   INTERACTION OF A PLUME WITH A CEILING

As seen above, the hot products generated by a fire source are transported upward due to a buoyancy force induced by the density difference with the surrounding environment. If the vertical convective flow impinges onto a horizontal surface (typically a ceiling), it undergoes a change in direction and starts to spread radially. The latter radial flow is referred to as a ceiling jet. If the latter is bounded by walls it is called a confined (as opposed to unconfined) ceiling jet.

The development of a ceiling jet during the early stages of a fire is generally tracked by smoke detectors mounted on the ceiling. These sensors can trigger then either a sound alarm (to indicate the need for evacuation) and/or more active measures to control the fire during its early stages. Such measures can be for instance sprinkler activation and/or smoke extraction via natural or mechanical vents. In order to evaluate the detection time based on the ceiling-jet layer, the latter must be well-characterized in terms of temperature and velocity fields.

4.5.1   DESCRIPTION OF A CEILING JET

As shown in Figure 4.9, the ceiling jet is mainly characterized by a horizontal velocity, u, and a temperature difference, ΔT. In many studies (e.g., [27,28]), these profiles are assumed to be self-similar and are expressed as Gaussian functions:

$$u(r,z) = u_{\max}(r)e^{-\left[(z-\delta_{u_{\max}})/(\ell_u - \delta_{u_{\max}})\right]^2} \quad \text{for} \quad z \geq \delta_{u_{\max}} \tag{4.100}$$

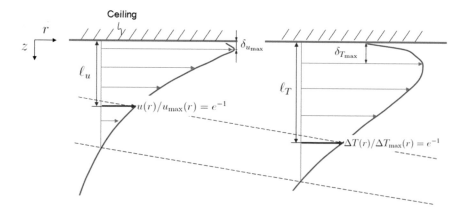

FIGURE 4.9   Schematic of a ceiling jet and its characteristic parameters.

and

$$\Delta T(r,z) = \Delta T_{\max}(r) e^{-\left[(z-\delta_{T\max})/(\ell_T-\delta_{T\max})\right]^2} \quad \text{for} \ \ z \geq \delta_{T\max}, \tag{4.101}$$

where r is the radial distance from the plume axis and z is the vertical distance from the ceiling. The variables $u_{\max}$ and $\Delta T_{\max}$ denote, respectively, the maximum horizontal velocity and temperature difference at a specific radial distance r. The variables $\delta_{u\max}$ and $\delta_{T\max}$ denote, respectively, the viscous boundary layer (defined as the distance from the ceiling to the position of the maximum velocity) and the thermal boundary layer (defined as the distance from the ceiling to the position of the maximum temperature). The variables ℓ_u and ℓ_T represent, respectively, the Gaussian momentum and thermal thicknesses (i.e., the loci of the points in the z direction where $u(r)/u_{\max}(r) = e^{-1}$ and $\Delta T(r)/\Delta T_{\max}(r) = e^{-1}$).

It is noted that Emmons provides in Ref. [29] an analysis based on a top hat velocity and temperature profile for the ceiling jet.

Based on the description provided above, a full characterization of the ceiling-jet velocity profile requires the knowledge of the radial profiles of:

- the Gaussian momentum thicknesses (i.e., ℓ_u and ℓ_T),
- the maximum gas velocities and excess temperature (i.e., $u_{\max}$ and $\Delta T_{\max}$), and
- the viscous and thermal boundary layer depths (i.e., $\delta_{u\max}$ and $\delta_{T\max}$).

Most of the studies and design calculations for ceiling jets are based on correlations for maximum excess temperature and velocity in the ceiling jet flow. However, often, fire detectors and sprinklers are placed at ceiling standoff distances that are outside of this region and therefore experience lower temperatures and lower velocities than predicted. This might result in a substantial increase in response time [30]. Thus, it is important to have a full characterization of the ceiling jet.

4.5.2 Alpert's Integral Model

A generalized theory for unconfined turbulent ceiling jets under steady-state conditions has been developed by Alpert [27]. In the developed model, the ceiling-jet properties are matched to the properties of an axisymmetric turbulent buoyant plume at the entrance of the turning region (see Figure 4.10). The turbulent buoyant plume properties are based on Morton's integral point source model [31] addressed at the start of this chapter. Point e in Figure 4.10 indicates the extent of the turning region in both radial and vertical directions. The coordinates of this point are expressed in terms of a:

- **Non-dimensional radial distance**

$$\frac{r_e}{H} = \frac{6}{5}\sqrt{\frac{3}{2}} E_p \left(1 + \frac{\sqrt{3}}{5} E_p\right)^{-1}. \tag{4.102}$$

- **Non-dimensional characteristic ceiling-jet thickness**

$$\frac{h_e}{H} = \frac{\sqrt{3}}{5} E_p \left(1 + \frac{\sqrt{3}}{5} E_p\right)^{-1}, \tag{4.103}$$

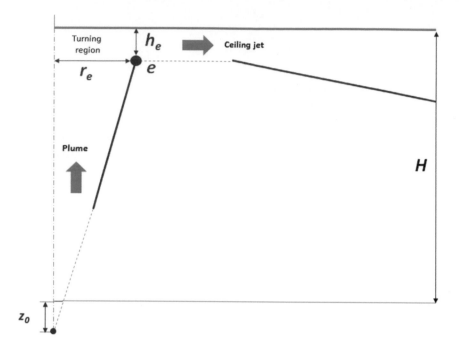

FIGURE 4.10   From a fire plume to a ceiling jet.

where E_p is the entrainment coefficient of the vertical plume.

The ceiling-jet properties are then solved using the conservation of mass, momentum and energy under the Boussinesq approximation.

The momentum thickness and maximum ceiling-jet velocity and temperature are expressed, respectively, as follows:

$$\frac{\ell_u}{H} = f_1\left(E_p, E, F, \frac{r}{H} \right), \tag{4.104}$$

$$\frac{u_{max}}{g\,H^{1/2}\,\dot{Q}^*} = f_2\left(E_p, E, F, K, \frac{r}{H} \right), \tag{4.105}$$

$$g\,\frac{\Delta T_{max}}{T_{max}} = f_3\left(\dot{Q}, E_p, E, F, K, \frac{r}{H} \right), \tag{4.106}$$

where E is the ceiling-jet entrainment coefficient, F the friction factor, K a coefficient of heat transfer to the ceiling and $\dot{Q}^*$ a non-dimensional heat release rate. The values of the plume entrainment coefficient, E_p, reported in Ref. [27] are between 0.116 and 0.14. An intermediate value of 0.12 was used. The ceiling-jet entrainment coefficient, E, has been assumed to be equal to E_p as reported in Ref. [27]. The friction factor, F, for typical ceiling materials F ranges between 0.001 and 0.01. However, Alpert noted that the ceiling-jet flow solutions are not sensitive to the exact value of F because $F \ll E$. The non-dimensional HRR is expressed as follows:

$$\dot{Q}^* = \frac{\dot{Q}}{\rho_{amb}T_{amb}c_p g^{1/2}H^{5/2}}. \tag{4.107}$$

Note that if only the convective part of the HRR is considered, Eq. (4.107) can be rewritten as follows:

$$\frac{u_{\max}}{(B_0/H)^{1/3}} = f_2\left(\frac{r}{H}\right), \tag{4.108}$$

where B_0 is the buoyancy flux of the fire source. This expression clearly shows that the maximum ceiling jet at a given radial position from the plume axis is mainly a function of the buoyancy flux at the source.

4.5.3 SIMPLIFIED CORRELATIONS

4.5.3.1 Alpert

It can be seen from Eqs. (4.107) and (4.108) that for constant coefficients E, E_p, and F, the horizontal ceiling-jet velocity and temperature are solely a function of $\dot{Q}$, H and r. Simplified expressions are provided in Ref. [32] for the maximum ceiling-jet gas velocity and temperature based on experimental data [27]. The fire sources examined in Ref. [27] are liquid pools (of ethanol and heptane) with characteristic diameters ranging from 0.305 to 3.66 m as well as wood cribs of 1.22 m × 1.22 m × 2.44 m height. The HRR varied between 27 and 13,500 kW. The ceiling height, H, varied between 2.44 and 15.9 m.

Alpert's correlations for unconfined ceiling jet are the following:

If $r > 0.18\,H$, then

$$u_{\max} = \frac{0.197\,\dot{Q}^{1/3}H^{1/2}}{r^{5/6}} \tag{4.109}$$

and

$$T_{\max} - T_{\mathrm{amb}} = \frac{5.38\left(\dot{Q}/r\right)^{2/3}}{H}. \tag{4.110}$$

The coefficient 0.18 corresponds to the value of $\bar{r}_e = r_e/H$ that could be calculated based on a constant plume entrainment coefficient E_p in Eq. (4.102).

Estimates of the maximum velocity and temperature at the turning region (i.e., $r < r_e$) are provided in the following expressions:

If $r \leq 0.18\,H$, then

$$u_{\max} = 0.946\left(\frac{\dot{Q}}{H}\right)^{1/3} \tag{4.111}$$

and

$$T_{\max} - T_{\mathrm{amb}} = \frac{16.9\,\dot{Q}^{2/3}}{H^{5/3}}. \tag{4.112}$$

4.5.3.2 Motevalli

In Ref. [28], a reduced-scale model was used to examine ceiling-jet properties. A premixed methane-air burner of 2.7 cm diameter was used to generate an HRR that varied between 0.5 and 2 kW. Two ceiling heights were tested: $H = 0.5$ m and $H = 1$ m. The obtained correlation for the maximum ceiling-jet velocity reads:

$$\frac{u_{\max}}{g\,H^{1/2}\,\dot{Q}^*} = 0.0415\left(\frac{r}{H}\right)^{-2} + 0.427\left(\frac{r}{H}\right)^{-1} + 0.281. \tag{4.113}$$

A similar expression has been derived for the temperature. However, two separate empirical correlations were obtained for $H = 0.5$ m and $H = 1$ m.

In addition to the maximum ceiling-jet velocity and temperature, empirical correlations were obtained in Ref. [28] for momentum (resp. thermal) thickness:

$$\frac{\ell_u}{H} = 0.205\left[1 - \exp\left(-1.75\,\frac{r}{H}\right)\right] \quad 0.26 \leq \frac{r}{H} < 1.5, \tag{4.114}$$

$$\frac{\ell_T}{H} = 0.112\left[1 - \exp\left(-2.24\,\frac{r}{H}\right)\right] \quad 0.26 \leq \frac{r}{H} < 1.5. \tag{4.115}$$

Furthermore, the momentum and thermal boundary layer thicknesses are given in Ref. [28] by:

$$\delta_{u_{\max}} = 0.0187\,r^{0.668} \quad 0.26 \leq \frac{r}{H} \leq 2.0, \tag{4.116}$$

$$\delta_{T_{\max}} = 0.0152\,r^{1.35} \quad 0.26 \leq \frac{r}{H} \leq 2.0. \tag{4.117}$$

4.5.3.3 Heskestad and Yao

In Ref. [33], the following correlations for the maximum velocity are provided (based on alcohol fire tests):

$$\frac{u_{\max}}{\sqrt{gH}\left(Q^*\right)^{1/3}} = 1.06(r/H)^{-0.69} \quad \text{for} \quad 0.17 < r/H < 4.0 \tag{4.118}$$

and

$$\frac{u_{\max}}{\sqrt{gH}\left(Q^*\right)^{1/3}} = 3.61 \quad \text{for} \quad r/H \leq 0.17. \tag{4.119}$$

4.5.4 ADDITIONAL CONSIDERATIONS

- **Weak vs strong plume:** The generalized theory and the correlations described above have been developed mainly for the case of *weak plume*-driven flow field. This implies that the flame height is much less than the ceiling height, H. If the flame height is comparable to the height of the ceiling or if the flame extends extensively beneath the ceiling, the plume is referred to as a *strong plume*. The density difference between the ceiling flow and the ambient air cannot be neglected anymore (i.e., non-Boussinesq plume). More details on strong plumes and flame impingement are provided in Chapter 3.

- **Steady-state vs time-dependent fires:** The main focus in this chapter has been on steady-state. In Ref. [30], Alpert states that the quasi-steady assumption may be used if the fire growth rate is sufficiently small. In the quasi-steady assumption the constant heat release rate, $\dot{Q}$, is replaced by an appropriate time-dependent $\dot{Q}(t)$. In doing so, it is implicitly assumed that the change in the heat release rate has an almost immediate effect on the flow field. This cannot be true in reality, given transport times, but as long as changes in HRR are slow, the effect is small.

- **Sloped ceilings:** As noted in Ref. [30], very few studies have addressed the effect of a slope on the ceiling. Small-scale experiments of Kung et al. [34] have shown that an increase in the slope significantly reduces the rate of velocity decrease in the upward direction. In the downward direction, the flow separated from the ceiling and turned upwards.

- **Confined configuration:** The ceiling-jet development has been discussed above essentially for the unconfined configuration. This is generally the case during the early stages of the fire as well as for large compartments (such as warehouses and industrial facilities with flat ceiling). In the later stages and also for small rooms, accumulated heat in the smoke layer will significantly affect the ceiling-jet velocities and temperatures. Furthermore, the presence of ceiling beams or corridor walls may create long channels that partially confine the flow [30].

4.5.5   SMOKE LAYER BUILD-UP IN A ROOM

When the ceiling jet encounters the bounding walls of a compartment, 'hot smoke' will accumulate in the room. Due to thermal stratification, a layer of lower density fluid (i.e., 'hot smoke') will rest above a layer of higher density fluid (i.e., 'fresh air'). More details on this 'two-layered' structure are addressed in Chapter 5.

4.6   BALCONY AND WINDOW SPILL PLUMES

4.6.1   BALCONY SPILL PLUMES

As stated in Ref. [26], balcony spill plumes are *representative of fires within an adjacent space, with the plume spilling beneath a protruding balcony before rising within the atrium or covered mall*. A sketch is provided in Figure 4.11a.

Many correlations exist (e.g., [26]) and the reader is referred to these sources to make a choice. However, from the configuration, it is clear that the plume resembles a line plume, not an axisymmetric plume. Therefore, Eq. (4.82) is the basis for the correlations. The effective width of the plume as it spills at the lower balcony edge is the characteristic length scale L and the convective power of the smoke flow at the balcony edge replaces $\dot{Q}$ in Eq. (4.82).

Whether or not the plume is free or adhered will depend on the geometry of the mall (flat or balconies) in the atrium (see Figure 4.11), as well as on the presence or not of a 'downstand'. The latter can break the momentum of the horizontally flowing smoke out of the fire compartment, increasing the likelihood of an adhered plume in the atrium, compared to the situation where the horizontal momentum pushes the smoke plume into the atrium.

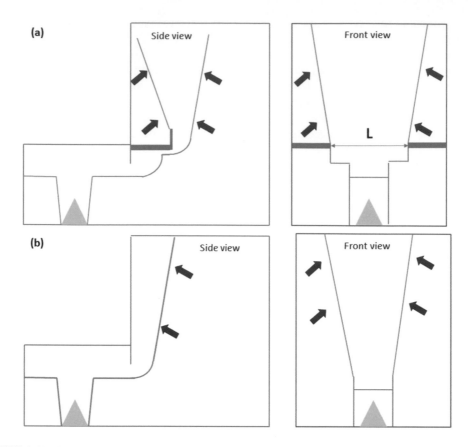

FIGURE 4.11 2-D sketches of (a) balcony spill plume and (b) an adhered plume.

4.6.2 Window Plumes

The term 'window plume' refers to the situation where flames emerge from the fire compartment into the atrium. This corresponds to post-flashover fully developed fire conditions (Section 5.3). As such, Eq. (4.3.13) recovers the basic expression with L the effective width, but the HRR is replaced by a factor proportional to the ventilation factor $A_w\sqrt{H_w}$ (with A_w the window area and H_w the window height), as explained in Section 5.3. Indeed, the effective HRR inside the fire compartment is limited by the ventilation (i.e., by the incoming air flow rate). Note that in such conditions, there would also still be combustion inside the atrium.

Whether or not the flames interact with the atrium wall, depends on the distance of the flames from the wall, and thus on the horizontal momentum of the combustible gases, flowing out of the window, as well as on the distance between the spill edge and the atrium wall (i.e., is there a balcony or not). In the case of interaction with a flat non-combustible atrium wall, the situation resembles what has been described in Section 3.5.2 (Eq. 3.59), namely a fire line plume interaction with a non-combustible wall, albeit that the inlet is now not a line plume with pure fuel low vertical momentum, but a hot line plume of a mixture of combustion products and combustible gases that is turning from a horizontal motion, with certain mass flow rate and momentum, into a vertical motion, air being entrained during this turning motion.

4.7 SCALING LAWS AND BUOYANT RELEASES

The simulation of a fire-induced smoke plume can be performed at a reduced scale using either an isothermal buoyant release (e.g., helium/air mixture in Ref. [35]) or a source of hot air (e.g., the Shabbir and George case [13]). The purpose is to essentially create a buoyancy force via a density difference with the surrounding environment. This can be done not only using aeraulic models (e.g., helium/air mixture or hot air) but also using hydraulic models such as in salt water experiments [36]. The advantages of a model scale approach are:

1. The low cost in comparison with large-scale experiments and thus the possibility to address a wider range of parameters and scenarios, and
2. The focus on fluid mechanics aspects without dealing with the complexity of other fire features like combustion and radiation.

The correspondence between a large-scale fire and a small-scale buoyant release is, however, a complex issue. Vauquelin [37] decomposed the problem into two steps:

1. In the first step, the large-scale fire source is replaced by a large-scale buoyant release with representative inlet density and flow rate. For example, a semi-empirical model for pool fires is presented in Ref. [38].
2. In the second step, scaling laws are applied to find plume parameters at a small scale that are representative of the large scale.

A theoretical development for Step 2 has been described in Ref. [37] and is recalled here for the sake of completeness.

The governing equations for the rising plume can be written for both Boussinesq and non-Boussinesq approaches in a single form as follows [37]:

- **Mass**

$$\frac{d}{dz}\left[\left(\frac{\bar{\rho}}{\rho_{amb}}\right)^{j} \bar{w}R^{2}\right] = 2a\left(\frac{\bar{\rho}}{\rho_{amb}}\right)^{j/2} \bar{w}R. \tag{4.120}$$

- **Momentum**

$$\frac{d}{dz}\left[\left(\frac{\bar{\rho}}{\rho_{amb}}\right)^{j} \bar{w}^{2}R^{2}\right] = \frac{\rho_{amb} - \bar{\rho}}{\rho_{amb}} gR^{2}. \tag{4.121}$$

- **Energy**

$$\frac{d}{dz}\left[\left(\rho_{amb} - \bar{\rho}\right)\bar{w}R^{2}\right] = 0, \tag{4.122}$$

where the exponent j introduced in these equations takes into account the presence of large density differences. For $j = 0$, equations are valid for Boussinesq plumes, while $j = 1$ has to be considered for the non-Boussinesq plumes.

When considering the momentum equation (Eq. 4.121) and subtracting the term

$$\frac{d}{dz}\left[\left(\frac{\rho_{amb}}{\rho_{amb}}\right)^{j} R^2 \bar{w}^2\right] = \frac{d}{dz}\left[R^2 \bar{w}^2\right] \tag{4.123}$$

from both sides, one obtains:

$$\frac{d}{dz}\left[\bar{w}^2 R^2\left(1 - j\frac{\Delta\bar{\rho}}{\rho_{amb}}\right)\right] = \frac{\Delta\bar{\rho}}{\rho_{amb}} gR^2, \tag{4.124}$$

where $\Delta\bar{\rho} = \rho_{amb} - \bar{\rho}$.

In order to write Eq. (4.124) in a non-dimensional form, the parameters of the problem are expressed using the following dimensionless variables:

- Vertical coordinate, $\tilde{z} = z/R_0$,
- Radius, $\tilde{R} = R/R_0$,
- Vertical velocity, $\tilde{w} = \bar{w}/\bar{w}_0$,
- Density difference, $\Delta\tilde{\rho} = \Delta\bar{\rho}/\Delta\bar{\rho}_0$

where the subscript 0 denotes the parameters at the source.

The momentum equation then becomes:

$$\frac{d}{d\tilde{z}}\left[\tilde{w}^2 \tilde{R}^2\left(1 - j\frac{\Delta\bar{\rho}_0}{\rho_{amb}}\Delta\tilde{\rho}\right)\right] = \frac{\Delta\bar{\rho}_0 gR_0}{\rho_{amb}\bar{w}_0^2}\Delta\tilde{\rho}\tilde{R}^2. \tag{4.125}$$

Two dimensionless groups are then defined from this equation:

$$Ri = \frac{\Delta\bar{\rho}_0 gR_0}{\rho_{amb}\bar{w}_0^2} \tag{4.126}$$

and

$$\frac{\Delta\bar{\rho}_0}{\rho_{amb}}. \tag{4.127}$$

One can note that the first parameter (in the form of a Richardson number) has to be preserved in all cases, whereas the second parameter (density ratio) must particularly be preserved for non-Boussinesq plumes since it is multiplied by j in the momentum equation.

Vauquelin [37] points out furthermore the importance of considering the Reynolds number and the difficulties to achieve strict conservation of both Reynolds and Froude numbers. As explained in Section 2.7.3, it is more important to preserve the Froude number in fire-related flows. However, as illustrated in Ref. [39], there are limits in the sense that the Reynolds number must not become too low at a reduced scale.

4.8   EXERCISES

4.8.1   ANALYTICAL SOLUTION FOR THE LINE PLUME PROBLEM

4.8.1.1   Problem Description

1. Perform the mass, momentum and energy balance for a line plume (similarly to the axisymmetric plume) and verify the set of Eqs. (4.74) to (4.76).
2. Provide the mathematical modelling details that led to the solution given in Eqs. (4.77) to (4.80) for the conservation equations (Eqs. 4.74 to 4.76) of a line plume under the Boussinesq approximation.

4.8.1.2   Suggested Solution

1. Conservation equations of mass momentum and energy:

- **Mass**
  The mass balance for a control volume (in a line plume) in steady-state conditions yields:

$$\left(2bL\bar{\rho}\bar{w}\right)_{z+\Delta z} = \left(2bL\bar{\rho}\bar{w}\right)_{z} + 2L\,\Delta z\,\rho_{amb}u. \tag{4.128}$$

In the differential form, the mass conservation reads:

$$\frac{d\left(2bL\bar{\rho}\bar{w}\right)}{dz} = 2L\,\rho_{amb}u. \tag{4.129}$$

Using the Boussinesq approximation (i.e., $\bar{\rho} \simeq \rho_{amb}$), assuming a non-stratified environment, (i.e., $d\rho_{amb}/dz = 0$) and dividing by $2L$ on both sides of Eq. (4.129), one obtains:

$$\frac{d\left(b\bar{w}\right)}{dz} = u. \tag{4.130}$$

- **Momentum**
  The momentum equation can simply be obtained by considering Eq. (4.17) for the axisymmetric plume and replacing the area πR^2 by $2bL$:

$$\frac{d\left(2bL\bar{\rho}\bar{w}^2\right)}{dz} = 2bL\rho_{amb}\bar{g'}. \tag{4.131}$$

Under the same assumptions used above for the mass balance, Eq. (4.131) becomes:

$$\frac{d\left(b\bar{w}^2\right)}{dz} = b\,\bar{g'}. \tag{4.132}$$

- **Energy (Buoyancy)**

The conservation of energy under the assumptions used for the axisymmetric plume is obtained by replacing the area πR^2 by $2bL$ for Eq. (4.22):

$$B = B_0 = 2bL\overline{w}\,\overline{g'}. \tag{4.133}$$

If one defines the buoyancy flux per unit length as $B' = B/L$, Eq. (4.133) becomes:

$$B' = B'_0 = 2b\overline{w}\,\overline{g'}. \tag{4.134}$$

2. Based on a dimensional analysis, the expressions for the plume half-width, average vertical velocity, air entrainment velocity and reduced gravity could be respectively written as:

$$b = az, \tag{4.135}$$

$$\overline{w} = c\left(B'_0\right)^{1/3}, \tag{4.136}$$

$$u = E\overline{w}, \tag{4.137}$$

$$\overline{g'} = d\left(B'_0\right)^{2/3} z^{-1}, \tag{4.138}$$

where E is the entrainment coefficient and the other coefficients (i.e., a, c and d) can be determined by inserting the obtained set of equations in the conservation equations (Eqs. 4.74 to 4.76). For example, the mass conservation becomes:

$$\frac{d\left[azc\left(B'_0\right)^{1/3}\right]}{dz} = Ec\left(B'_0\right)^{1/3}, \tag{4.139}$$

which leads to:

$$a = E. \tag{4.140}$$

Applying the same procedure to the momentum and energy equations, respectively, gives:

$$c^2 = d \tag{4.141}$$

and

$$2acd = 1. \tag{4.142}$$

Combining the last three equations gives:

$$a = E, \qquad (4.143)$$

$$c = (2E)^{-1/3}, \qquad (4.144)$$

$$d = (2E)^{-2/3}. \qquad (4.145)$$

Replacing the coefficients a, c and d by their values in Eqs. (4.135) to (4.138) gives the desired solution.

4.8.2 Design of a Reduced-Scale Helium/Air Mixture Experiment of a Car Fire in a Tunnel

4.8.2.1 Problem Description

You are required to design a car fire experiment in a reduced-scale tunnel using the release of a helium/air mixture through a circular injection. The scale reduction is $\alpha = 1/20 = 0.05$. The average dimensions of a car are 4.5 m by 1.75 m. The estimated convective heat release rate and smoke flow rate for a burning car are, respectively, $\dot{Q}_{conv} = 4$ MW and $\dot{q}_s = 20$ m^3/s (as reported by Ref. [35] from Ref. [40]). Assume throughout the problem standard atmospheric conditions ($T_{amb} = 293$ K and $\rho_{amb} = 1.205$ kg/m^3). The specific heat of air is taken as $c_p = 1$ kJ/kg K and the gravitational acceleration is $g = 9.81$ m/s^2. The density of helium at 293 K is $\rho_{He} = 0.179$ kg/m^3.

4.8.2.1.1 Full-scale Parameters

1. Calculate the buoyancy flux at the source, B_0.
2. Calculate the reduced density difference at the fire source, $\Delta\rho_s = \rho_{amb} - \rho_s$, and deduce the smoke density, ρ_s, and temperature T_s.
3. Calculate the hydraulic diameter, D_h, based on the dimensions of the car.
4. Calculate the average vertical velocity of gases at the source, w_s, assuming an axisymmetric plume.
5. Calculate the Reynolds number at the source, Re. Take the kinematic viscosity for air as: $\nu_{air} = 5.5 \times 10^{-5}$ m^2/s.

4.8.2.1.2 Reduced-Scale Parameters

1. Calculate the diameter of the circular fire source.
2. Calculate the average velocity and the volume flow rate of the helium/air mixture at the injection.
3. Calculate the required volume flow rates of air and helium at the injection in order to comply with the scaling laws.
4. Calculate the Reynolds number at the source, Re$_{model}$. Take the kinematic viscosity for the mixture of air/helium as: $\nu_{mixture} = 3.75 \times 10^{-5}$ m^2/s.

4.8.2.2   Suggested Solution

4.8.2.2.1   Full-Scale Parameters

1. The buoyancy flux is calculated as follows:

$$B = \frac{g}{\rho_{\text{amb}} T_{\text{amb}} c_p} \dot{Q}_{\text{conv}} \simeq 111 \text{ m}^4/\text{s}^3. \tag{4.146}$$

2. According to Eq. (4.4), the buoyancy flux at the source can be written as follows:

$$B_0 = \pi R_0^2 w_0 \frac{\Delta \rho_s}{\rho_{\text{amb}} g}. \tag{4.147}$$

The term $\pi R_0^2 w_0$ represents the volume flow rate of smoke at the source and thus can be replaced by $\dot{q}_s$. The buoyancy flux is then rewritten as follows:

$$B_0 = \dot{q}_s \frac{\Delta \rho_s}{\rho_{\text{amb}}} g. \tag{4.148}$$

The reduced density difference at the fire source is:

$$\frac{\Delta \rho_s}{\rho_{\text{amb}}} = \frac{B_0}{\dot{q}_s g} \simeq 0.567, \tag{4.149}$$

which corresponds to a smoke density of

$$\rho_s = \rho_{\text{amb}} \left(1 - \frac{\Delta \rho_s}{\rho_{\text{amb}}} \right) \simeq 0.522 \text{ kg/m}^3 \tag{4.150}$$

and a smoke temperature of

$$T_s = \frac{\rho_{\text{amb}} T_{\text{amb}}}{\rho_s} \simeq 623\,\text{K} \simeq 403°\text{C} \tag{4.151}$$

3. The hydraulic diameter is calculated as follows:

$$D_{h,s} = \frac{2 L_{\text{car}} W_{\text{car}}}{L_{\text{car}} + W_{\text{car}}} = 2.52\,\text{m}, \tag{4.152}$$

which corresponds to a radius

$$R_s = 1.26\,\text{m}. \tag{4.153}$$

4. The average vertical velocity of gases at the source is:

$$w_s = \frac{\dot{q}_s}{\pi R_s^2} \simeq 4 \text{ m/s}.$$ (4.154)

5. Note that the kinematic viscosity for air, $v_{air} = 5.5 \times 10^{-5} \text{ m}^2/\text{s}$, is taken at the calculated temperature of $T_s = 623$ K. The Reynolds number is calculated as follows:

$$\text{Re} = \frac{w_s D_{h,s}}{v_{air}} \simeq 180,000.$$ (4.155)

4.8.2.2.2 Reduced-Scale Parameters

1. The diameter of the circular fire source is:

$$\left(D_s \right)_{\text{model}} = \alpha \left(D_{h,s} \right)_{\text{full-scale}} = 0.126 \text{ m}$$ (4.156)

and corresponds to a radius

$$\left(R_s \right)_{\text{model}} = 0.063 \text{ m}.$$ (4.157)

2. The average velocity is calculated as follows:

$$\left(w_s \right)_{\text{model}} = \alpha^{1/2} \left(w_s \right)_{\text{full-scale}} \simeq 0.894 \text{ m/s}.$$ (4.158)

The corresponding volume flow rate is calculated as follows:

$$\left(\dot{q}_s \right)_{\text{model}} = \pi \left(R_s^2 \right)_{\text{model}} \left(w_s \right)_{\text{model}} \simeq 0.011 \text{ m}^3/\text{s}.$$ (4.159)

3. The volume fraction of helium is calculated as follows:

$$\chi_{\text{He}} = \frac{\rho_{\text{amb}} - \rho_s}{\rho_{\text{air}} - \rho_{\text{He}}} \simeq 0.666.$$ (4.160)

The volume fraction of air is thus $\chi_{air} = 0.334$. The corresponding volume flow rates for helium and air are:

$$\dot{q}_{\text{He}} = \chi_{\text{He}} \left(\dot{q}_s \right)_{\text{model}} \simeq 0.073 \text{ m}^3/\text{s},$$ (4.161)

$$\dot{q}_{\text{air}} = \chi_{\text{air}} \left(\dot{q}_s \right)_{\text{model}} \simeq 0.034 \text{ m}^3/\text{s}.$$ (4.162)

4. Note that the kinematic viscosity for the mixture of air/helium, $v_{\text{mixture}} = 3.75 \times 10^{-5} \text{ m}^2/\text{s}$, is calculated based on the mass proportions of air and helium in the mixture and the kinematic viscosity of each (air and helium) at ambient temperature, T_{amb}. The Reynolds number for the reduced scale is calculated as follows:

$$\text{Re}_{\text{model}} = \frac{(w_s)_{\text{model}} (D_s)_{\text{model}}}{v_{\text{mixture}}} \simeq 3,000. \tag{4.163}$$

REFERENCES

1. O. Vauquelin and Y. Wu (2006) "Influence of tunnel width on longitudinal smoke control", *Fire Safety Journal*, Vol. 41, pp. 420–426.
2. B.R. Morton (1959) "Forced plumes", *Journal of Fluid Mechanics*, Vol. 5, pp. 151–163.
3. B. Karlsson and J.G. Quintiere (2000) *Enclosure Fire Dynamics*. CRC Press: Boca Raton, FL.
4. E.E. Zukoski, T. Kubota, and B. Cetegen (1981) "Entrainment in fire plumes", *Fire Safety Journal*, Vol. 3, pp. 107–121.
5. D. Drysdale (2011) *An Introduction to Fire Dynamics*, 3rd Ed. Wiley: Hoboken, NJ.
6. J.G. Quintiere (2006) *Fundamentals of Fire Phenomena*. John Wiley & Sons Ltd: Hoboken, NJ.
7. G. Heskestad (1983) "Virtual origins of fire plumes", *Fire Safety Journal*, Vol. 5, pp. 109–114.
8. B.M. Cetegen, E.E. Zukoski and T. Kubota (1984) "Entrainment into the near and far field of fire plumes", *Combustion Science and Technology*, Vol. 39, pp. 305–331.
9. G. Cox and R. Chitty (1985) "Some source-dependent effects of unbounded fires", *Combustion and Flame*, Vol. 60, pp. 219–232.
10. G. Heskestad (2008) "Fire plumes, flame height and air entrainment". In: P.J. DiNenno, D. Drysdale, C.L. Beyler, et al. (eds.) *SFPE Handbook of Fire Protection Engineering*, 3rd Ed, pp.2-1–2-17. National Fire Protection Association: Quincy, MA.
11. A. Shabbir and W.K. George (1994) "Experiments on a round turbulent buoyant plume", *Journal of Fluid Mechanics*, Vol. 275, pp. 1–32.
12. J. Rouse, C.S. Yih and H.W. Humphrey (1952) "Gravitational convection from a boundary source", *Tellus*, Vol. 4, p. 201.
13. A. Shabbir and W.K. George (1994) "Experiments on a round turbulent buoyant plume", *Journal of Fluid Mechanics*, Vol. 275, pp. 1–32.
14. M. Van Reeuwijk, P. Salizzoni, G.R. Hunt and J. Craske (2016) "Turbulent transport and entrainment in jets and plumes; a DNS study", *Physical Review Fluids*, Vol. 1, p. 074301.
15. H. Nakagome and M. Hirata (1976) The structure of turbulent diffusion in an axisymmetric thermal plume, *In Proceedings of 1976 ICHMT Seminar on Turbulent Buoyant Convection*, pp. 361–372, Hemisphere.
16. P.N. Papanicolaou and E.J. List (1988) "Investigations of round vertical turbulent buoyant jets", *Journal of Fluid Mechanics*, Vol. 195, pp. 341–391.
17. C.J. Chen and W. Rodi (1980) *Vertical Turbulent Buoyant Jets*. Elsevier Science & Technology, New York.
18. B.J. McCaffrey (1979) Purely buoyant diffusion flames: Some experimental results, NBSIR 79–1910, National Bureau of Standards.
19. S.-L. Lee and H.W. Emmons (1961) "A study of natural convection above a line fire", *Journal of Fluid Mechanics*, Vol. 11, pp. 353–368.
20. R. Sangras, Z. Dai and G.M. Faeth (1998) "Mixing structure of plane self-preserving buoyant turbulent plumes", *ASME Journal of Heat Transfer*, Vol. 120, pp. 1033–1041.
21. M.-C. Lai and G.M. Faeth, "Turbulent structure of vertical adiabatic wall plumes", *ASME Journal of Heat Transfer*, Vol. 109, pp. 663–670.
22. R. Sangras, Z. Dai and G.M. Faeth (1999) "Buoyant turbulent jets and flames: I. Adiabatic wall plumes", Annual report, Grant No. 60NANBD0081.

23. R. Sangras, Z. Dai and G.M. Faeth (1999) "Mixture fraction statistics of plane self-preserving buoyant turbulent adiabatic wall plumes", *Journal of Heat Transfer*, Vol. 121, pp. 838–843.
24. R. Sangras, Z. Dai and G.M. Faeth (2000) "Velocity statistics of plane self-preserving buoyant turbulent adiabatic wall plumes", *Journal of Heat Transfer*, Vol. 122, pp. 693–700.
25. M.-C. Lai, S.-M. Jeng and G.M. Faeth (1986), "Structure of turbulent adiabatic wall plumes", *ASME Journal of Heat Transfer*, Vol. 108, pp. 827–834.
26. J.A. Milke and F.W. Mowrer (1993) A design algorithm for smoke management systems in atria and covered malls, Report No. FP93-04, Department of Fire Protection Engineering, University of Maryland, College Park, MD.
27. R.L. Alpert (1975) "Turbulent ceiling-jet induced by large-scale fires, combustion", *Science and Technology*, Vol. 11, pp. 197–213.
28. V. Motevalli and C.H. Marks (1991) "Characterizing the unconfined ceiling jet under steady-state conditions: A reassessment", *Fire Safety Science-Proceedings of the Third International Symposium*, pp. 301–312.
29. H.W. Emmons (1991) "The ceiling jet in fires", *Proceedings of the Third International Symposium on Fire Safety Science*, pp. 249–260.
30. R.L. Alpert (2002) "Ceiling jet flows", in: P.J. Di Nenno (Ed.), *SFPE Handbook of Fire Protection Engineering*, 3rd Ed, Chapter 2.2. National Fire Protection Association: Quincy, MA.
31. B.R. Morton, G.I. Taylor and J.S. Turner, (1956) "Turbulent gravitational convection from maintained and instantaneous sources." *Proceedings of the Royal Society A*, Vol. 236, p. 1.
32. R.L. Alpert (1972) "Calculation of response time of ceiling-mounted fire detectors", *Fire Technology*, Vol. 8, pp. 181–195.
33. G. Heskestad and C. Yao (1971) "A new approach to development of installation standards for fire detectors", Technical Proposal No. 19574, prepared for the Fire Detection Institute by Factory Mutual Research Corporation.
34. H.C. Kung, R.D. Spaulding and P. Stavrianidis (1991) "Fire induced flow under a sloped ceiling, fire safety science", *Fire Safety Science, Proceedings of the Third International Symposium*, 3, 271.
35. O. Vauquelin (2008) "Experimental simulations of fire-induced smoke control in tunnels using an air-helium reduced scale model: Principle, limitations, results and future", *Tunneling and Underground Space Technology*, Vol. 23, pp. 171–178.
36. X. Yao and A.W. Marshall (2006) "Quantitative slat-water modeling of fire induced flow", *Fire Safety Journal*, Vol. 41, pp. 497–508.
37. O. Vauquelin, G. Michaux and C. Lucchesi (2009) "Scaling laws for a buoyant release used to simulate fire-induced smoke in laboratory experiments", *Fire Safety Journal*, Vol. 44, pp. 665–667.
38. O. Megret and O. Vauquelin (2000) "A model to evaluate fire tunnel characteristics", *Fire Safety Journal*, Vol. 34, pp. 393–401.
39. N. Tilley, P. Rauwoens, D. Fauconnier, and B. Merci (2013) "On the extrapolation of CFD results for smoke and heat control in reduced-scale set-ups to full scale: Atrium configuration", *Fire Safety Journal*, Vol. 59, pp. 160–165.
40. D. Lacroix (1998) "The new PIARC report on fire and smoke control in road tunnels", *Third International Conference on Safety in Road and Rail Tunnels*, ITC, Nice, France.

5 Fire and Smoke Dynamics in Enclosures

5.1 SOME FUNDAMENTALS ON FLOWS THROUGH OPENINGS

A fundamental principle is that a fluid flow is driven by a pressure difference, flowing from higher to lower pressure. Bernoulli's equation, Eq. (2.74), reveals the relationship between velocities and pressure differences along a streamline:

$$p + \frac{1}{2}\rho v^2 + \rho g z = \text{Constant.} \tag{5.1}$$

This equation is only valid for constant density flows and in the absence of losses in the flow (Section 2.4). Obviously, density cannot be assumed constant in case of fire in an enclosure. Yet, this is a classical assumption, necessary to develop analytical solutions. It must be appreciated, though, that this assumption is never fulfilled in reality. Yet, the analytical solutions provide a first estimate and reveal the important parameters for the problem at hand.

Equation (2.79), mentioned in Section 2.4, allows to estimate flows through openings:

$$\dot{V} = C_d A \sqrt{\frac{2\Delta p}{\rho}}. \tag{5.2}$$

Equation (5.2) expresses that the volume flow rate through an opening is proportional to the discharge coefficient of the opening, the cross-sectional area of the opening and the square root of the pressure difference over the opening, and it is inversely proportional to the density of the fluid flowing through the opening. The discharge coefficient, in fact, takes into account the vena contracta effect (see Section 2.4).

The mass flow rate is readily retrieved by multiplication with the mass density:

$$\dot{m} = C_d A \sqrt{2\rho\Delta p}. \tag{5.3}$$

The key question, therefore, becomes: what is the local pressure difference over the opening? For that purpose, the law of hydrostatics is applied, Eq. (2.82):

$$p = p_{ref} - \rho_{amb} g \left(z - z_{ref} \right). \tag{5.4}$$

In Eq. (5.4), p_{ref} is the reference pressure at reference height z_{ref} and z denotes the height.

The flows through openings can now be derived from these equations. Again, it must be noted that the law of hydrostatics must be used with care, as there is motion. In fact, the implicit assumption made when studying flows through openings in the context of fires in enclosures is that the flows are weak, compared to the flows induced by the static pressure.

DOI: 10.1201/9781003204374-5

This, however, is not guaranteed at all. One illustration has been given in Ref. [5], showing that for a sufficiently high smoke extraction rate at the ceiling in an atrium, the induced air supply rate could strongly disturb the smoke layer, essentially due to a very large vortex flow. The position and size of the vents for the supply rate also have an impact on the flow pattern. The reader is referred to Ref. [5] for more details, but it is important to note that the smoke dynamics and flows through openings are not only governed by static pressure differences but also, and often even more strongly, by the air flow inside the enclosure. For small compartments, though, it makes sense to perform the analysis based on the equations above and the pressure differences are induced by the fire.

In addition to the equations above, a very important observation is that pressure differences in the context of fires in enclosures are typically very small (well below 1 kPa), compared to atmospheric pressure (100 kPa). Moreover, as explained in Chapter 4, smoke can be treated as hot air, as far as fluid dynamics is concerned. These two observations lead to the fact that the ideal gas law can be applied, using the gas constant R for air (around 287 J/(kg K)). Using $p_{atm} = 100$ kPa, this reads:

$$\rho = \frac{p}{RT} \approx \frac{p_{atm}}{T} \approx \frac{348}{T}.$$

(5.5)

In Eq. (5.5), absolute temperatures (in K) must be used. Equation (5.5) illustrates that the local mass density is inversely proportional to the local temperature. The equation is assumed valid inside the smoke as well as in ambient conditions. It is not valid inside the flames because there the local gas composition can deviate too strongly from air. Yet, even inside the flames, the inversely proportional relation of mass density with temperature holds. Only the factor 348 can be different.

In the analyses below, density ratios will appear. Equation (5.5) shows that mass density ratios are equivalent to temperature ratios:

$$\frac{\rho_1}{\rho_2} = \frac{T_2}{T_1}.$$

(5.6)

As in Eq. (5.5), absolute temperatures (in K) must be used.

5.2 GROWING FIRE

During the early stages after ignition, several evolutions of the fire are possible. One possibility is that the fire dies. This can be 'fuel-controlled', namely if the energy lost from the flames is larger than the energy required for further pyrolysis of virgin material (see Section 3.6 on flame spread), or because the originally burning object completely burns out, but the heat released is not sufficient for fire spread to other combustible material. The fire can also die in a 'ventilation-controlled' manner, namely if there is already lack of oxygen to consume all the pyrolysis gases released during early stages after ignition. This can happen in airtight compartments (Figure 5.1).

Another possibility, though, is that the fire grows. This happens if there is a positive feedback loop: the heat from the flames causes more material to pyrolyse, providing more pyrolysis gases and therefore larger flames (cf. Section 3.5), which provide more heat to be transferred to the not-yet burning material, providing more pyrolysis gases, etc. The fire can spread as flame spread over surfaces (cf. Section 3.6) or by distant ignition of objects through radiation. The reader is referred to, e.g., Refs. [1–4] for more details. In certain conditions, the fire can grow to become a fully

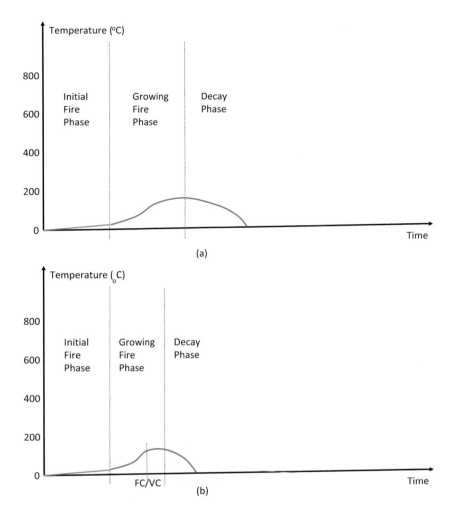

FIGURE 5.1 Evolution of average temperature in an enclosure in case of a fire that does not develop to flash-over due to lack of fuel (a) or lack of oxygen (b). FC/VC refers to the transition from 'fuel-controlled' (FC) to 'ventilation-controlled' (VC). Note: temperature values are merely indicative.

developed fire after flash-over (Figure 5.2), as discussed in Section 5.3. The upper layer refers to the smoke layer underneath the ceiling in an enclosure (see Section 5.2.1).

It is noteworthy that in a fuel-controlled growing fire, the fuel governs the fire dynamics because essentially there is more than enough oxygen to consume the pyrolysis gases released by the fuel. This corresponds to situations where the 'Global Equivalence Ratio' (GER) is much lower than 1, or stated in another manner, globally the conditions are very lean. The equivalence ratio has been defined in Section 2.2.1, Eq. (2.12). In Ref. [6], the GER is defined as 'the ratio of the mass of gas in the upper layer derived from the fuel divided by that introduced from air, normalized by the stoichiometric ratio'.

In an enclosure, the GER can be defined as follows:

$$\Phi = \frac{s\dot{m}_f}{\dot{m}_{ox}}. \tag{5.7}$$

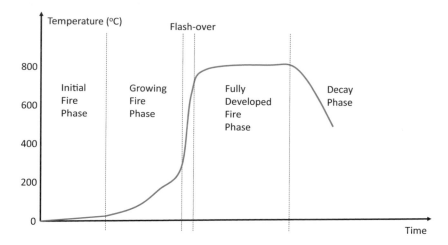

FIGURE 5.2   Evolution of average temperature in an enclosure in case of a fire that develops to flash-over in 'fuel-controlled' conditions. Note: temperature values are merely indicative.

In Eq. (5.7), $\dot{m}_f$ is the amount of combustible gases released per unit time (in kg/s) inside the enclosure, $\dot{m}_{ox}$ is the amount of oxidizer (typically oxygen from air) entering the enclosure per unit time (in kg/s) and s is the stoichiometric oxidizer to fuel ratio:

$$s = \left(\frac{\dot{m}_{ox}}{\dot{m}_f} \right)_{st}. \tag{5.8}$$

Conditions where $\Phi \ll 1$ correspond to situations where there is supply of plenty of oxidizer and the fire is fuel-controlled. Note that while in principle complete combustion is possible under these conditions, the GER does not provide information on the local conditions near the fire, so that incomplete combustion is still possible. This is almost always the case in fires.

The growth of a fire can also be limited by the supply of oxygen, though. Indeed, if $\Phi > 1$, complete combustion is impossible. Then the fire dynamics is no longer governed by the fuel only, e.g., temperature build-up inside the enclosure can be limited if combustible gases leave the enclosure before having reacted with oxygen. This is also discussed in Section 5.2.1.

Finally, it is noted that fire growth can be complex, with different stages, in large compartments. This is briefly discussed in Section 5.4.

5.2.1   FIRE SOURCE

5.2.1.1   Fuel-Controlled Growing Fire

Consider first the situation where there is always sufficient oxygen available for complete combustion of the combustible gases, released by the growing fire. Figure 5.2 sketches the mean temperature evolution in an enclosure where the fire grows into a fully developed fire (Section 5.3). Focus remains on the growth phase in the present section. Note that with a fuel-controlled fire, all the heat released by the fire is in principle released within the enclosure. Indeed, all the combustible gases are supposed to react completely within the enclosure, due to the abundance of oxygen available. Depending on the geometry and heat transfer processes, this can be an onerous situation for the structure in which the fire exists. This corresponds to a 'regime II' diagram in Thomas' terminology [7,8].

In fire safety engineering calculations, it is common practice to primarily characterize the fire source as a combination of heat release rate (HRR) and area. The implicit simplifying assumption is that the concept of a 'free fire plume', as discussed in Section 3.5.1, applies. While such a situation can occur during early stages of a fire and/or in large enclosures (with large floor area and high ceiling), this idealized condition is often not met in reality in an enclosure, for a number of reasons.

First of all, as indicated in Table 3.2, the average flame height can be quite high, compared to the height of the enclosure. Considering, e.g., a fire of 1 MW, an average flame height in the order of 2 m is easily found. It must be appreciated that this implies much higher instantaneous flame heights, due to the turbulent nature of the fire plume (see Section 3.5 for more detail). In other words, unless the ceiling is sufficiently high, interaction between the fire plume and the ceiling of the enclosure is very likely. This would cause a substantial deviation in behaviour from the free fire plume. Flame impingement onto the ceiling has been discussed in Section 3.5.3.

Moreover, it is typical for a smoke layer to develop inside an enclosure. Indeed, as explained in Section 4.2, the hot smoke plume rises and will, in case of an enclosure, interact with the ceiling to form a ceiling jet (Section 4.5). The horizontal flow underneath the ceiling will deflect downwards upon interaction with the side walls. The smoke is pushed downwards due to the horizontal momentum, but it enters a region with cold air, so buoyancy pushes the smoke upwards again. As such, a hot upper smoke layer is formed, above a cold bottom layer of air. The smoke plume transfers mass and energy from the cold bottom layer into the hot upper layer. For such a situation to persist, buoyancy must keep the smoke upwards. In other words, the fire HRR must be sufficiently high and/or the heat loss from the smoke to the enclosure must be sufficiently small (i.e., the enclosure must not be too large).

Under such conditions, it is very well possible that the upper part of the fire plume is within the smoke layer (unless the ceiling is sufficiently high). Figure 5.3 sketches this situation. This has a few important consequences. One aspect concerns the change in entrainment into the fire plume. Indeed, the upper part is no longer surrounded by (presumably) cold air, but by hot smoke. This modifies the buoyant force (Section 2.6), and consequently the turbulence (Section 3.5.1.3), substantially. Another aspect concerns the fact that the upper part of the fire plume does not entrain air, but hot smoke and combustion products. This corresponds to vitiated conditions, as explained

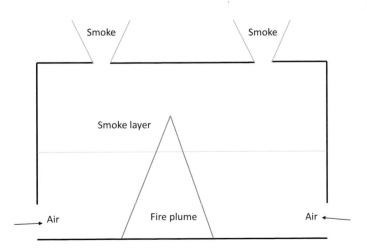

FIGURE 5.3 Sketch of a fire plume where the upper part is inside a smoke layer in an enclosure. Note that horizontal vents have been drawn, but this is not essential.

in Section 3.5.4.3. All these aspects make clear that the simplification of a fire source into a combination of fire HRR and area is not always justified for an enclosure fire.

There are still other aspects, jeopardizing the assumption of the free fire plume, e.g., there can be interaction with walls. As described in Section 3.5.2, the presence of the wall affects the entrainment process, resulting in strongly different flame heights, compared to the free fire plumes.

Entrainment can also be strongly affected by the air flow itself, for obvious reasons. The air flow can be induced by the fire itself or can be caused by external factors. This depends on the pressure differences involved. Consider first the fire-induced air flow, i.e., the situation where fire-induced pressure differences dominate all other static pressure differences. If the fuel composition is known, the required amount of air for complete combustion of 1 kg fuel can be calculated (see Section 2.2.1). The implicit assumption is then that all the released combustible gases burn completely inside the enclosure, which need not be the case (see, e.g., Section 5.1.1.2). For many organic fuels, it is found that per kg air consumed, about 3 MJ of energy is released by complete combustion, Eq. (2.16a) (e.g., [3]). In other words, 1 kg/s air can sustain a 3 MW fire. However, it must be appreciated that the fire-induced air flow rate is much larger than the air flow rate consumed by the combustion reactions. Indeed, air will also entrain into the smoke plume or even simply flow into and out of the enclosure, without really participating in any entrainment process. In any case, it is clear that a few kg/s air is attracted into the enclosure by a fire with an order of magnitude of 1 MW. Depending on the size of the enclosure, this can cause air velocities in the order of a few m/s. This can strongly affect the entrainment process into the fire plume.

From a fluid mechanics point of view, the induction of the air flow by the fire has been explained in Section 3.5. Essentially, due to the combustion reaction, the temperature in the flames is much higher than in ambient air. The resulting decrease in mass density causes the hot gases to rise, due to the buoyancy force. This creates an 'under-pressure' in the fire plume, compared to ambient, and this pressure difference sets the ambient air into motion.

Particularly complex flow patterns can occur if the fire-induced air flow rate approaches the fire plume from one side in a not-too-large enclosure. In Ref. [9], e.g., CFD (Computational Fluid Dynamics) results illustrate the impact of such a fire-induced air flow rate, schematically reproduced in Figure 5.4. The fire plume is tilted backwards, and in horizontal planes, the air flow

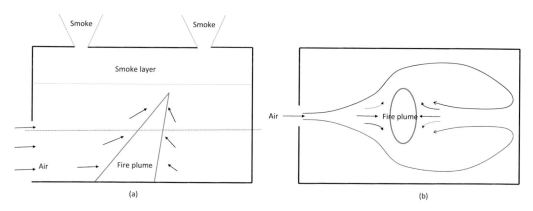

FIGURE 5.4 Schematic reproduction of the effect of fire-induced air flow rate approaching the fire plume from one side in a not-too-large enclosure (similar to the set-up in Ref. [9]). (a) Vertical plane, showing backward plume tilting due to the fire-induced air flow; dashed line: horizontal plane, shown in right figure. (b) Horizontal plane (top view), showing the stretching of the plume from a round into an elliptic shape due to the induced air flow. Note that horizontal vents have been drawn, but this is not essential.

stretches the plume from a round into an elliptic shape. This increases the area for entrainment. Moreover, the additional shear stress also increases turbulence and thus the turbulent entrainment. These factors obviously affect the flame height.

Finally, it is noted that also external factors can drive the air flow rate. External factors include the possible stack effect (particularly in high-rise buildings), HVAC (Heating, Ventilation, Air-Conditioning) imposed flows, wind, moving elevators, positive pressure ventilation, etc. These are covered in more detail in Chapter 6. It suffices here to mention that this can also affect the entrainment into the fire plume inside the compartment substantially.

5.2.1.2 Ventilation-Controlled Growing Fire

In Section 5.1.1.1, it was mentioned that about 1 kg/s air supply is required to sustain a 3 MW fire. If, however, the supply of air – be it in a natural way as to flow through vents caused by an under-pressure in the enclosure, or in a mechanical way by a fan – is insufficient to sustain the growing fire, the fire growth becomes ventilation-controlled, rather than fuel-controlled. Figure 5.5 sketches the mean temperature evolution. Due to the limited ventilation, temperatures remain lower than the ones sketched in Figure 5.2. Indeed, part of the combustible gases no longer find oxygen to react with inside the enclosure; so, they either burn outside the enclosure or simply leave the enclosure without reacting. The latter is more likely because the temperature of the smoke leaving the enclosure is typically too low for (spontaneous) ignition outside the enclosure if the fire was still in its growth phase. External flaming occurs in post-flash-over conditions (see Section 5.3).

In ventilation-controlled conditions during the fire growth phase, there is 'excess fuel', very much as in the case of a fully developed fire after flash-over (Section 5.3) [3]. Excess fuel can be the fuel that has not reacted at all, or combustion products from incomplete combustion, which are as a consequence still combustible. If the fuel ignites with oxygen outside the enclosure, this obviously imposes a thermal attack onto the structure and its surroundings. If not, then the out-flowing gases can still be very toxic (e.g., containing high amounts of carbon monoxide). In any case, however, the incomplete combustion and resulting lower temperatures inside the enclosure,

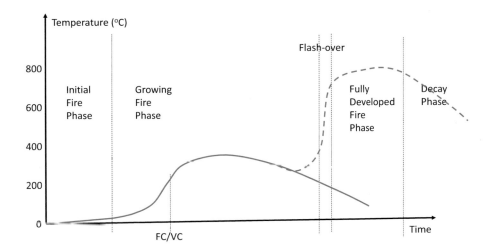

FIGURE 5.5 Evolution of average temperature in an enclosure in case of a ventilation-controlled growing fire. Solid line: flash-over and fully developed fire conditions are not met. Dashed line: evolution to ventilation-induced flash-over by sudden increase in air supply rate. Note: temperature values are merely indicative.

represent in principle a less onerous situation for the enclosure structure itself than if all the fuel were consumed completely inside the enclosure.

The dashed line in Figure 5.5 shows the possible evolution towards a 'ventilation-induced' flash-over and subsequent fully developed fire. Such an evolution is possible if there is a sudden increase in air supply rate (e.g., by breakage of a window or by the opening of a door). For obvious reasons, this is a potentially very hazardous fire evolution for firefighters. Ventilation-induced flash-over is relatively similar to 'backdraft' (Section 5.6). However, whereas a backdraft involves pressure build-up and a fireball, a ventilation-induced flash-over is rather a continued fire development (growth) upon the additional supply of air.

5.2.2 SMOKE DYNAMICS

An obvious driving force for the smoke dynamics is the fire source itself. As explained in Chapter 4, different types of smoke plumes can occur.

The classical free smoke plume concept with the according correlations (see Section 4.2) is often used in fire safety engineering calculations, particularly for smoke and heat control calculations. However, just like the free fire plume does not often appear in an enclosure, also the free smoke plume above a fire plume (as sketched in Figure 5.6) does not always exist in reality (except during the early stages of a fire and in enclosures with large floor area and high ceiling). Indeed, if the fire plume is already distorted, this also holds for the smoke plume. Even if the fire plume resembles a free fire plume, complex induced flows inside the enclosure may still strongly affect the entrainment, thereby tilting or stretching the smoke plume. It is also possible that there is no real smoke plume, as sketched in, e.g., Figure 5.3, namely if the flames reach up to the smoke layer. It is important to realize that the absence of a free smoke plume strongly jeopardizes many two-zone model calculations (see Section 5.2.5 for more explanation on two-zone models), because typically one of the correlations as mentioned in Section 4.2 is implemented to quantify the smoke mass flow rate at a certain height.

The spill plume (Section 4.6) does occur more often in practice. The classical example is the spill plume in an atrium, emerging from an adjacent room where the fire is positioned. However, also in such circumstances, care must be taken in quantifying the entrainment, as induced air

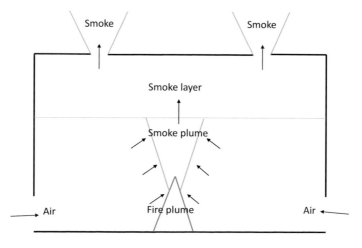

FIGURE 5.6   Sketch of a free fire plume and a free smoke plume in an enclosure. This idealized situation does not occur often in reality. Note that horizontal vents have been drawn, but this is not essential.

flow rates can strongly affect the smoke dynamics. This has been illustrated in, e.g., Ref. [5] (see Section 5.1): for a sufficiently high smoke extraction rate at the ceiling, the induced air supply rate could strongly disturb the smoke layer, essentially due to a very large vortex flow. The position and size of the vents for the supply rate also have an impact on the flow pattern.

5.2.3 Flows Through Openings

5.2.3.1 Horizontal Openings

Consider a situation as sketched in Figure 5.7, namely a smoke layer underneath a horizontal ceiling, from which smoke leaves through a horizontal opening in the ceiling. A uniform smoke temperature is assumed in order to allow for the development of an analytical solution. Steady-state conditions are assumed. Strictly speaking, this does not apply to 'growing' fires, but the true meaning here is that the analysis applies to non-flash-over fires. In case of a growing fire, quasi steady state is assumed for the instantaneous situations.

Outside the enclosure, the law of hydrostatics, Eq. (5.4), reveals a linear decrease in static pressure. Inside the enclosure, the classical assumption is that the temperature equals ambient temperature, underneath the smoke layer. This implies that the mass density in that region equals the ambient mass density, since pressure differences are negligible compared to atmospheric pressure. Using all these assumptions leads to a static pressure line inside the enclosure that is parallel to the static pressure line outside. Note that no wind is considered here. The possible effect of wind is discussed in Section 6.4.

Consider first the situation where the area for supply of intake air is infinitely large. In other words, the pressure difference over the inlet vents equals zero. This implies that the static pressure lines inside and outside the enclosure coincide. The static pressure difference over the opening approximately equals the static pressure difference between points 2_{in} and 2_{out} and is readily calculated from Eq. (5.4):

$$\Delta p = \left(\rho_{amb} - \rho\right)gD. \tag{5.9}$$

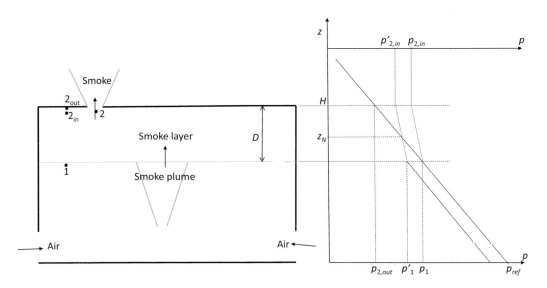

FIGURE 5.7 Sketch of a smoke layer underneath a horizontal ceiling. Smoke leaves the smoke layer through a horizontal opening in the ceiling. Static pressure lines are sketched at the right hand side.

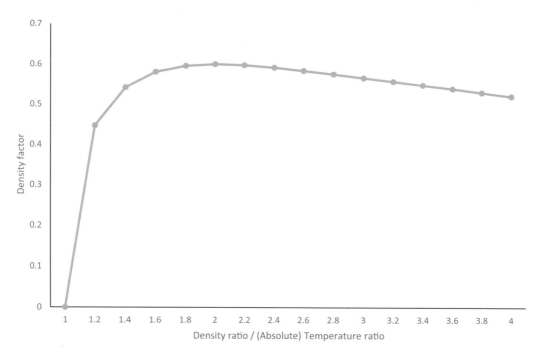

FIGURE 5.8　Mass flow rate through a horizontal opening: non-monotonic dependence on the smoke mass density. The plot shows the density factor $\rho\sqrt{\left(\dfrac{\rho_{amb}}{\rho}-1\right)}$, for $\rho_{amb}=1.2$ kg/m^3.

In Eq. (5.9), D is the smoke layer depth (see Figure 5.7). Using Eq. (5.3), the mass flow rate through the vent can be calculated:

$$\dot{m}=C_d A\sqrt{2\rho\left(\rho_{amb}-\rho\right)gD}\;=\;\rho C_d A\sqrt{2\left(\frac{\rho_{amb}}{\rho}-1\right)gD}.\tag{5.10}$$

Equation (5.10) reveals that the mass flow rate through the opening increases linearly with the opening cross-section area A and with the discharge coefficient of the opening. It also increases proportional to the square root of the smoke layer thickness, given a certain mass density of the smoke.

It is instructive now to examine the factor $\rho\sqrt{\left(\dfrac{\rho_{amb}}{\rho}-1\right)}$ for realistic temperature ratios. Approximating ambient temperature around 300 K, a temperature ratio of 4 (implying a temperature of 1,200 K) seems an upper bound. This then also applies to the density ratio (Eq. 5.6). Figure 5.8 reveals the non-monotonic behaviour. For lower smoke temperatures, the mass flow rate through the opening decreases substantially, illustrating that, in principle, if natural ventilation is applied (see Section 5.1.4), higher smoke temperatures make the ventilation more efficient. However, the curve levels off, decreasing beyond a temperature ratio of 2 (corresponding to a temperature of 600 K). The reason is as follows: a higher temperature ratio implies a higher smoke temperature, leading to a stronger buoyancy force. This leads to an increase in volume flow rate through the opening. This can be understood from Figure 5.7: the orange part of the static pressure line (i.e., the static pressure line inside the smoke layer) becomes more vertical. However, the higher smoke temperature also implies a lower smoke mass density, as can be understood from the ideal gas law, Eq. (2.10), with approximately constant pressure. Indeed, the global pressure level remains approximately equal to atmospheric pressure, because pressure differences are much smaller than the absolute pressure level. As a consequence, the increase in volume flow rate becomes less and less pronounced (because the static pressure line inside the smoke layer never becomes completely vertical and the static pressure line outside remains unchanged), and the reduction in mass density leads to an effective reduction in mass flow rate for sufficiently high smoke temperature (i.e., higher than 600 K), despite the increase in volume flow rate. It is noteworthy that the density factor becomes essentially constant (around 0.55), for high-enough smoke temperatures (above 150 °C).

In reality, obviously, the openings for the incoming air are not infinitely large. As a consequence, there will be a pressure loss for the incoming air as it flows through the inlet openings. This is illustrated in Figure 5.7 by a shift of the static pressure line inside the enclosure to the left (i.e., to lower pressure values). The consequence is that the pressure difference over the outlet opening, all other parameters kept constant, decreases. In other words, the flow through the ceiling vent decreases due to the reduction in opening for the incoming air. This is logical: for the flow, coming from outside and leaving the enclosure through the ceiling vent, the two openings are a series of pressure losses. Before elaborating on this, the neutral plane height is introduced first. The neutral plane height denotes the vertical position (z_N in Figure 5.7) where the pressure inside the enclosure equals the outside pressure. This position can be determined from a mass balance, as follows.

The flow through the outlet opening is given by Eq. (5.10), replacing D by $(H - z_N)$, with H the height of the compartment:

$$\dot{m}_{out} = C_{d,out} A_{out} \sqrt{2\rho\left(\rho_{amb} - \rho\right)g\left(H - z_N\right)} . \tag{5.11}$$

The pressure difference over the inlet opening is also related to the position of the neutral plane height, still assuming that the static pressure line inside the enclosure underneath the smoke layer is parallel to the ambient static pressure line:

$$\Delta p_{in} = \left(\rho_{amb} - \rho\right)g\left(D - \left(H - z_N\right)\right). \tag{5.12}$$

From this pressure difference, the inlet mass flow rate is determined as follows:

$$\dot{m}_{in} = C_{d,in} A_{in} \sqrt{2\rho_{amb}\left(\rho_{amb} - \rho\right)g\left(D - \left(H - z_N\right)\right)} . \tag{5.13}$$

In order to obtain steady-state conditions, Eqs. (5.11) and (5.13) must be equal. Note that the fuel mass flow rate is ignored. This is justified, given typical stoichiometric air-to-fuel ratios (see Section 2.2.1), but can become questionable in ventilation-controlled conditions.

A typical value for the discharge coefficient is around 0.6 for an opening with sharp edges, with slight dependence on the Reynolds number (slightly higher values for lower Reynolds numbers). Assuming, for the sake of simplicity, equal discharge coefficients for the inlet and outlet openings, the end result is:

$$z_N = \frac{A_{out}^2 \rho H + A_{in}^2 \rho_{amb}\left(H - D\right)}{A_{out}^2 \rho + A_{in}^2 \rho_{amb}}. \tag{5.14}$$

With this information, the mass flow rate (Eq. 5.11) can be computed, given a certain smoke layer thickness and temperature:

$$\dot{m}_{out} = \rho C_{d,out} A_{out} \sqrt{\frac{2gD\left(\dfrac{T}{T_{amb}} - 1\right)}{1 + \dfrac{T_{amb}}{T}\left(\dfrac{C_{d,out} A_{out}}{C_{d,in} A_{in}}\right)^2}} \, . \tag{5.15}$$

It is important to reflect on the inlet opening area. Equation (5.14) reveals that in the limit to infinitely large inlet opening, the position of the neutral plane coincides with the bottom side of the smoke layer, in line with the analysis above. For the limit of A_{in} going to zero, on the other hand, the neutral plane height is at the ceiling level. This implies that there is no pressure difference over the opening, and thus, there is no flow through the opening. This is a very important result for natural ventilation because it shows that the outlet vent, regardless of its size, becomes completely useless if no openings are in place for replacement air. The reason is easy to understand: initially, the fire causes an expansion of gases and an outflow of smoke. This outflow of mass, not compensated by any inflow, leads to a reduction in pressure inside the enclosure. After a while, the fire may die due to lack of oxygen (see above) or a pulsating fire (Section 5.5) may establish if the vent acts as an air supply opening due to under-pressure in the enclosure and the fire can grow again. In any case, no stable situation as sketched in Figure 5.7 can be obtained.

To summarize: the smaller the opening area for the inlet air, the higher the position of the neutral plane height and the lower the mass flow rate through the outlet vent, keeping all other parameters unchanged.

To conclude this section, it is noted that it an implicit assumption in all the reasonings above is that the velocity in point 1 is equal to zero. This need not be the case. In fact, the outlet vent can be positioned above the smoke plume. In that case, an upward velocity will lead to an increase in total pressure at the inner side of the outlet vent, as can easily be seen from Bernoulli's equation, Eq. (5.1). As expected, the outlet vent would become more effective when it is positioned directly above the smoke plume. Obviously, it cannot be predicted where exactly the smoke plume will be in reality; so, in that sense, the assumption of zero upward velocity at the bottom side of the smoke layer is a conservative approach.

Finally, it is recalled that the analysis corresponds to a strong simplification of reality: no flow fields are considered, there are no changes in the fire and smoke dynamics, and smoke temperatures are assumed to be uniform. Still, it provides an order of magnitude analysis and reveals the important parameters for the problem at hand.

Example 5.1

In steady-state conditions, a 2 m thick smoke layer is established underneath a flat ceiling, in which a horizontal ventilation opening of 3 m^2 is present. The compartment height is 5 m. The compartment floor area is 15 m×10 m. The average smoke temperature is 150 °C. The total opening for intake of fresh air is 4 m^2. The discharge coefficient for all openings is 0.6. The ambient temperature is 20 °C, and the outside pressure is 1,00,000 Pa. Estimate the position of the neutral plane height and the mass flow rate through the vent in the ceiling. What is the corresponding convective HRR of the fire, ignoring heat transfer to the structure?

In Eq. (5.14), $A_{out} = 3$ m^2, $\rho = 0.823$ kg/m^3, $H = 5$ m, $A_{in} = 4$ m^2, $\rho_{amb} = 1.189$ kg/m^3 and $D = 2$ m, so that $z_N = 3.56$ m. Equation (5.15) yields: $\dot{m}_{out} = 4.6$ kg/s.

Ignoring heat transfer to the structure, this corresponds to a convective HRR of the fire, equal to $\dot{Q}_{conv} = \dot{m}_{out}c_p(T - T_{amb}) = 597$ kW. Indeed, this is the part of the fire HRR that is removed by the smoke flowing out of the compartment (and replaced by air flowing into the compartment).

Exercises

For the setting of the example, calculate the position of the neutral plane height, the mass flow rate through the vent in the ceiling and the corresponding convective HRR of the fire (ignoring heat transfer to the structure), if:

1. The inlet opening area is only $2\,m^2$. Explain the effect of reduction of inlet area. (Answer: $z_N = 4.22$ m; $\dot{m}_{out} = 3$ kg/s; $\dot{Q}_{conv} = 390$ kW. The neutral plane height increases due to larger pressure loss over the inlet opening (shift of the inner pressure line in Figure 5.7 to the left); this leads to a reduced pressure difference over the outlet opening and thus reduced mass flow rate; with the same smoke temperature, this corresponds to a lower convective HRR.)

2. The smoke layer thickness is 3 m. Explain the effect. (Answer: $z_N = 2.84$ m; $\dot{m}_{out} = 5.63$ kg/s; $\dot{Q}_{conv} = 731$ kW. The neutral plane height decreases as the bottom of the smoke layer decreases; this leads to an increased pressure difference over the outlet opening and thus increased mass flow rate; with the same smoke temperature, this corresponds to a higher convective HRR.)

3. The smoke layer temperature is 250°C. (Answer: $z_N = 3.48$ m; $\dot{m}_{out} = 4.7$ kg/s; $\dot{Q}_{conv} = 1,081$ kW. The neutral plane height decreases because the pressure line in the smoke layer is steeper due to the higher smoke temperature; this leads to an increased pressure difference over the outlet opening and slightly increased mass flow rate because the smoke temperature is still below 600 K (Figure 5.8); the higher smoke temperature and mass flow rate correspond to a higher convective HRR.)

4. The smoke layer temperature is 450°C. (Answer: $z_N = 3.37$ m; $\dot{m}_{out} = 4.26$ kg/s; $\dot{Q}_{conv} = 1,831$ kW. The neutral plane height decreases because the pressure line in the smoke layer is steeper due to the higher smoke temperature; this leads to an increased pressure difference over the outlet opening, but not to an increased mass flow rate, because the smoke temperature is above 600 K (Figure 5.8) so that the mass density drops more than the volume flow rate increases; the higher smoke temperature leads to a higher convective HRR, despite the lower mass flow rate.)

5.2.3.2 Vertical Openings

Figure 5.9 sketches the situation, equivalent to Figure 5.7, but for vertical openings, rather than horizontal openings. A smoke layer is assumed present underneath a horizontal ceiling. Smoke leaves the compartment through the upper part of a vertical opening (e.g., a door opening or a window). A uniform smoke temperature is assumed in order to allow for the development of an analytical solution. Note that the simplification is made that the smoke plume is not affected by the flows into and out of the compartment. In Section 5.2.2, it has been explained that this is a very strong simplification of reality. Steady-state conditions are assumed. Strictly speaking, this does not apply to 'growing' fires, but the true meaning here is that the analysis applies to non-flash-over fires. In case of a growing fire, quasi steady state is assumed for the instantaneous situations.

Outside the enclosure, the law of hydrostatics, Eq. (5.4), still applies, i.e., a linear decrease in static pressure with height. Inside the enclosure, the classical assumption is that the temperature equals ambient temperature below the smoke layer. This implies that the mass density in that region equals the ambient mass density because pressure differences are negligible compared to atmospheric pressure. Using all these assumptions leads to a static pressure line inside the

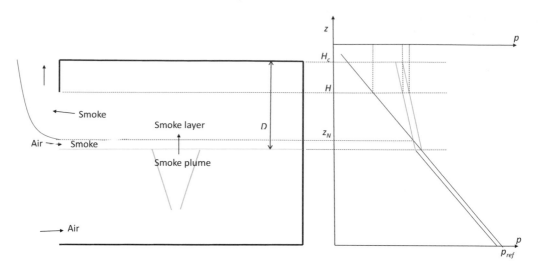

FIGURE 5.9 Sketch of a smoke layer underneath a horizontal ceiling. Smoke leaves the smoke layer through a vertical opening in the side wall. Static pressure lines are sketched at the right hand side. The outflow of smoke is sketched for the situation where there is a non-zero pressure difference over the opening for the air intake (i.e., the opening for air intake is not infinitely large).

enclosure that is parallel to the static pressure line outside. Note that no wind is considered in the present section. The possible effect of wind is discussed in Section 6.7.

Consider first the situation where the area for supply of intake air is infinitely large. In other words, the pressure difference for the air intake equals zero. This implies that the static pressure lines inside and outside the enclosure coincide.

In contrast to the analysis for the horizontal opening (Section 5.2.3.1), the pressure difference over the vertical opening for the smoke outflow is not uniform. On the contrary, the pressure difference evolves as follows:

$$\Delta p(z) = (\rho_{amb} - \rho) g (z - z_N). \tag{5.16}$$

For the special case of an infinitely large opening area for air intake, the neutral plane height coincides with the bottom of the smoke layer. In all other cases, i.e., in reality, the neutral plane height is at a (slightly) higher position than the bottom of the smoke layer. This can easily be understood from the right part of Figure 5.9; the change in slope in the static pressure line is due to the lower mass density inside the smoke layer. This mass density is always lower than the mass density of ambient air because the smoke has a higher temperature (Eq. 5.6). Thus, for steady conditions to be possible, the situation must be as sketched in Figure 5.9; if the static pressure inside the enclosure were completely to the right (respectively, left) of the static pressure outside, there would be outflow (respectively, inflow) through the entire opening. The corresponding loss (respectively, gain) in mass inside the enclosure would then bring back the static pressure line inside the enclosure to the left (respectively, right), as sketched in Figure 5.9. As long as the static pressure line in the lower layer inside the enclosure coincides with the pressure line outside, there is no inflow into the bottom layer (except for the limiting case of infinite opening area, from which the analysis started). In other words, for steady-state conditions, the situation as sketched in Figure 5.9 is the only possibility and the position of the neutral plane height is by definition higher than the bottom of the smoke layer.

From Eq. (5.16), the mass flow rate of smoke out of the opening can be calculated as an integration:

$$\dot{m}_{\text{out}} = C_{d,\text{out}} \int_{z_N}^{H} \rho v(z) W(z) dz. \tag{5.17}$$

This is a generalization of the approach used to derive result Eq. (5.11). Note that H now denotes the opening height. In Eq. (5.17), W is the opening width (which can in principle vary with height). The velocity at height z reads (Eqs. 5.1 and 5.16):

$$v(z) = \sqrt{\frac{2\Delta p(z)}{\rho}} = \sqrt{2\left(\frac{\rho_{\text{amb}}}{\rho} - 1\right) g(z - z_N)}. \tag{5.18}$$

As a consequence, Eq. (5.17) can be elaborated as follows:

$$\dot{m}_{\text{out}} = C_{d,\text{out}} \int_{z_N}^{H} \rho \sqrt{2\left(\frac{\rho_{\text{amb}}}{\rho} - 1\right) g(z - z_N)} W(z) dz. \tag{5.19}$$

For constant opening width, Eq. (5.19) simplifies to read:

$$\dot{m}_{\text{out}} = \frac{2}{3} C_{d,\text{out}} \rho W \sqrt{2\left(\frac{\rho_{\text{amb}}}{\rho} - 1\right) g} (H - z_N)^{3/2} = \frac{2}{3} C_{d,\text{out}} \rho W \sqrt{2\left(\frac{T}{T_{\text{amb}}} - 1\right) g} (H - z_N)^{3/2}. \tag{5.20}$$

In the latter equation, Eq. (5.6) has been applied.

Equation (5.20) reveals that the mass flow rate through the opening increases linearly with the opening width W and with the discharge coefficient of the opening. The dependence is not linear with respect to the height of the opening. This point will also be addressed in Section 5.3.3.2, where the relation between intake of fresh air and the 'ventilation factor', $AH^{1/2}$, is made (with A the area of the vertical opening). It can be understood from Section 5.2.3.1, though, where Eq. (5.11) revealed a linear dependence on the opening area and a dependence, proportional to the square root of the smoke layer thickness, given a certain mass density of the smoke. For a vertical opening, the thickness of the smoke layer plays a double role: it determines the 'height' of the area of the opening through which smoke flows out (linear dependence), and it determines the magnitude of the pressure difference (square root dependence), hence the exponent 3/2.

It is noteworthy that also the factor $\rho \sqrt{\left(\frac{\rho_{\text{amb}}}{\rho} - 1\right)}$ appears in expression (5.20), just as in Eq. (5.11). This implies that the non-monotonic behaviour, explained in Figure 5.8, is present again.

As mentioned above, in reality, the openings for the incoming air and for the outflowing smoke are not infinitely large. As a consequence, there is a pressure loss for the incoming air as it flows through the inlet openings. This is illustrated in Figure 5.9 by a shift of the static pressure line inside the enclosure to the left (i.e., to lower pressure values). Calling that pressure difference Δp_{in}, Eq. (5.12) now becomes:

$$\Delta p_{\text{in}} = \left(\rho_{\text{amb}} - \rho\right) g\left(z_N - (H_c - D)\right). \tag{5.21}$$

The mass flow rate of air into the enclosure reads:

$$\dot{m}_{air,in} = C_{d,in} \int_0^{H_c-D} \rho_{amb} \sqrt{2\left(1 - \frac{\rho}{\rho_{amb}}\right)g\left(z_N - (H_c - D)\right)} W dz$$

$$+ C_{d,in} \int_{H_c-D}^{z_N} \rho_{amb} \sqrt{2\left(1 - \frac{\rho}{\rho_{amb}}\right)g\left(z_N - z\right)} W dz \qquad (5.22)$$

or (for constant opening width W)

$$\dot{m}_{air,in} = C_{d,in} \rho_{amb} \sqrt{2\left(1 - \frac{\rho}{\rho_{amb}}\right)g\left(z_N - (H_c - D)\right)} W\left(H_c - D\right)$$

$$+ C_{d,in} \int_{H_c-D}^{z_N} \rho_{amb} \sqrt{2\left(1 - \frac{\rho}{\rho_{amb}}\right)g\left(z_N - z\right)} W dz. \qquad (5.23)$$

One could argue that the density to be used in the second term of Eq. (5.22) or (5.23) should not be ρ_{amb}, but ρ (or a value in between), because the smoke that intends to leave the enclosure is pushed back in (because ambient pressure is higher than the pressure inside) and there will be some mixing in reality (which is much more complex and lively than the steady simplification in the analysis above). Yet, the second term is often small, and in fact, often ignored in two-zone modelling (see Section 5.2.5).

The exact position of the neutral plane is determined by setting the mass flow rate into the enclosure, Eq. (5.22), equal to the mass flow rate out of the enclosure, Eq. (5.22), as explained in Section 5.2.3.1. Note that again the fuel mass flow rate is ignored, as mentioned in Section 5.2.3.1.

It is noted that in the case at hand, a reduction in the opening area directly affects both the inflow of ambient air and the outflow of hot smoke. The first effect would be an increase in pressure drop over the opening for the intake of fresh air. This implies a shift to the left of the line of static pressure inside the enclosure and a shift upwards of the neutral plane height, as explained in Section 5.2.3.1. This would imply a reduction in the mass flow rate, flowing out of the compartment, as seen from Eq. (5.20). Another possibility, though, is that the neutral plane height position remains unchanged and the smoke layer thickness D increases. What exactly will happen, depends on the fire source as well (e.g., the change in fire size depending on the ventilation conditions).

Moreover, a reduction in the width of the opening does not have the same impact as the reduction in the height of the opening. Indeed, there is a linear dependence on the width, while the dependence on height is through an exponent of 1.5, as explained above.

It is very important to appreciate that neither the height of the compartment nor the smoke layer depth by itself is the characteristic length scale in the expression for the mass flow rate of smoke, emerging from the enclosure. Indeed, Eq. (5.20) contains neither H_c nor D, but rather $(H - z_N)$ as a characteristic length scale. This is an important difference from Eq. (5.10), which contains the smoke layer thickness D as a characteristic length scale. (Note that for a horizontal opening in the ceiling, $H_c = H$ and $(H - z_N) \approx D$ in Eq. (5.11).) The reason is obvious: in the configuration sketched in Figure 5.9, the smoke cannot flow out of the compartment until it is below the upper point of the opening (H). This also implies that the position of the opening itself will determine the position of the neutral plane height. In other words, if an opening (e.g., a window) is positioned higher or lower, the neutral plane height will move along with the opening.

Finally, it is recalled that the analysis corresponds to a strong simplification of reality: no flow fields are considered, there are no changes in the fire and smoke dynamics, and smoke temperatures are assumed to be uniform. Still, it provides an order of magnitude analysis and reveals the important parameters for the problem at hand.

5.2.4 NATURAL AND MECHANICAL VENTILATION

The results obtained in Section 5.2.3 are very valuable with respect to smoke and heat control in case of fire (Chapter 6).

First of all, from a qualitative point of view, the static pressure lines shown in Figures 5.7 and 5.9 reveal that openings for the intake of ambient air are best put at lower positions, where there is a natural inflow. Similarly, vents for natural extraction of smoke are best positioned high in the enclosure (e.g., the ceiling), where the difference between the pressure inside and the pressure outside (still ignoring wind) is at its maximum. Also, in case of mechanical extraction by means of a fan instead of natural extraction through a vent, it is advantageous to have the extraction fan at high positions. Indeed, in that case, the natural pressure difference assists the mechanical fan. This can be understood from fan characteristics, as schematically shown in Figure 5.10. On the horizontal axis, the pressure difference is shown that needs to be overcome by the fan (e.g., due to friction losses in a ductwork through which the smoke is pushed, see Chapter 6). The higher this counter-pressure, the lower the volume flow rate extracted by the fan (shown on the vertical axis). If there is now a natural pressure difference to assist the fan, the volume flow rate extracted by the fan will automatically increase for a given fan.

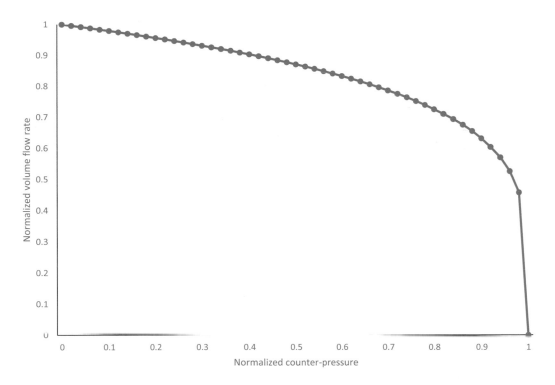

FIGURE 5.10 Schematic representation of extraction fan characteristics: volume flow rate as a function of the counter-pressure the fan has to overcome.

Next, it is instructive to re-examine Eq. (5.10) from the point of view of the effectiveness of a vent for smoke extraction by means of natural ventilation. The vent will be more effective when more smoke is extracted, i.e., if the smoke mass flow rate is higher. For a situation as sketched in Figure 5.7 in (quasi) steady state, the mass flow rate through the vent, given by Eq. (5.10), equals the mass flow rate of smoke entering the smoke layer from the bottom. The latter mass flow rate is typically determined by the fire and smoke dynamics. Recall the assumption of uniform smoke temperature, so the possibility of flames entering the smoke layer is not considered here. Assuming now that the smoke mass flow rate is determined by the fire and smoke dynamics, Eq. (5.10) reveals that for a given smoke temperature, a smaller opening area A_{out} will lead to a higher value of the smoke layer thickness D. Given the non-linear dependence on D, Eq. (5.10) reveals that a reduction in A_{out} by a factor of 2 would, keeping everything else unchanged, lead to an increase in smoke layer thickness by a factor of 4. The opposite is also true: an increase in A_{out} leads to a reduction in smoke layer thickness D. The limit is clear: for an infinite opening area, there is no smoke inside. However, other effects may also occur if A_{out} becomes too large. This is addressed by the end of this subsection, in the context of plug-holing.

It must be recalled that also the inlet opening area has a strong impact, as explained in Section 5.2.3. Indeed, in fact, the end result was expression (5.15) for the flow through horizontal vents. For the limit of large A_{in} values, Eq. (5.10) is recovered, but the other interesting limit is the limit of Eq. (5.15) for small values of A_{in}. After elaboration, this reads:

$$\dot{m}_{\text{out}} = \rho_{\text{amb}} C_{d,\text{in}} A_{\text{in}} \sqrt{2gD\left(1 - \frac{T_{\text{amb}}}{T}\right)}. \tag{5.24}$$

In other words, if A_{in} becomes small, this becomes the limiting factor, regardless of the size of the outlet vent. And again, keeping all other parameters unchanged, a reduction in A_{in} by a factor of 2 implies an increase in smoke layer thickness by a factor of 4. This clearly illustrates the huge impact of the inlet opening area.

It is, in this sense, also instructive to examine the ratio of expressions (5.10) and (5.15), because this illustrates the impact of the inlet opening area on the mass flow rate through the outlet vent (compared to the theoretical situation of an infinitely large inlet opening area), and thus quantifies the reduction in 'effectiveness' of the outlet vent, keeping all other parameters unchanged. Calling Eq. (5.10) the maximum possible mass flow rate, given a certain outlet vent area, smoke layer thickness and smoke layer temperature, the effectiveness is obtained by dividing Eq. (5.15) by (5.10):

$$\text{Effectiveness} = \frac{\dot{m}_{\text{out}}}{\dot{m}_{\text{out,max}}} = \frac{\rho C_{d,\text{out}} A_{\text{out}} \sqrt{\dfrac{2gD\left(\dfrac{T}{T_{\text{amb}}} - 1\right)}{1 + \dfrac{T_{\text{amb}}}{T}\left(\dfrac{C_{d,\text{out}} A_{\text{out}}}{C_{d,\text{in}} A_{\text{in}}}\right)^2}}}{\rho C_{d,\text{out}} A_{\text{out}} \sqrt{2\left(\dfrac{\rho_{\text{amb}}}{\rho} - 1\right)gD}} = \sqrt{\frac{1}{1 + \dfrac{T_{\text{amb}}}{T}\left(\dfrac{C_{d,\text{out}} A_{\text{out}}}{C_{d,\text{in}} A_{\text{in}}}\right)^2}}. \tag{5.25}$$

Equation (5.25) quantifies what mass flow rate of smoke effectively flows through the outlet vent, compared to what would flow through the same vent, keeping all other parameters unchanged, but with infinite area for the inlet air. Figure 5.11 illustrates the effectiveness of a few values of the ratio T_{amb} / T. For the limit of A_{in} going to infinity, i.e., for the limit $\left(\dfrac{C_{d,\text{out}} A_{\text{out}}}{C_{d,\text{in}} A_{\text{in}}}\right)^2$ going to zero,

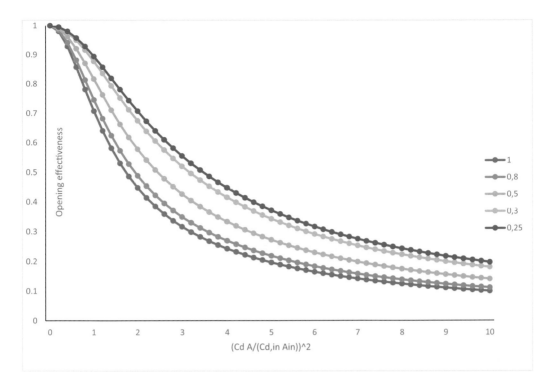

FIGURE 5.11 Illustration of the impact of the inlet opening area on the effectiveness of a horizontal outlet vent in a ceiling for smoke extraction for a few values of the ratio T_{amb}/T (indicated by the line legend). Visualization of Eq. (5.25).

the effectiveness of the outlet vent opening goes to 100%, regardless of the smoke temperature. This effectiveness decreases rapidly with decreasing A_{in}. The decrease is more rapid for lower smoke temperature. Note, however, that even for a smoke temperature of 600 K (curve '0.5' in Figure 5.11), the effectiveness of the outlet vent has already dropped down to about 82% for the situation $A_{in} = A_{out}$ (for curve '1', this is even as low as 71%). The graphs in Figure 5.11 reveal that the openings for inlet air are always important for natural ventilation.

Now consider mechanical ventilation in the form of mechanical smoke extraction and natural intake of fresh air. Essentially, the outlet vent is replaced by a mechanical fan. This fan will extract a certain mass flow rate of smoke, depending on its rpm (revolutions per minute) and the pressure difference over the fan (Figure 5.10). This extraction leads to a reduction of the pressure inside the enclosure. Static pressure lines still resemble the lines as sketched in Figure 5.7, but the fundamental difference is now that the pressure difference at the ceiling (which is typically where the extraction points are positioned, as explained above) is not the driving force for the smoke to flow out of the enclosure. Rather, the mechanical fan imposes a reduction of pressure inside the enclosure, thereby also affecting the flow rate of air into the enclosure. Indeed, as the fan extracts more strongly, the static pressure line inside the enclosure will shift to the left, leading to a stronger intake of fresh air. Indeed, keeping everything else unchanged, the intake air volume flow rate is proportional to the square root of the pressure difference over the inlet opening (Eq. 5.2). Figure 5.12 illustrates this. It also illustrates that the neutral plane height, and typically also the position of the bottom of the smoke layer, move to a higher position as the fan extracts more strongly (or, equivalently, the pressure inside the enclosure decreases). As a closing

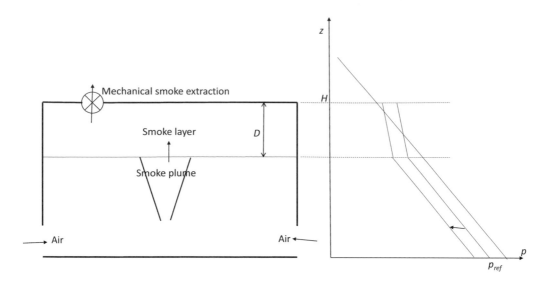

FIGURE 5.12 Illustration of the effect of mechanical smoke extraction. An increase in extraction rate shifts the static pressure line inside the enclosure to the left (indicated by the black arrow on the right figure). The result is an increase in air intake and the neutral plane height moving upwards. The smoke layer thickness has not been varied in the figure, but normally the smoke layer thickness also decreases with increase in smoke extraction rate. See, however, e.g., Ref. [5], illustrating that this is not always the case.

comment for this discussion, it is mentioned that in some situations, there is a potential danger in mechanical smoke extraction, namely for under-ventilated fires. If under such circumstances the increase in air intake due to the mechanical smoke extraction provides more oxygen to the fire, the fire dynamics can be affected and the fire can become stronger than without the mechanically imposed smoke extraction. On the other hand, if the fire is already fuel-controlled, the mechanically imposed smoke extraction will remove more smoke and, along with the smoke, heat, from the enclosure, which is beneficial in that this will slow down the development or growth of the fire. Clearly, this all depends on the configuration at hand.

Regardless of the fire dynamics, mechanical smoke extraction can also strongly affect the flow field inside the enclosure. This has clearly been illustrated in, e.g., Ref. [5]. It must be appreciated that the effect of mechanical ventilation quickly overwhelms the impact of the fire on the smoke dynamics. This statement also holds for natural ventilation in high-rise buildings. This will be discussed in Chapter 6, Sections 6.4 and 6.5.

Finally, it is important to stress that so far, it has implicitly been assumed that only smoke is flowing through the outlet vents, or only smoke is extracted by the mechanical fans. This need not be the case. Consider first mechanical extraction. If the extraction rate becomes too large, there is a risk of so-called 'plug-holing': if the smoke layer is too shallow, air will be extracted directly from below the smoke layer, rather than smoke. Obviously, also smoke will still be extracted, but clearly, the effectiveness of the extraction fan decreases if the phenomenon of plug-holing occurs. The important question is the quantification of 'too large' and 'too shallow'. In Ref. [10], the following expression is provided for the maximum mass flow rate, to be extracted through a single exhaust, based on Ref. [11]:

$$\dot{m}_{\text{crit}} = C\beta d^{5/2} \left(1 - \frac{T_{\text{amb}}}{T}\right)^{1/2} \left(\frac{T_{\text{amb}}}{T}\right)^{1/2}. \tag{5.26}$$

In Eq. (5.26), C is a constant (3.13 if SI units are used, 0.354 if US units are used) and d is the depth of the smoke layer below the lowest point of the exhaust (which can be placed vertically in a wall). This equals D in case of an exhaust in the ceiling as sketched in Figure 5.12. The factor β equals 2.8 for exhausts 'far away' from walls and equals 2 if the exhaust in a ceiling is close to a wall or if the exhaust in a wall is close to a ceiling [10]. Equation (5.26) can be rewritten in terms of maximum volume flow rates as follows:

$$\dot{V}_{crit} = \frac{\dot{m}_{crit}}{\rho} = \frac{\dot{m}_{crit}}{\rho_{amb}}\frac{T}{T_{amb}} = C'\beta d^{5/2}\left(T_{amb}\left(T - T_{amb}\right)\right)^{1/2}. \tag{5.27}$$

In Eq. (5.27), using the ideal gas law, C' can be computed from C : $C' = 0.00887$ (SI units) or 0.537 (US units). Equation (5.27) can now also be written in an inverse manner, leading to the minimum smoke layer depth d_{crit} where plug-holing is still avoided, for a given volume flow rate through the exhaust:

$$d_{crit} = \frac{\dot{V}^{2/5}}{\left(C'\beta\right)^{2/5}\left(T_{amb}\left(T - T_{amb}\right)\right)^{1/5}}. \tag{5.28}$$

It is noted that plug-holing is also possible for natural ventilation [10].

Whereas Eqs. (5.26)–(5.28) are useful in practice, more systematic research is required to solidify the theoretical basis of the plug-holing phenomenon. In Ref. [12], e.g., a 'Froude' number is defined as follows:

$$Fr = \frac{\dot{V}}{\left(g\left(\frac{\rho_{amb} - \rho}{\rho_{amb}}\right)\right)^{1/2} d^{5/2}}. \tag{5.29}$$

Reference is made to Ref. [13] to define a critical value $Fr_{crit} = 1.6$, beyond which plug-holing would occur. This statement is claimed to be valid for mechanical and natural ventilation in Ref. [12]. However, it is instructive to compare definition (5.29) to the original definition, Eq. (2.93):

$$Fr = \frac{\rho v^2}{\Delta\rho gL}. \tag{5.30}$$

Appreciating that for natural ventilation, the volume flow rate $\dot{V}$ can be written as $\dot{V} = Av$, with v the (average) outflow velocity through the vent and accepting that a square root has been taken in Eq. (5.29), compared to Eq. (5.30), there are two important things to note:

- In Eq. (5.30), the non-dimensional parameter A/d^2 has been introduced.
- The density ratio in Eq. (5.29) reads $\rho_{amb}/\Delta\rho$, while it reads $\rho/\Delta\rho$ in Eq. (5.30). The latter seems more logical because the density should relate to the flow through the vent (*in casu* smoke from the upper layer).

It is interesting to note that in Ref. [14], the Froude number is defined as Eq. (5.30), using originally the diameter of the vent as characteristic length scale, but relating it to the smoke layer thickness at critical conditions (in fact, Ref. [14] deals with water layers of different density, but this can be translated into a smoke layer upon a cold air layer; it also deals with openings in a vertical wall, rather than a horizontal ceiling, but this is of secondary importance here). In other words, at least the density ratio in Eq. (5.29) from Ref. [12] seems odd. This is confirmed when the critical volume flow rate for plug-holing is calculated from $Fr_{crit} = 1.6$, using Eq. (5.29):

$$\dot{V}_{crit} = 1.6\sqrt{g}\sqrt{1 - \frac{T_{amb}}{T}}d^{5/2} = \frac{1.6\sqrt{g}}{T_{amb}}\sqrt{\frac{\rho}{\rho_{amb}}}\sqrt{T_{amb}(T - T_{amb})}d^{5/2}.$$ Interestingly, using $T_{amb} = 288\,\text{K}$,

this yields exactly the same result as Eq. (5.27), with $\beta = 2$, were it not for the density factor $\sqrt{\rho/\rho_{amb}}$.

In other words, Eq. (5.29) should probably be updated into $Fr = \dfrac{\dot{V}}{\left(g\left(\dfrac{\rho_{amb} - \rho}{\rho}\right)\right)^{1/2} d^{5/2}}$.

In Ref. [15], the expression reads: $\dot{V}_{crit} = 1.33\dfrac{\sqrt{g}}{T}\sqrt{T_{amb}(T - T_{amb})}d^{5/2}$. Interestingly, with

$T = 481\,\text{K}$ in Ref. [15], this yields $\dot{V}_{crit} = 0.0087\sqrt{T_{amb}(T - T_{amb})}d^{5/2}$. This is again close to Eq. (5.27), but different by about 10%. However, it is important to note that it is also reported in Ref. [15] that the multiplication factor in large-scale experiments is higher than 0.0087, which was obtained in small-scale experiments. Deviations of, on average, 17%, up to 49% in 1 test, have been observed. This is non-negligible.

This illustrates the need for systematic research into the phenomenon of plug-holing, so the phenomenon can be quantified more precisely.

5.2.5 ZONE MODELLING

In this section, the basic concept of zone modelling for pre-flash-over fires is briefly described. Many zone models exist (see, e.g., [16]), all with specific features. It is not the intention to provide an overview of the existing models here. Rather, the fundamental joint features are briefly addressed.

For pre-flash-over fires, the basic configuration is as sketched in Figures 5.7, 5.9 and 5.12. A first basic assumption is that the enclosure can be subdivided into two zones: a hot upper layer (containing smoke) and a cold bottom layer of air. This assumption is reflected in the name 'two-zone modelling'. The fire plume and smoke plume are not considered to belong to any of those two zones but serve as a 'pump' for heat and mass transfer from the bottom layer to the upper layer.

The hot upper layer volume and temperature can evolve in time, but the temperature in each layer is assumed uniform and the interface between both is assumed to be perfectly horizontal. Obviously, this cannot be true in reality. Temperatures will always be higher near the fire and smoke plume, and there will be a vertical temperature gradient inside the smoke layer as well. Yet, if the fire and smoke plume can be considered a 'local' effect and there is strong turbulent mixing inside the smoke layer, the assumption of the uniform temperature inside the upper layer can be reasonable.

The interface will also not be perfectly horizontal in reality. There will always be fluctuations due to the turbulent flows. Yet, the situation is 'stable' in the sense that hot, lighter fluid, is 'resting' on top of the cold, denser fluid. Indeed, buoyancy tries to maintain this situation, so there will be much less mixing across the interface than in the rising fire and smoke plume. This is in line with what was mentioned in Section 4.2 for the smoke plume and Section 4.5 for the ceiling jet.

Starting from these basic assumptions, the fundamental conservation laws, as described in Section 2.3, are applied to two control volumes in Figure 5.7 (or 5.9 or 5.12), namely once to the hot upper layer and once to the cold bottom layer.

Before expressing the conservation of mass and energy, it is essential to appreciate that the conservation of total momentum (Section 2.3.2) cannot be expressed, starting from the detailed Navier–Stokes equations, Eq. (2.38), since only two large zones are considered. This is a fundamental difference from CFD, described in Chapter 8. In two-zone models, the 'flows' are simplified as follows: a correlation (e.g., one of the expressions given in Section 4.2) is introduced to determine the mass flow rate of the smoke plume, entering the hot upper layer at the height of the interface, $z_{int}(t)$, at time t; the inflow and outflow through openings are determined from expressions like Eqs. (5.11), (5.13), (5.15), (5.20), (5.22), etc. Particularly, the correlations for the plume can be questionable, as mentioned in Sections 5.2.1 and 5.2.2, depending on the size and geometry of the enclosure, the size and position of the fire, and the size and position of the ventilation openings. Sensitivity to these aspects cannot be (accurately) accounted for in two-zone models, in contrast to what is possible in CFD simulations (Chapter 8).

Now conservation of mass and energy will be discussed, starting from a prescribed fire. The fire is typically determined as a combination of HRR, $\dot{Q}(t)$ (in W) and geometrical size (horizontal dimensions or hydraulic diameter). Thus, a fuel mass flow rate $\dot{m}_F(t)$ (in kg/s) is released per unit time.

The following notations are introduced: ul, upper layer; bl, bottom layer; $\dot{m}_{in}$, mass flow rate of gases (in kg/s) per unit time from the outside into the enclosure (typically this is air); $\dot{m}_{out}$, mass flow rate of gases (in kg/s) per unit time from the enclosure to the outside (typically this is smoke, but it can also be air); m, mass (in kg); E, energy (in J); and he, heat exchange.

The conservation of mass then reads:

$$\text{For the bottom layer}: \frac{dm_{bl}(t)}{dt} = -\dot{m}_{entr}(t) + \dot{m}_{in,bl}(t) - \dot{m}_{out,bl}(t). \tag{5.31}$$

$$\text{For the upper layer}: \frac{dm_{ul}(t)}{dt} = \dot{m}_F(t) + \dot{m}_{entr}(t) + \dot{m}_{in,ul}(t) - \dot{m}_{out,ul}(t). \tag{5.32}$$

The term $\dot{m}_{entr}(t)$ expresses the amount of mass per unit time that is entrained into the smoke plume. This quantity, leaving the bottom layer and flowing into the upper layer with the smoke plume, depends on the instantaneous height of the interface, $z_{int}(t)$, and the fire, through one of the correlations mentioned in Section 4.2. Note that no transport times are taken into account in two-zone models, which is yet another important difference from the CFD approach. Also, note that the fire plume itself, which has a different entrainment from the smoke plume (see Chapter 3), is typically not considered in two-zone models.

As mentioned, the mass flow rate entrained into the plume, $\dot{m}_{entr}(t)$, enters the upper layer. The sum of the fuel mass flow rate and the entrained mass flow rate provides the total mass flow rate of smoke into the upper layer, again depending on the instantaneous height of the interface:

$$\dot{m}_p(t) = \dot{m}_p(z_{int}(t)) = \dot{m}_F(t) + \dot{m}_{entr}(z_{int}(t)). \tag{5.33}$$

Note that the correlations of Section 4.2 actually apply to $\dot{m}_p$, rather than $\dot{m}_{entr}$. Indeed, empirical correlations cannot distinguish between m_F and $\dot{m}_{entr}$, and in the point source model, there is no clear fuel mass flow rate $\dot{m}_F$. In practice, this distinction is not made and the sum, Eq. (5.33), is imposed by means of a correlation, expressing the mass flow rate into the hot upper layer. Making an error in the cold bottom layer by assuming that there $\dot{m}_{entr}(t)$ is also obtained from the same correlation as $\dot{m}_p(t)$, i.e., by assuming that $\dot{m}_{entr}(t) = \dot{m}_p(t)$, is not dramatic because the energy content of that mass flow rate is small.

The other terms in Eqs. (5.31) and (5.32) refer to flows through openings. Note that no mass transfer is considered through the interface (outside the smoke plume), as mentioned above.

Conservation of energy requires somewhat more discussion. Consider first the fire source. The HRR consists of a convective part and a part that is emitted as radiation:

$$\dot{Q}(t) = \dot{Q}_{conv}(t) + \dot{Q}_{rad}(t). \tag{5.34}$$

Only the convective part, $\dot{Q}_{conv}(t)$, travels as heat with the smoke plume into the upper layer. The radiative part is emitted. Part of it, $\dot{Q}_{rad,abs,ul}(t)$, is absorbed in the hot upper layer (smoke), part is radiated to the ambient (e.g., through door openings or windows) and part is absorbed by the enclosure (in the ceiling, walls and floor). The latter part will induce a temperature rise in the structure and thus affect the convective heat transfer between the hot upper layer (and cold bottom layer) and the structure. Indeed, for a given temperature difference between the gas phase and the structure, the convective heat exchange reads:

$$\dot{Q}_{conv,gas-structure}(t) = hA_{he}\left(T_{gas}(t) - T_{structure}(t)\right). \tag{5.35}$$

In Eq. (5.35), h is the convection coefficient (in W/(m² K)). Typically, correlations for natural convection are applied, or h is assumed constant (within the range of $h = 5$–20 W/(m² K)). There is a heat loss from the gas if $T_{gas}(t) > T_{structure}(t)$, which is typical for the hot upper layer, and vice versa (which is typical for the cold bottom layer). A_{he} is the area (in m²) available for convective heat transfer between the gas and the structure.

Note that radiation cannot be accounted for in a very accurate manner in two-zone models, as average temperatures are assumed. Thus, it is common practice to prescribe the fraction of $\dot{Q}(t)$ that is radiated. This fraction is fuel-dependent but should also depend on ventilation conditions, as these affect the completeness of the combustion.

Taking the ambient temperature, T_{amb}, as reference temperature, there is also a change in energy in each layer if there is an outflow or inflow of gases with a temperature different from T_{amb}:

$$\dot{Q}_{conv,in/out}(t) = \dot{m}_{in/out}c_p\left(T_{gas}(t) - T_{amb}\right). \tag{5.36}$$

The specific heat, c_p, in principle depends on the gas flowing in or out. In practice, the value for air at ambient temperature, $c_p = 1$ kJ/$(kg\,K)$, is typically used. The conservation of energy now reads:

- For the bottom layer:

$$\frac{dE_{bl}(t)}{dt} = -\dot{m}_{entr}c_p\left(T_{bl}(t) - T_{amb}\right) + \dot{m}_{in,bl}c_p\left(T_{gas}(t) - T_{amb}\right)$$
$$- \dot{m}_{out,bl}c_p\left(T_{bl}(t) - T_{amb}\right) - h_{bl}A_{he,bl}\left(T_{bl}(t) - T_{structure,bl}(t)\right). \tag{5.37}$$

- For the upper layer:

$$\frac{dE_{ul}(t)}{dt} = \dot{Q}_{conv}(t) + \dot{Q}_{rad,abs,ul}(t) + \dot{m}_{entr}c_p\left(T_{bl}(t) - T_{amb}\right) + \dot{m}_{in,ul}c_p\left(T_{gas}(t) - T_{amb}\right)$$
$$- \dot{m}_{out,ul}c_p\left(T_{ul}(t) - T_{amb}\right) - h_{bl}A_{he,ul}\left(T_{ul}(t) - T_{structure,ul}(t)\right). \tag{5.38}$$

In Eqs. (5.37) and (5.38), $T_{gas}(t)$ is the instantaneous temperature of the gases entering from the outside into the bottom and upper layers, respectively. Typically, this is equal to T_{amb}. It is also typical that $\dot{m}_{in,ul}c_p\left(T_{gas}(t)-T_{amb}\right)$ equals zero for the upper layer.

There is no radiation absorption term in Eq. (5.37) because air is transparent for radiation.

With respect to the equation of state, all gases are treated as ideal gases and pressure in the energy equation is assumed constant. Pressure differences over openings are taken into account to determine the flow through these openings, though, as mentioned above.

The solution procedure is then as follows. At time $t = 0$, there is no smoke yet, so that the interface height equals the height of the compartment: $z_{int}(0) = H_c$. Given a fire with a certain size and HRR, the conservation equations, Eqs. (5.31), (5.32), (5.37) and (5.38), are solved to determine the change in z_{int} and the temperatures of the upper layer and the bottom layer at time $t = \Delta t$, where Δt is a time step, chosen by the user. In other words, solutions are calculated for discrete times $t_n = n\Delta t$. Note that $T_{ul}(t)$ determines the instantaneous mass density $\rho_{ul}(t)$, so that the volume of the upper layer (and thus the interface height) can be determined as follows:

$$V_{ul}(t) = \frac{m_{ul}(t)}{\rho_{ul}(t)}. \tag{5.39}$$

With the new interface height and temperatures, the different mass flow rates (into and out of the enclosure and into the smoke plume) are calculated and starting from the situation at time $t = \Delta t$, the equations are solved again, to determine the situation at $t = 2\Delta t$. This procedure is repeated until the end time $t_{end} = n_{end}\Delta t$, which is chosen by the user.

Note that in the procedure as described, no transport times are taken into account. In other words, the smoke is assumed to reach the bottom of the smoke layer immediately. This is not realistic, particularly during the early stages in an enclosure with a high ceiling. This is a direct consequence of not solving the momentum equations in two-zone models, in contrast to CFD (Chapter 8). Also, the ceiling-jet phenomenon (Section 4.5) is typically not taken into account.

In general, it can be stated that compared to CFD calculations, two-zone model calculations are extremely fast, but the level of accuracy to be expected is moderate. Two-zone model calculations can be a very handy tool to obtain the first order of magnitude approximations and to perform a priori sensitivity studies, but whenever more detail or accuracy is needed, CFD calculations prevail, as explained in Chapter 8.

5.3 FULLY DEVELOPED FIRE

As mentioned in Section 5.2, if the fire keeps growing, more and more heat is released per unit time. As such, the temperature in the upper layer underneath the ceiling can become so high that radiation from the smoke layer causes all combustible items within the enclosure to release combustible pyrolysis gases. This is called 'flash-over': all combustible materials participate in the fire. Different definitions exist for the term 'flash-over' [3], but typically a temperature value of around 600°C for the smoke layer is used as the 'flash-over' threshold value. This evolution is called a 'fuel-controlled' flash-over (Figure 5.2). Clearly, this only occurs if a sufficient amount of energy is released sufficiently rapidly inside the enclosure, such that the smoke layer temperature can evolve to a sufficiently high value.

Another possibility is that the fire gets to a stage where the fuel mass release rate becomes so high that the air supply rate becomes insufficient to consume all the released pyrolysis gases. The fire evolves from 'fuel-controlled' to 'ventilation-controlled' in the sense that the ventilation rate will determine the evolution of the HRR of the fire. Indeed, the oxygen available determines how fast and, more importantly, how complete the combustion reactions can be (see Section 2.2.1). However, the pyrolysis rate primarily depends on the heat transfer to the fuel. Therefore, as long as there is sufficient heat transfer towards the fuel, the pyrolysis process continues. Subsequently, if there is a sudden additional supply of oxygen (e.g., by breaking a window or by opening a door), the combustion reactions can become (more) complete again, leading to a very rapid increase in HRR and temperature inside the enclosure. This is called 'ventilation-induced' flash-over (Figure 5.5) [17]. Similar to the notion that a sufficient HRR is required to come to (fuel-controlled) flash-over, a minimum critical ventilation rate is required to obtain flash-over conditions.

After flash-over, the fire becomes 'fully developed' in that all combustible materials inside the enclosure participates in the pyrolysis process.

It is recalled that the present section mainly deals with not-too-large compartments, where the fire drives the flows into, out of and inside the compartment. Section 5.4 briefly addresses large compartments, with momentum-driven flows (affected by the fire, of course).

5.3.1 Fire Source

In case of post-flash-over fully developed fire, the situation can be fuel-controlled or ventilation-controlled. An important parameter is the 'Opening Factor' (OF):

$$OF = \frac{A\sqrt{H}}{A_T}. \tag{5.40}$$

The origin of the ventilation factor $A\sqrt{H}$ is explained in more detail in Section 5.3.3, Eq. (5.60). It relates to the fact that in fully developed fire conditions inside the enclosure, the flow rate through a single vertical opening of the enclosure is proportional to the ventilation factor of that opening. A_T is the total area for heat transfer in the compartment.

In the case of fuel-controlled conditions (regime II [7,8] in Figure 5.13), in principle all the fuel burns inside the enclosure. This corresponds to large values of OF and can involve complex flow phenomena inside the enclosure. In any case, the assumption of uniform temperatures is not guaranteed and there is much scatter in the experimental data (indicated by the use of a dashed line in Figure 5.13) [7,8].

In the case of ventilation-controlled conditions, not all the fuel finds oxygen inside the enclosure. This corresponds to a 'regime I' [7,8] fire in Figure 5.13. The 'excess fuel' burns outside the compartment. Indeed, in contrast to a ventilation-controlled growing fire (Section 5.2.1.2), the combustion products and combustible gases leaving the enclosure are now at such high temperature that auto-ignition (or piloted ignition by an external ignition source) occurs. Flames are therefore emerging from the enclosure openings, which affects the smoke dynamics outside the enclosure (Section 5.3.2), but which also poses a strong external thermal attack onto the structure by radiation (and convection). External fire spread can be a consequence hereof. If S is the stoichiometric ratio in that S kg air is required for complete combustion of 1 kg fuel (Section 2.2.2.1), the excess fuel factor can be defined as follows [3]:

$$f_{ex} = 1 - \frac{\dot{m}_{air}/S}{\dot{m}_F}. \tag{5.41}$$

As long as there is sufficient air supply ($\dot{m}_{air}$, kg/s) for complete combustion of all the fuel mass released per unit time ($\dot{m}_F$, kg/s), $f_{ex} \leq 0$. This corresponds to fuel-controlled conditions. As soon

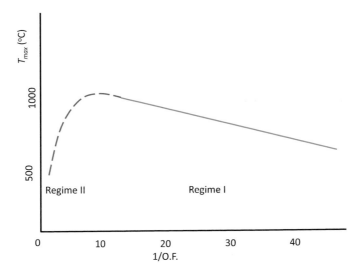

FIGURE 5.13 Sketch of dependence of maximum temperature inside enclosures for regime I and regime II fires [7,8]. The horizontal axis concerns the inverse of the opening factor as defined in Eq. (5.40). The dashed line for regime II fires (and the transition zone between regimes I and II) refers to the high degree of uncertainty and non-uniformity of temperatures for regime II fires.

as the air supply becomes insufficient, though, ventilation-controlled conditions are met, corresponding to $f_{ex} > 0$. This need not imply that pure fuel will leave the enclosure, but at least a certain mass flow rate of combustible gases does. In the limit of no air supply at all, $f_{ex} = 1$. The larger f_{ex}, the more fuel emerges from the enclosure and the larger the external flames will be (and, therefore, the larger the external thermal attack onto the structure and its surroundings). However, for the fire compartment itself, a larger f_{ex} value corresponds to less onerous conditions because less energy is released within the compartment itself.

In post-flash-over conditions, all combustible items are pyrolysing, so that the assumption of fire and smoke plumes is clearly not valid. A commonly made assumption for fully developed fires after flash-over is instead that the temperature inside the enclosure would be uniform [18]. This, however, is not true, even for small compartments [19]. In Ref. [19], it is explained that this over-simplification of reality might affect assessment analyses of the impact of a fully developed fire on the structure. This is beyond the scope of this book, but the non-homogeneity in temperature also implies a non-uniform mass density inside the enclosure. Yet, the assumption of uniformity is adhered to in order to allow for the development of analytical solutions, as presented in Section 5.3.3. It also allows for zone modelling as described in Section 5.3.5.

5.3.2 Smoke Dynamics

Inside the fire compartment, not much needs to be said on the smoke dynamics for regime I fires. The enclosure is in principle engulfed in flames, so there is no separate smoke dynamics to be considered. Outside of the enclosure, though, smoke dynamics can be important. This was referred to as 'window plumes' in Section 4.6.2. The ejecting flames provide buoyancy for the fire and smoke plume, both of which typically interact with a wall (e.g., inside an atrium) or a façade (outside), which by itself can be combustible. Moreover, there is interaction with flow fields induced by (mechanical) ventilation (inside) or with wind (outside). Needless to say, this leads to much more complex flow fields and different entrainment rates than those described in Chapter 4.

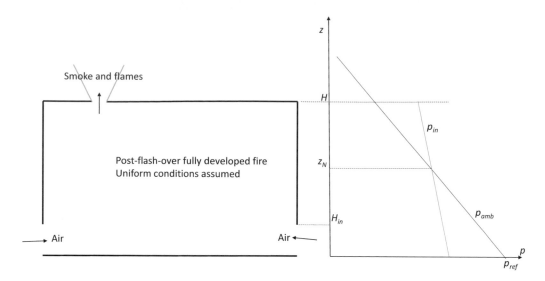

FIGURE 5.14   Sketch of post-flash-over conditions with smoke and flames leaving through a horizontal opening in the ceiling. Static pressure lines are sketched at the right hand side. Uniform conditions are assumed within the enclosure.

A review of recent research is provided in Ref. [20], but much more research is required to build up understanding. CFD (see Chapter 8) can be expected to play an important role in this.

For regime II fires, the fluid dynamics inside the compartment can also be much more complex, with strong excess air flows inside the enclosure. Again, CFD can help gain understanding as to how this affects fire and smoke dynamics. This is beyond the scope of this book and the reader is invited to follow up on research in this field.

5.3.3   FLOWS THROUGH OPENINGS

As mentioned in Section 5.3.1, the basic assumption for the calculation of flows through openings in post-flash-over conditions is that the temperature inside the enclosure is uniform. Although this is an over-simplification of reality [19], the assumption is adopted here, in order to allow for the development of analytical solutions. Steady-state conditions are assumed. No wind is considered in the present section. The possible effect of wind is discussed in Section 6.7.

5.3.3.1   Horizontal Openings

Consider a situation as sketched in Figure 5.14, which resembles Figure 5.7, but now uniform conditions are assumed within the enclosure. It is recalled that this is never really true, but it can be a reasonable estimate for the flow dynamics for a 'regime I' fire (see Section 5.3.1). The entire reasoning, explained in Section 5.2.3.1, is not repeated here. Only the main results are provided.

The static pressure difference over the outlet opening reads:

$$\Delta p(z) = (\rho_{\mathrm{amb}} - \rho)g(H - z_N).$$ (5.42)

The mass flow rate through the vent then reads:

$$\dot{m}_{\mathrm{out}} = C_{d,\mathrm{out}} A_{\mathrm{out}} \sqrt{2\rho(\rho_{\mathrm{amb}} - \rho)g(H - z_N)}\,.$$ (5.43)

The discussion in Section 5.2.3.1, given for Eq. (5.11), prevails, as does Figure 5.8.

The position of the neutral plane height is determined by a mass balance, as explained in Section 5.2.3.1. The mass flow rate for the intake air reads:

$$\dot{m}_{\text{air,in}} = C_{d,\text{in}} \int_0^{H_{\text{in}}} \rho_{\text{amb}} \sqrt{2\left(1 - \frac{\rho}{\rho_{\text{amb}}}\right) g(z_N - z)} W(z)\, dz. \tag{5.44}$$

For not-too-large values of H_{in}, this can be approximated as follows:

$$\dot{m}_{\text{air,in}} \approx \rho_{\text{amb}} C_{d,\text{in}} A_{\text{in}} \sqrt{2\left(1 - \frac{\rho}{\rho_{\text{amb}}}\right) g\left(z_N - \frac{H_{\text{in}}}{2}\right)} \approx \rho_{\text{amb}} C_{d,\text{in}} A_{\text{in}} \sqrt{2\left(1 - \frac{\rho}{\rho_{\text{amb}}}\right) g z_N}. \tag{5.45}$$

In Eq. (5.45), A_{in} is the total area of the openings for air intake. Setting Eq. (5.43) equal to Eq. (5.45) yields:

$$\frac{H - z_N}{z_N} = \frac{T}{T_{\text{amb}}} \left(\frac{C_{d,\text{in}} A_{\text{in}}}{C_{d,\text{out}} A_{\text{out}}}\right)^2. \tag{5.46}$$

Elaboration yields:

$$z_N = \frac{H}{1 + \dfrac{T}{T_{\text{amb}}} \left(\dfrac{C_{d,\text{in}} A_{\text{in}}}{C_{d,\text{out}} A_{\text{out}}}\right)^2}. \tag{5.47}$$

Equation (5.47) reveals that in the limit of A_{in} going to zero (note that small values of H_{in} have been assumed), the neutral plane height is at the ceiling level, as in Section 5.2.3.1. This implies that there is no pressure difference over the opening, and thus, there is no flow through the opening.

Using Eq. (5.47) in Eq. (5.43) yields the mass flow rate:

$$\dot{m} = \rho C_{d,\text{out}} A_{\text{out}} \sqrt{\frac{2gH\left(\dfrac{T}{T_{\text{amb}}} - 1\right)}{1 + \dfrac{T_{\text{amb}}}{T}\left(\dfrac{C_{d,\text{out}} A_{\text{out}}}{C_{d,\text{in}} A_{\text{in}}}\right)^2}}. \tag{5.48}$$

Note that this is exactly the same expression as Eq. (5.15), replacing the thickness of the smoke layer, D, in Eq. (5.15), by the compartment height (which is now filled with smoke).

Example 5.2

A steady fully developed post-flash-over fire exists in a 5 m high compartment (with floor area 15 m×10 m), with a horizontal ventilation opening of 3 m² in the ceiling. The temperature inside the compartment is assumed uniform, equal to 900°C. The total opening for intake of fresh air is 4 m². The discharge coefficient for all openings is 0.6. Ambient temperature is 20°C, and the outside pressure is 1,00,000 Pa. Estimate the position of the neutral plane height and the mass flow rate through the vent in the ceiling. What is the corresponding convective HRR of the fire, ignoring heat transfer to the structure?

In Eq. (5.47), $H = 5$ m, $T = 1173.5$ K, $A_{in} = 4$ m², $T_{amb} = 293.15$ K and $A_{out} = 3$ m², so that $z_N = 0.62$ m. Equation (5.48) yields: $\dot{m} = 5.1$ kg/s .

Ignoring heat transfer to the structure, this corresponds to a convective HRR of the fire, equal to $\dot{Q}_{conv} = \dot{m}c_p(T - T_{amb}) = 4.48$ MW. This is part of the fire HRR, removed by the hot gases flowing out of the compartment (and replaced by air flowing into the compartment).

Exercises

For the setting of the example, calculate the position of the neutral plane height, the mass flow rate through the vent in the ceiling and the corresponding convective HRR of the fire (ignoring heat transfer to the structure), if:

1. The inlet opening area is only 2 m². Explain the effect of reduction of inlet area. (Answer: $z_N = 1.8$ m, $\dot{m} = 2.9$ kg/s, $\dot{Q}_{conv} = 2.55$ MW. The neutral plane height increases due to the larger pressure loss over the inlet opening (shift of the inner pressure line in Figure 5.14 to the left); this leads to a reduced pressure difference over the outlet opening and thus reduced mass flow rate; with the same compartment temperature, this corresponds to a lower convective HRR.)

2. The compartment temperature is 800°C. (Answer: $z_N = 0.66$ m, $\dot{m} = 5.4$ kg/s, $\dot{Q}_{conv} = 4.2$ MW. The neutral plane height slightly increases because the pressure line in the compartment is less steep due to the lower temperature; this leads to a decrease in pressure difference over the outlet opening, but the mass flow rate increases due to an increase in mass density of the hot gases flowing out, Figure 5.8; the convective HRR is lower, due to the lower temperature.)

3. The ambient temperature is −10°C. (Answer: $z_N = 0.56$ m, $\dot{m} = 5.26$ kg/s, $\dot{Q}_{conv} = 4.88$ MW. The neutral plane height decreases because the pressure line outside is less steep due to the lower ambient temperature; this leads to an increased pressure difference over the outlet opening, and thus to an increased mass flow rate, given the same temperature inside the compartment; this results in a higher convective HRR.)

5.3.3.2 Vertical Openings

Figure 5.15 sketches a situation for two small vertical openings in a side wall.

The situation is, in fact, very similar to the situation sketched in Figure 5.14. Indeed, for not-too-large openings, the most important parameter is the height difference between the positions of the openings, indicated as   in Figure 5.15. For larger openings, the analysis is given below, because then the variation of the pressure difference over the vertical opening must be considered. For small openings, an approximation as in Eq. (5.45) can be made:

$$\dot{m}_{out} = \rho C_{d,out} A_{out} \sqrt{2\left(\frac{\rho_{amb}}{\rho} - 1\right)g(z_{out} - z_N)} , \tag{5.49}$$

$$\dot{m}_{air,in} \approx \rho_{amb} C_{d,in} A_{in} \sqrt{2\left(1 - \frac{\rho}{\rho_{amb}}\right)g(z_N - z_{in})}. \tag{5.50}$$

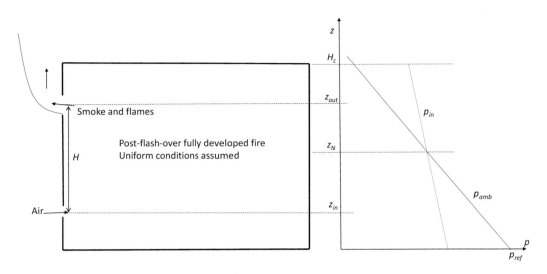

FIGURE 5.15 Sketch of post-flash-over conditions with air entering and smoke and flames leaving through small vertical openings in the side wall. Static pressure lines are sketched at the right hand side. Uniform conditions are assumed within the enclosure.

Setting Eq. (5.50) equal to Eq. (5.51) yields the position of the neutral plane, from which the mass flow rate can be computed as follows, using $H = z_{out} - z_{in}$:

$$\dot{m} = \rho C_{d,out} A_{out} \sqrt{\frac{2gH\left(\dfrac{T}{T_{amb}} - 1\right)}{1 + \dfrac{T_{amb}}{T}\left(\dfrac{C_{d,out}A_{out}}{C_{d,in}A_{in}}\right)^2}} \;. \tag{5.51}$$

Not surprisingly, this is exactly the same result as Eq. (5.48). Note that:

- The compartment height is not in expression (5.51). It is now replaced by the vertical distance between the vertical openings, because that distance determines the pressure differences, driving the flow through the openings.
- The area and discharge coefficient of both openings, for inlet of intake air and for outlet of smoke and flames, determine the mass flow rate. This agrees with the viewpoint of a series of flow resistances.
- The factor $\rho\sqrt{\dfrac{\rho_{amb}}{\rho} - 1}$ appears again, so the non-monotonic dependence on mean temperature inside the enclosure applies as sketched in Figure 5.8.

Figure 5.15 is not the most common situation, though. A more common situation is sketched in Figure 5.16, with a single vertical opening in a side wall. In this case, the integration must be performed as in Section 5.2.3.2.

At each height z the local pressure difference is found from Eq. (5.16). Thus, the mass flow rates read:

$$\dot{m}_{out} = C_{d,out} \int_{z_N}^{z_{top}} \rho \sqrt{2\left(\frac{\rho_{amb}}{\rho} - 1\right)g(z - z_N)}\,W(z)\,dz, \tag{5.52}$$

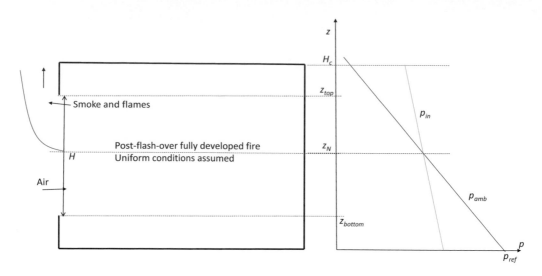

FIGURE 5.16   Sketch of post-flash-over conditions with air entering and smoke and flames leaving through a vertical opening in the side wall. Static pressure lines are sketched at the right hand side. Uniform conditions are assumed within the enclosure.

$$\dot{m}_{air,in} = C_{d,in} \int_{z_{bottom}}^{z_N} \rho_{amb} \sqrt{2\left(1 - \frac{\rho}{\rho_{amb}}\right) g (z_N - z)} W dz. \tag{5.53}$$

For constant opening width, this yields:

$$\dot{m}_{out} = \frac{2}{3} C_{d,out} \rho W \sqrt{2\left(\frac{T}{T_{amb}} - 1\right) g} \left(z_{top} - z_N\right)^{3/2}, \tag{5.54}$$

$$\dot{m}_{air,in} = \frac{2}{3} C_{d,in} \rho_{amb} W \sqrt{2\left(1 - \frac{T_{amb}}{T}\right) g} \left(z_N - z_{bottom}\right)^{3/2}. \tag{5.55}$$

Setting Eq. (5.54) equal to Eq. (5.55) yields the position of the neutral plane height:

$$\frac{z_N - z_{bottom}}{z_{top} - z_N} = \left(\frac{C_{d,out}}{C_{d,in}}\right)^{2/3} \left(\frac{T_{amb}}{T}\right)^{1/3}. \tag{5.56}$$

Assuming that the discharge coefficient does not vary strongly, this simplifies to:

$$\frac{z_N - z_{bottom}}{z_{top} - z_N} = \left(\frac{T_{amb}}{T}\right)^{1/3}. \tag{5.57}$$

As $z_{top} - z_{bottom} = H$, the position of the neutral plane height can also be expressed as follows:

$$\frac{z_N - z_{bottom}}{H} = \frac{1}{1 + \left(\dfrac{T}{T_{amb}}\right)^{1/3}}. \tag{5.58}$$

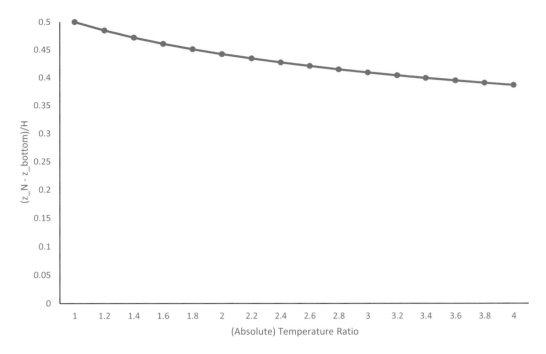

FIGURE 5.17 Evolution of the position of the neutral plane height in case of uniform temperature inside an enclosure with a vertical opening in a side wall, expressed as the ratio of the neutral plane height to the opening height, as function of the temperature ratio (Eq. 5.58).

In the limit that $T = T_{amb}$, Eq. (5.58) shows that the neutral plane is at half height. The higher the temperature inside the enclosure, the lower the neutral plane height. This is logical: smoke has a lower mass density than air, so it requires a larger fraction of the opening area for the same mass flow rate (which is the basic assumption for steady-state conditions to apply). Figure 5.17 shows the evolution of the position of the neutral plane height as function of the temperature ratio.

Combining the above equations yields the following end expression for the mass flow rate:

$$\dot{m}_{out} = \frac{2}{3} C_{d,out} \rho \sqrt{2\left(\frac{T}{T_{amb}} - 1\right) g} WH^{3/2} \left(\frac{1}{1 + \left(\frac{T_{amb}}{T}\right)^{1/3}}\right)^{3/2}. \tag{5.59}$$

The important observation in Eq. (5.59) is again the presence of the ventilation factor:

$$WH^{3/2} = A\sqrt{H}. \tag{5.60}$$

This shows that the mass flow rate is proportional to the opening width and the square root of the opening height or, equivalently, linearly proportional to the opening width and proportional to the opening height with exponent 3/2, in line with the discussion in Section 5.2.3.2. This shows that a reduction in opening height has a stronger effect on the mass flow rate than a similar reduction in opening width.

It is also noteworthy again that the compartment height itself does not appear in any of the expressions, neither for the position of the neutral plane height (Eq. 5.58), nor for the mass flow rate (Eq. 5.59).

Example: A steady fully developed post-flash-over fire exists in a 5 m high compartment (with floor area 15 m × 10 m), with a vertical door opening of 3 m high and 1 m wide in one side wall as the only ventilation opening. The temperature inside the compartment is assumed uniform, equal to 900°C. The discharge coefficient of the door opening is 0.6. Ambient temperature is 20°C, and the outside pressure is 1,00,000 Pa. Estimate the position of the neutral plane height and the mass flow rate through the door opening. What is the corresponding convective HRR of the fire, ignoring heat transfer to the structure?

In Eq. (5.58), $z_{bottom} = 0$ m (door opening starts at floor level), $H = 3$ m, $T = 1173.15$ K and $T_{amb} = 293.15$ K, so that $z_N = 1.16$ m. Equation (5.59) yields, with $H = 3$ m and $W = 1$ m: $\dot{m}_{out} = 2.28$ kg/s. Ignoring heat transfer to the structure, this corresponds to a convective HRR equal to $\dot{Q}_{conv} = \dot{m}_{out} c_p (T - T_{amb}) = 2$ MW. Indeed, this is the part of the fire HRR that is removed by the hot gases flowing out of the compartment (and replaced by air flowing into the compartment).

Exercises

For the settings of the previous example, calculate the position of the neutral plane height, the mass flow rate through the door opening and the corresponding convective HRR (ignoring heat transfer to the structure), if:

1. The door height is only 2 m. (Answer: $z_N = 0.77$ m, $\dot{m}_{out} = 1.24$ kg/s, $\dot{Q}_{conv} = 1.1$ MW. The neutral plane height is lower due to the decrease in z_{top} (Figure 5.16): the reduction in opening height implies a reduced pressure difference over the opening, and in combination with the reduced opening area, this leads to a reduced mass flow rate; with the same compartment temperature, this corresponds to a lower convective HRR.)
2. The door height is only 2 m, but the width is 1.5 m (so that the opening area is again 3 m²). (Answer: $z_N = 0.77$ m, $\dot{m}_{out} = 1.86$ kg/s, $\dot{Q}_{conv} = 1.6$ MW. The neutral plane height is identical to Exercise 1; the pressure difference over the opening is identical to Exercise 1. However, the 50% increase in opening area results in a 50% increase in mass flow rate. With the same compartment temperature, this corresponds to a 50% increase in convective HRR, compared to Exercise 1.) Note that the mass flow rate is lower than in the example, despite identical opening area (and identical settings, other than the opening height and width), illustrating the importance of the opening height.

5.3.4   Natural and Mechanical Ventilation

Ventilation systems are usually not expected to operate during fully developed fires, as usually they are not designed to withstand the high temperatures related to the fully developed fire conditions. If they do operate, however, the same principles apply as described in Section 5.2.4.

5.3.5   Zone Modelling

In the assumption of uniform temperature inside the enclosure, there is no distinction between a hot upper layer and a cold bottom layer, as described in Section 5.2.5. As such, there is no 'two-zone' modelling, but 'single-zone' modelling. The enclosure itself is now the control volume.

Conservation of mass reads:

$$\frac{dm_{encl}(t)}{dt} = \dot{m}_F(t) + \dot{m}_{in}(t) - \dot{m}_{out}(t). \tag{5.61}$$

In Eq. (5.61), $\dot{m}_{in}(t)$ is the instantaneous mass flow rate of gases (typically air, in kg/s) into the enclosure and $\dot{m}_{out}(t)$ is the instantaneous mass flow rate of gases (typically combustion products, in kg/s) out of the enclosure. Depending on the configuration of the ventilation openings, they are determined by Eqs. (5.43) and (5.44), (5.49) and (5.50), or (5.54) and (5.55).

For the conservation of energy, again the fire source can be decomposed into a convective part and a part that is emitted as radiation, Eq. (5.34). Only the convective part, $\dot{Q}_{conv}(t)$, is kept directly in the gas phase inside the enclosure. The radiative part is emitted. Part of it, $\dot{Q}_{rad,abs,encl}(t)$, is absorbed in the gas phase, part is radiated to the ambient (e.g., through door openings or windows) and part is absorbed by the enclosure (in the ceiling, walls and floor). The latter part will induce a temperature rise of the structure and thus affect the convective heat transfer between the gas phase and the structure. Indeed, for a given temperature difference between the gas phase and the structure, Eq. (5.35) remains valid. The convection coefficient for natural convection typically remains within the range of $h = 5 - 20$ W/(m^2 K). There is a heat loss from the gas if $T_{gas}(t) > T_{structure}(t)$, which is the typical situation in the fully developed fire phase. A_{he} is the area (in m^2) available for convective heat transfer between the gas and the structure.

It is recalled that radiation cannot be accounted for in a very accurate manner in zone models, as average temperatures are assumed. Thus, it is common practice to prescribe the fraction of $\dot{Q}(t)$ that is radiated. This fraction is fuel-dependent, but should also depend on ventilation conditions, as these affect the completeness of the combustion.

Taking the ambient temperature, T_{amb}, as a reference the conservation of energy can be written as follows:

$$\frac{dE_{encl}(t)}{dt} = \dot{Q}_{conv}(t) + \dot{Q}_{rad,abs,encl}(t) + \dot{m}_{entr}c_p\left(T_{bl}(t) - T_{amb}\right) + \dot{m}_{in}c_p\left(T_{gas}(t) - T_{amb}\right)$$

$$- \dot{m}_{out}c_p\left(T_{encl}(t) - T_{amb}\right) - hA_{he}\left(T_{encl}(t) - T_{structure}(t)\right). \tag{5.62}$$

In Eq. (5.62), $T_{encl}(t)$ is the gas temperature inside the enclosure. It is typical for the gas temperature of the gas entering the enclosure, $T_{gas}(t)$, to be equal to T_{amb}.

With respect to the equation of state, all gases are treated as ideal gases and pressure in the energy equation is assumed constant. Pressure differences over openings are taken into account to determine the flow through these openings, though, as mentioned above.

5.4 LARGE COMPARTMENTS

In large compartments, the simplifying assumption of uniform conditions (be it in the entire compartment, Section 5.3, or in a smoke layer, Section 5.2) is no longer justified. The spatial distribution of the energy inside the compartment and the flow field must be considered to obtain a realistic view on the compartment fire dynamics in such circumstances [21]. This is a recent lively research area, where many developments can be expected in the near future. Only some basis aspects are mentioned here.

It is useful to revisit Section 3.6.1, where flame spread over surfaces was discussed for different conditions (counter-current vs. concurrent, and thermally thin vs. thermally thick). Flame spread was expressed through a moving pyrolysis front, and because in counter-current situations, the flame size on average does not change much in steady conditions, the velocity with which the flame tip moves is the same as the propagation speed of the pyrolysis front. For concurrent flame spread, it was mentioned that the flame size grows in time (see Section 3.6.1.3), so that the propagation velocity of the flame tip exceeds that of the pyrolysis front.

In Refs. [21–23], another propagation velocity is considered, namely that of the 'burn-out' front, v_{BO}, which essentially expresses the motion of the 'back end' of the pyrolysing zone at the surface, rather than the 'front end' (x_p in Section 3.6.1). This has led to the distinction of three regimes, based on the ratio v_f/v_{BO}:

- *Travelling fire*: $\dfrac{v_F}{v_{BO}} \approx 1$;

- *Growing fire*: $\dfrac{v_F}{v_{BO}} > 1$;

- *Fully developed fire*: $\dfrac{v_F}{v_{BO}} \to \infty$.

In Refs. [21–23], horizontal fire spread was considered. It remains to be investigated what the impact of concurrent flows might be.

In general, it must be appreciated that in the discussions in Sections 5.2 and 5.3, no impact of the flow field is considered. Flows are assumed to be driven by the overall static pressure lines and are only considered to estimate the flow rates through openings in the compartment boundaries. Reality is far more complex and the flow field inside the compartment will inevitably strongly affect the fire and smoke dynamics. To name a few aspects:

- Entrainment into the fire and smoke plume is directly affected, and the assumption that the entrainment velocity is proportional to the buoyancy-induced upward velocity may no longer be valid;
- The shape of the fire plume can be affected (e.g., tilting), directly affecting the heat transfer by convection and radiation;
- The flame temperatures can be affected, directly affecting the heat transfer by convection and radiation;
- The formation of soot in the flame region can be affected, directly affecting the heat transfer by radiation;
- The flow through openings can be strongly affected by wind. This effect can also be important inside the compartment.

In large compartments, the flow field can have a significant impact on the fire spread, as it will affect the shape and orientation of the fire plume. It could, therefore, be argued that 'travelling' fires should be called 'moving fires', to indicate that the fire cannot 'decide' by itself where it travels as the air momentum, which depends on the ventilation settings, will also play a (possibly dominant) role. Detailed CFD simulations are expected to shed light on this in the foreseeable future.

5.5   FIRES IN WELL-CONFINED ENCLOSURES

In well-confined enclosures, such as passive houses or compartments in nuclear power plants [24], leakage plays an important role: in the absence of 'regular' openings, it is the leakage that determines the inflow into and outflow out of the compartment. Thus, it is important to know the total area of leakage and, ideally, how the leakage area is spread over the geometry of the compartment. Indeed, as made clear above, the position of the openings, in particular with respect to the neutral plane height, is important. Flows through leakage can be laminarized, due to the typically small length scale in the flow Reynolds number, Eq. (2.48). This is not discussed further here. The reader is referred to specialized literature.

An important feature of fires in well-confined enclosures is the relatively strong pressure build-up. This can be understood from the ideal gas law:

$$pV = mRT. \tag{5.63}$$

In the case of absolute tightness, the mass   inside the compartment and the volume V of the compartment are constant. Assuming little variation in the gas constant R (in J/(kg K)), Eq. (5.63) reveals a linear increase of the pressure with absolute temperature. Whereas in reality the tightness will not be absolute, it is clear that a much stronger pressure rise is to be expected than in the case of 'regular' openings in the compartment boundaries. This can cause important problems in, e.g., passive housing, where the ventilation openings are deliberately kept very small for daily energy savings. Indeed, the pressure rise inside can rapidly lead to circumstances where it is impossible to open doors inwards, so that evacuation can become impossible.

It is noted that the limited amount of ventilation openings can also cause relatively rapid under-ventilation (and possible extinction) of the fire, upon consumption of the initially present oxygen inside the compartment (see Section 5.2).

In the case of mechanical ventilation, combined with airtight enclosures, very complex fire dynamics and pressure evolutions can be met. An in-depth discussion is provided in Ref. [25]. Systematic research is required to further quantify the different effects, but it is instructive to discuss the energy equation, because that allows understanding the pressure evolution. In Ref. [25], the equation is provided (see also, e.g., [2,24]):

$$\frac{V}{\gamma - 1} \frac{d}{dt} p = \dot{Q}_f - \dot{Q}_w - \dot{Q}_v. \tag{5.64}$$

where p is the compartment pressure (Pa); V is the compartment volume (m³); γ is the gas isen-tropic coefficient; $\dot{Q}_f, \dot{Q}_v$ and $\dot{Q}_w$ represent the fire heat release rate (HRR) (w), the heat loss (which could be a net gain, though) through ventilation flows (w) and the wall (boundary) heat losses (w), respectively. The ventilation flow contains the inevitable leakage flows as well.

One possible phenomenon is then a pulsating fire at low frequency, which can occur in case of limited oxygen supply. This supply can be natural (through openings) or mechanical (in, e.g., nuclear power plants). The mechanism for natural air supply is as follows. If the fire grows, the temperature inside the enclosure increases if the HRR is higher than the heat loss rate from the enclosure. Consequently, for a given amount of mass inside the enclosure, the pressure inside the enclosure increases and a large fraction of the vents is occupied as exhaust for the hot gases. Even the entire area can be used as exhaust. As a consequence, less (or no) area is available for supply of fresh oxygen. The fire becoming ventilation-controlled (cf. Section 5.2), its HRR will drop and temperatures will decrease again. This leads to a reduction in pressure inside the enclo-sure, so that a larger fraction of the vents is used as air supply. Even the entire area can be used as inlet for fresh air. The fire can then grow again, leading to a pulsating cycle that repeats itself, until the fire dies due to lack of fuel. Figure 5.18 sketches this behaviour.

Such observation was also described in, e.g., Ref. [24]. In their set-up, in the context of nuclear power plants, both supply of air and extraction of hot gases occur by mechanical ventilation. If the fire grows, the temperature inside the enclosure increases if the HRR is higher than the heat loss rate from the enclosure. Consequently, the pressure inside the enclosure increases and air supply becomes less efficient as a consequence of the fan characteristics (Figure 5.10: increase in counter-pressure leads to a reduction in the volume flow rate supplied by the fan). As a conse-quence, there is less/too little supply of fresh oxygen to maintain the fire. If the pressure inside

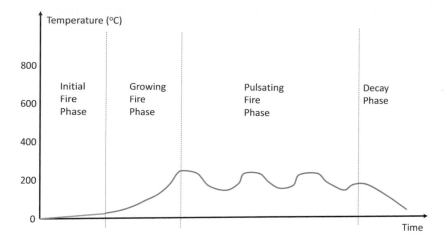

FIGURE 5.18 Evolution of average temperature in an enclosure in case of a pulsating fire. Note: temperature values are merely indicative.

the enclosure exceeds a critical value, the ducts meant as air supply even act as exhaust for the hot gases and there is no oxygen supply to the fire. This corresponds to a normalized counter-pressure higher than 1 in Figure 5.10. The fire being or becoming ventilation-controlled due to the lack (or absence) of oxygen supply (cf. Section 5.2), its HRR drops and temperatures decrease. This leads to a reduction in pressure inside the enclosure, so that more oxygen is supplied by the fans. The fire can grow again, leading to a pulsating cycle that repeats itself, until the fire dies due to lack of fuel (Figure 5.18).

Clearly the pulsating fire does not reach a true steady state. The fire and smoke dynamics are too complex to justify the simplifications made in Section 5.2. Detailed experiments and numerical simulations are required to obtain more profound understanding of all the phenomena involved. The reader is referred to Ref. [25] for a more complete discussion on this. An example of CFD simulation results is given in Section 8.9.5 as well.

5.6 BACKDRAFT

The evolution towards backdraft resembles the evolution towards a ventilation-induced flash-over (Sections 5.2.1.2). Specific for backdraft is that, due to the relative lack in oxygen supply, combustible gases are collected in the upper layer inside the enclosure. This corresponds to high GER (Eq. 5.7) values. In particular, large amounts of CO are produced under these circumstances, and besides being toxic, it is a combustible gas. The mixture becomes flammable and subsequently too rich to burn (Section 3.1). If there is now a sudden additional supply of oxygen (e.g., by a breaking window or by opening a door), this oxygen will mix with the hot gases to form a combustible mixture. This flammable mixture can ignite by auto-ignition (the auto-ignition temperature for CO is around 610°C) or by piloted ignition, e.g., by contact with flames or hot surfaces. Once ignited, very rapid premixed combustion (Section 3.2) occurs, and due to expansion and the induced additional turbulence, the flammable mixture of hot gases is expelled out of the enclosure through the openings. This is typically accompanied by pressure build-up inside the enclosure. The backdraft phenomenon causes impressive flames ejecting out of the enclosure openings, because the hot combustible gases are pushed through the openings. The combination of fierce turbulence and the mixing with ambient air causes indeed very rapid combustion (premixed and non-premixed). This relates to the 'fireball' phenomenon, which is associated with the backdraft phenomenon.

Such conditions are extremely hazardous and a potential danger for firefighters, particularly if they break a window or open a door to reach the fire scene (as such events can provide the oxygen needed for the mixture to become flammable).

The simplifications made in the previous sections no longer apply to describe the fire and smoke dynamics related to backdraft. The reader is referred to specialized literature on this topic.

REFERENCES

1. B. Karlsson and J.G. Quintiere (2000) Enclosure Fire Dynamics. CRC Press LLC: Boca Raton, FL.
2. J.G. Quintiere (2006) *Fundamentals of Fire Phenomena*. John Wiley & Sons, Ltd: Hoboken, NJ.
3. D. Drysdale (2011) *An Introduction to Fire Dynamics,* 3rd Ed. John Wiley & Sons, Ltd: Hoboken, NJ.
4. J.G. Quintiere (2002) "Compartment fire modeling", in M.J. Hurley et al. (Eds.), *The SFPE Handbook of Fire Protection Engineering*, Chapter 3-5. Springer: New York, pp. 3-162–3-170.
5. N. Tilley and B. Merci (2013) "Numerical study of smoke extraction for adhered spill plumes in atria: impact of extraction rate and geometrical parameters", *Fire Safety Journal,* Vol. 55, pp. 106–115.
6. W.M. Pitts (1995) "The global equivalence ratio concept and the formation mechanisms of carbon monoxide in enclosure fires", *Progress in Energy and Combustion Science*, Vol. 21, pp. 197–237.
7. P.H. Thomas, A.J. Heselden and M. Law, "Fully Developed Compartment. Fires - Two Kinds of Behavior", Fire Research Tech. Paper no. 18 (1967).
8. J.L. Torero, A.H. Majdalani, C. Abecassis-Empis and A. Cowlard (2014) IAFSS NZ paper.
9. B. Merci and K. Van Maele (2008) "Numerical simulations of full-scale enclosure fires in a small compartment with natural roof ventilation", *Fire Safety Journal*, Vol. 43(7), pp. 495–511.
10. J.H. Klote and J.A. Milke (2002) Principles of Smoke Management, ASHRAE.
11. CIBSE (1995) Relationships for smoke control calculations, TM19, London: Chartered Institute of Building Services Engineers.
12. L.Y. Cooper (2002) "Smoke and heat venting", in M.J. Hurley et al. (Eds.), *SFPE Handbook of Fire Protection Engineering*, 3rd Edition, Chapter 3-9. Springer: New York.
13. I.R. Wood (1968) "WoodSelective withdrawal from a stably stratified fluid", *Journal of Fluid Mechanics*, Vol. 32, pp. 209–223.
14. I.R. Wood (2001) "Extensions to the theory of selective withdrawal", *Journal of Fluid Mechanics*, Vol. 448, pp. 315–333.
15. D. Spratt and A.J.M. Heselden (1974) "Efficient extraction of smoke from a thin layer under a ceiling", Fire Research Note 1001.
16. S.M. Olenick and D.J. Carpenter (2003) "An updated international survey of computer models for fire and smoke", *SFPE Journal of Fire Protection Engineering*, Vol. 13(2), pp. 87–110.
17. K. Lambert and S. Baaij (2011) Brandverloop – Technisch bekeken, tactisch toegepast, Sdu Uitgevers.
18. P.H. Thomas (1983) "Modeling of compartment fires", *Fire Safety Journal*, Vol. 5(3–4), pp. 181–190.
19. J. Stern-Gottfried, G. Rein. L.A. Bisby and J.L. Torero (2010) "Experimental review of the homogeneous temperature assumption in post-flashover compartment fires", *Fire Safety Journal*, Vol. 45, pp. 249–261.
20. L. Hu et al., PECS review paper (submitted).
21. V. Gupta (2021) Open-plan compartment fire dynamics, PhD Thesis, The University of Queensland.
22. J. P. Hidalgo, A. Cowlard, C. Abecassis-Empis, C. Maluk, A. H. Majdalani, S. Kahrmann, R. Hilditch, M. Krajcovic, J L. Torero (2017) "An experimental study of full-scale open floor plan enclosure fires", *Fire Safety Journal,* Vol. 89, pp. 22–40.
23. J. P. Hidalgo, T. Goode, V. Gupta, A. Cowlard, C. Abecassis-Empis, J. Maclean, A. I. Bartlett, C. Maluk, J. M. Montalvá, A. F.Osorio, J L. Torero (2019) "When is the fire spreading and when it travels? Numerical simulations of compartments with wood crib fire loads", *Fire Safety Journal*, Vol. 108, p. 102827
24. H. Prétrel, W. Le Saux and L. Audoin (2012) "Pressure variations induced by a pool fire in a well-confined and force-ventilated compartment", *Fire Safety Journal*, Vol. 52, pp. 11–24.
25. B. Merci, J. Li, G. Maragkos (in press, corrected proof) *Proceedings of the Combustion Institute*, Vol. 39. https://doi.org/10.1016/j.proci.2022.06.011

6 Driving Forces in Smoke and Heat Control

6.1 THE IMPORTANCE OF FLUID MECHANICS

In case of fire, smoke and heat are produced. Particularly, smoke is known as a major 'killer'. As such, it is important to try and control the smoke and heat. For that purpose, there is a need for understanding where the smoke and heat flow and how they are diluted, given the driving forces. Local concentrations and temperatures, as well as heat fluxes, are then compared to criteria, quantifying conditions that are 'tenable' for people ('tenability criteria'). Such criteria can refer to the depth of the smoke layer (e.g., the bottom of the smoke layer must remain higher than 2 m above the floor), to the smoke temperature (e.g., the average smoke temperature must remain below 150 °C), to the radiative heat flux from the smoke layer (e.g., the radiation flux from the smoke layer must remain below 2 kW/m²), or to toxicity (e.g., the CO concentration at 2 m above the floor must remain below 5,000 ppm). A range of values for tenability criteria are found in the literature and the numbers provided here are merely indicative (albeit in the correct order of magnitude).

Yet, the primary question in the context of SHC is essentially binary – is there any smoke at a certain position at a given instant in time, or not? In other words, the implicit assumption is that any smoke is a threat to life or health (or for the property in terms of smoke damage), regardless of its composition. In that sense, physics becomes predominant over chemistry, because physics will determine the transport and dilution of smoke and heat (see Eqs. (2.67), respectively (2.41)), while chemistry is mainly important in the context of toxicity.

Several textbooks are available on the topic of SHC (e.g., [1,2]). Also, design guidelines exist (e.g., [3,4]). A recent textbook on fire toxicity is Ref. [5]. The present chapter is neither written in the light of design guidelines for SHC systems nor in the light of a discussion on tenability criteria. Rather, it is the intention to combine aspects described in the previous chapters to discuss the driving forces for (primarily) smoke in terms of fluid mechanics.

6.2 BUOYANCY – THE STACK EFFECT

In the absence of fire, there can be a natural stack effect in a building, due to differences between the temperature inside the building (T_{in}) and ambient temperature (T_{amb}). Figure 6.1 sketches the situation for a building with uniformly distributed openings (or leakage) over its height, in the situation where $T_{in} > T_{amb}$ (top) and vice versa (bottom). The former is called the 'natural' stack effect, while the latter is called the 'reverse' stack effect. Note that the sketch refers to a strong simplification of reality:

- Uniform temperatures are assumed.
- The openings are assumed uniformly distributed over the building height.
- Steady-state conditions are assumed.
- There is no wind.

The position of the neutral plane height is determined as described in Section 5.3.3, as essentially uniform temperatures are assumed. The static pressure lines are as illustrated in Figure 6.1

DOI: 10.1201/9781003204374-6

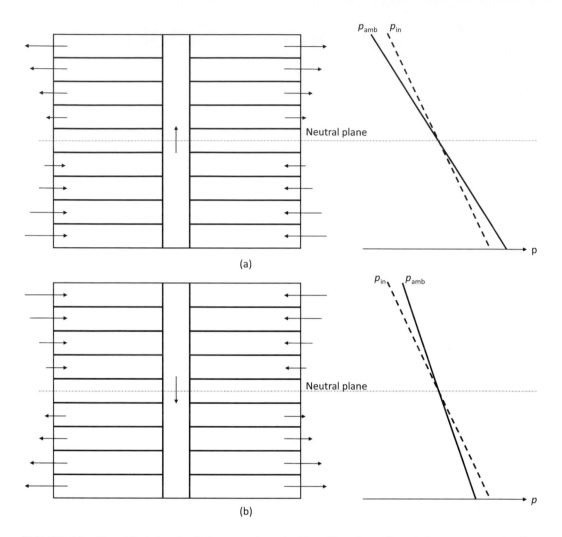

FIGURE 6.1 Simplified sketch of the natural stack effect ($T_{in} > T_{amb}$ (a)) and the reverse stack effect ($T_{in} < T_{amb}$ (b)). There is no wind and openings are assumed uniformly distributed over the building height. Steady-state conditions and uniform temperatures are assumed. Arrows indicate velocities. Static pressure lines are straight.

(see also Figure 2.8). Note that typical velocities in the natural stack effect are low, so dynamic pressures and friction losses related to the flow are neglected in the present analysis.

It is recalled that the situation, as sketched in Figure 6.1, exists in the absence of fire. In other words, during the early stages of a fire, when the fire itself is still small in terms of size and heat release rate, the flow momentum induced by the natural (or reverse) stack effect is the dominant driving force for the smoke flow, as long as no other driving force is activated. Note that the discussion still assumes the absence of wind. The effect of wind is discussed briefly in Section 6.5.

Whereas the principle of the natural and reverse stack effect is simple, it is instructive to discuss the order of magnitude of the induced pressure differences. From the basic law of hydrostatics, Eq. (2.82), the difference between the static pressure inside and the ambient pressure reads:

$$\Delta p = p_{in} - p_{amb} = \left(\rho_{amb} - \rho_{in} \right) g \left(z - z_n \right). \tag{6.1}$$

TABLE 6.1

Difference per Metre Distance from the Neutral Plane between the Pressure Inside the Building and Ambient Pressure, as a Function of the Temperature (Assumed Uniform) Inside the Building and Ambient Temperature

T_{in}	16	17	18	19	20	21	22	23
T_{amb}								
−40	2.875563	2.916825	2.957803	2.998502	3.038922	3.079068	3.118941	3.158546
−30	2.264929	2.306191	2.34717	2.387868	2.428288	2.468434	2.508307	2.547912
−20	1.702536	1.743798	1.784777	1.825475	1.865895	1.906041	1.945915	1.985519
−10	1.182884	1.224147	1.265125	1.305823	1.346244	1.38639	1.426263	1.465868
0	0.70128	0.742543	0.783521	0.824219	0.86464	0.904786	0.944659	0.984264
10	0.253693	0.294955	0.335934	0.376632	0.417052	0.457198	0.497072	0.536676
15	0.041548	0.082811	0.123789	0.164487	0.204908	0.245054	0.284927	0.324532
20	−0.16336	−0.1221	−0.08112	−0.04042	0	0.040146	0.080019	0.119624
25	−0.36139	−0.32013	−0.27915	−0.23846	−0.19804	−0.15789	−0.11802	−0.07841
30	−0.5529	−0.51164	−0.47066	−0.42996	−0.38954	−0.34939	−0.30952	−0.26991
40	−0.91756	−0.8763	−0.83532	−0.79462	−0.7542	−0.71405	−0.67418	−0.63458
50	−1.25965	−1.21839	−1.17741	−1.13671	−1.09629	−1.05615	−1.01627	−0.97667

Distance is considered positive if the floor is above the neutral plane and negative if the floor is below the neutral plane. Temperatures are expressed in °C. Pressure differences are expressed in Pa/m. Atmospheric pressure is assumed to be 1,01,325 Pa.

In Eq. (6.1), z_n denotes the vertical position of the neutral plane height. Eq. (6.1) reveals that:

- If $T_{in} > T_{amb}$ $\left(\rho_{amb} > \rho_{in}\right)$, i.e., in case of the natural stack effect, $p_{in} > p_{amb}$ above the neutral plane. Thus, there is flow out of the building above the neutral plane and vice versa below the neutral plane.
- If $T_{in} < T_{amb}$ $\left(\rho_{amb} < \rho_{in}\right)$, i.e., in case of the reverse stack effect, $p_{in} > p_{amb}$ below the neutral plane. Thus, there is flow out of the building below the neutral plane and vice versa above the neutral plane.

Both for the natural and reverse stack effect, the pressure differences increase linearly with the distance from the neutral plane and they become larger as the density (or temperature) inside the building deviates more strongly from the ambient density (or temperature). Table 6.1 quantifies this in terms of pressure difference per metre from the neutral plane height, given a certain ambient temperature and a certain temperature inside the building and assuming standard atmospheric pressure (1,01,325 Pa).

Table 6.1 contains some extreme circumstances (e.g., outside temperature of −40 °C or +50 °C), but it reveals that even for 'moderate' conditions, the pressure differences due to the stack effect can become quite substantial, particularly for high-rise buildings. To give an example: for $T_{amb}=0°C$ and $T_{in}=21°C$, the pressure difference per metre distance from the neutral plane is 0.905 Pa/m. So for a 200 m high building, approximating the neutral plane position at mid-height of the building, the pressure difference at the top and floor level mounts up to 90 Pa. For larger temperature differences (e.g., $T_{amb}=-20°C$), this mounts up to even 195 Pa, as can be seen from Table 6.1. This is quite substantial, compared to other typical pressure differences, as explained below.

Note that the information in Table 6.1 can also readily be obtained by reformulating Eq. (6.1) in terms of temperature, using the ideal gas law (Eq. 2.10) with gas constant $R = 287.1$ J/(kg K) for air:

$$\Delta p = p_{in} - p_{amb} = 3,462 \left(\frac{1}{T_{amb}} - \frac{1}{T_{in}} \right)(z - z_n). \tag{6.2}$$

Note that the factor 3,462 is not dimensionless in Eq. (6.2). SI units (Pa, K and m) must be used. If pressure is expressed in H_2O temperature in °R and height in ft, Eq. (6.1) becomes:

$$\Delta p = p_{in} - p_{amb} = 7.64 \left(\frac{1}{T_{amb}} - \frac{1}{T_{in}} \right)(z - z_n). \tag{6.3}$$

6.3　FIRE-INDUCED BUOYANCY

The local situation for a growing fire, in terms of pressure differences between the room of the fire and an adjacent shaft or room or ambient, corresponds to the sketches as presented in Figure 5.9 (or Figure 5.7).

In case of a fire in the growing phase, hot smoke is flowing out of the room of the fire and the pressure difference can be calculated from Eq. (5.16). Approximating the vertical position of the neutral plane to be close to the bottom of the smoke layer, the maximum pressure difference is found at the top of the opening between the room of the fire and the adjacent shaft or room or ambient. Using the notation of Figure 5.9 (i.e., H_c is the height of the room of the fire, H is the height of the opening and D is the thickness of the smoke layer), the maximum pressure difference reads:

$$\Delta p_{max} = \left(\rho_{amb/in} - \rho_s \right) g \left(H - (H_c - D) \right). \tag{6.4}$$

In Eq. (6.4), $\rho_{amb/in}$ is the short notation to state that depending on the situation, the density in the adjacent shaft or room $\left(\rho_{in} \right)$ needs to be used, or ambient density $\left(\rho_{amb} \right)$. ρ_s denotes the density of the smoke. Assuming standard atmospheric pressure (1,01,325 Pa) and using the ideal gas law (Eq. 2.10), with gas constant $R = 287.1$ J/(kg K) for air, as well as for smoke, Eq. (6.4) can be reformulated in terms of temperature again, like Eq. (6.2) or (6.3):

$$\Delta p_{max} = 3,462 \left(\frac{1}{T_{amb/in}} - \frac{1}{T_s} \right) \left(H - (H_c - D) \right). \tag{6.5}$$

Note that the factor 3,462 is to be used with SI units (Pa, K and m). If pressure is expressed in in H_2O, temperature in °R and height in ft, the factor becomes 7.64, as in Eq. (6.3).

Table 6.2 provides some values for the factor $3,462 \left(\frac{1}{T_{amb/in}} - \frac{1}{T_s} \right)$, given a certain smoke temperature and a certain temperature in the adjacent shaft or room or ambient temperature, and assuming standard atmospheric pressure (1,01,325 Pa). Only positive values are found, as long as the smoke temperature is higher than $T_{amb/in}$. The values in Table 6.2 are also obviously much higher than in Table 6.1, because smoke temperatures can be high, but a key aspect is that the height $\left(H - (H_c - D) \right)$ is typically small, except for very high fire compartments. Take as an example a 3 m high room $\left(H_c \right)$ with a 2 m high door $\left(H \right)$ to the adjacent shaft or room or ambient and a 1.5 m thick smoke layer $\left(D \right)$. This yields: $H - (H_c - D) = 2 - (3 - 1.5) = 0.5$ m, whereas distances

TABLE 6.2

Values for the Factor $3{,}462\left(\dfrac{1}{T_{amb/in}} - \dfrac{1}{T_s}\right)$ (in Pa/m) of Eq. (6.5), as a Function of the Smoke Temperature (Assumed Uniform) and the Temperature in the Adjacent Shaft or Room or Ambient Temperature

T_s	50	100	150	200	250	300	350	400
$T_{amb/in}$								
−40	4.135213	5.570651	6.66687	7.531409	8.230695	8.807975	9.292617	9.705265
−30	3.52458	4.960017	6.056236	6.920775	7.620061	8.197341	8.681984	9.094631
−20	2.962187	4.397624	5.493844	6.358382	7.057668	7.634948	8.119591	8.532238
−10	2.442535	3.877973	4.974192	5.838731	6.538016	7.115297	7.599939	8.012587
0	1.960931	3.396369	4.492588	5.357127	6.056412	6.633693	7.118335	7.530983
10	1.513344	2.948781	4.045001	4.909539	5.608825	6.186105	6.670748	7.083395
15	1.301199	2.736637	3.832856	4.697395	5.39668	5.973961	6.458603	6.871251
20	1.096291	2.531729	3.627948	4.492487	5.191773	5.769053	6.253695	6.666343
25	0.898256	2.333694	3.429913	4.294452	4.993737	5.571017	6.05566	6.468307
30	0.706753	2.142191	3.23841	4.102948	4.802234	5.379514	5.864157	6.276804
40	0.342092	1.77753	2.873749	3.738288	4.437573	5.014854	5.499496	5.912144
50	0	1.435438	2.531657	3.396196	4.095481	4.672761	5.157404	5.570051

Temperatures are expressed in °C. Atmospheric pressure is assumed to be 101325 Pa.

in Section 6.1.1 could be more than 100 times larger. In other words, it must not be forgotten that the eventual pressure difference is a product of values as given in Table 6.1 or 6.2, with a length scale.

In case of a fully developed fire, the situation is as sketched in Figure 5.16. Using the notation of Section 5.3.3.2, but keeping for the density of the hot gases flowing out of the room of the fire, Eq. (6.4) becomes:

$$\Delta p_{max} = \left(\rho_{amb/in} - \rho_s\right)g\left(z_{top} - z_n\right). \tag{6.6}$$

Using Eq. (5.58), knowing that the opening height $H = z_{top} - z_{bottom}$, yields:

$$z_{top} - z_n = H - \left(z_n - z_{bottom}\right) = H\frac{\left(\dfrac{T_s}{T_{amb/in}}\right)^{1/3}}{1+\left(\dfrac{T_s}{T_{amb/in}}\right)^{1/3}}. \tag{6.7}$$

Substituting Eq. (6.7) into Eq. (6.6) and assuming standard atmospheric pressure (1,01,325 Pa) and using the ideal gas law (Eq. 2.10), with gas constant $R = 287.1$ J/(kg K) for air, as well as for the hot gases, Eq. (6.6) becomes:

$$\Delta p_{max} = 3{,}462\left(\frac{1}{T_{amb/in}} - \frac{1}{T_s}\right)\frac{\left(\dfrac{T_s}{T_{amb/in}}\right)^{1/3}}{1+\left(\dfrac{T_s}{T_{amb/in}}\right)^{1/3}}H. \tag{6.8}$$

TABLE 6.3

Values for the Maximum Pressure Difference per Metre Opening Height (in Pa/m), Eq. (6.8), as a Function of the Temperature of the Emerging Hot Gases (Assumed Uniform) and the Temperature in the Adjacent Shaft or Room or Ambient Temperature

T_s	550	600	650	700	750	800	850	900
$T_{amb/in}$								
−40	6.423631	6.62007	6.799674	6.964736	7.117134	7.258425	7.389911	7.512691
−30	6.021415	6.214322	6.390655	6.552675	6.702232	6.84086	6.969844	7.090264
−20	5.653286	5.842892	6.016171	6.175349	6.322253	6.458397	6.585045	6.703263
−10	5.315214	5.501728	5.672145	5.828663	5.973084	6.106901	6.231362	6.347519
0	5.003774	5.187384	5.355113	5.509133	5.651222	5.782856	5.905264	6.019487
10	4.716038	4.896913	5.062112	5.21378	5.353675	5.483254	5.603732	5.716135
15	4.580257	4.759823	4.923811	5.074354	5.2132	5.341796	5.46135	5.572884
20	4.449485	4.627779	4.79059	4.94004	5.077866	5.205506	5.324164	5.434852
25	4.323459	4.500516	4.662183	4.81057	4.947404	5.074116	5.191902	5.301768
30	4.201936	4.377789	4.538343	4.685696	4.821565	4.947374	5.064312	5.173379
40	3.971497	4.145038	4.303453	4.448819	4.582836	4.706911	4.822219	4.929753
50	3.756516	3.92786	4.084246	4.227727	4.359986	4.482415	4.596179	4.702258

Temperatures are expressed in °C. Atmospheric pressure is assumed to be 1,01,325 Pa

Table 6.3 provides some values of maximum pressure difference per metre height of the opening H, given a certain temperature of the hot gases emerging and a certain temperature in the adjacent shaft or room or ambient temperature, and assuming standard atmospheric pressure (1,01,325 Pa). Only positive values are found, as the temperature of the hot gases in a fully developed fire is (much) higher than $T_{amb/in}$. The values in Table 6.3 are lower than those in Table 6.2 because

the factor $\dfrac{\left(\dfrac{T_s}{T_{amb/in}}\right)^{1/3}}{1+\left(\dfrac{T_s}{T_{amb/in}}\right)^{1/3}}$ has been included in Table 6.3. However, the key aspect is again that the

height of the opening H is typically small, except for very high fire compartments.

As a closing note of this section, it is mentioned that in situations where the hot gases emerge into an adjacent room or shaft, the temperature T_{in} will rise. This will reduce the buoyancy-induced pressure difference (Eq. 6.5 or 6.8), but, more importantly, it will affect the stack effect as described in Section 6.1.1. If the starting situation is a natural stack effect $(T_{in} > T_{amb})$, then the fire will strengthen this effect, because the temperature difference $(T_{in} - T_{amb})$ will increase (Eq. 6.2 or 6.3). If the initial conditions are a reverse stack effect $(T_{in} < T_{amb})$, the situation can evolve towards the situation of a natural stack effect $(T_{in} > T_{amb})$, due to the temperature increase inside the building caused by the fire. However, this is not necessarily the case and will depend on the fire Heat Release Rate (HRR) and the size of the building. In any case, it is important to consider both buoyancy effects and it is important to realize that the fire is not necessarily the dominant driving force for the flow of smoke inside a building. In the case of reverse stack effect as initial conditions, smoke can effectively flow downwards, particularly during (but not necessarily limited to) the early stages of a fire.

6.4 THE EFFECT OF WIND

The effect of wind is often very important. At the same time, it is very hard to predict. Particularly in the built environment, flow patterns due to wind can be very complex, with strong interaction between different buildings. Wind comfort is a research discipline by itself, of particular relevance to architects, where computational techniques become more and more successful [6]. This is beyond the scope of the book at hand. It suffices here to say that even for an isolated building, it is impossible to predict what exactly will be the wind conditions in terms of wind velocity and wind direction at the moment of the fire. Average wind fields do not provide information on that. In other words, any smoke control system design will need to foresee some safety margin to overcome wind-induced pressure.

Indeed, wind induces pressure differences due to the flow itself. The pressure differences relate to the square of the wind velocity v_{wind} through a wind coefficient C_w:

$$\Delta p_{\text{wind}} = \frac{1}{2} \rho_{\text{amb}} C_w v_{\text{wind}}^2. \tag{6.9}$$

The wind coefficient, C_w, is positive in case of over-pressure and negative in case of under-pressure. Indeed, Figure 6.2 illustrates that both are possible. The illustration concerns a simple cube-shaped isolated building, with wind approaching perpendicular to one side of the building (from left to right in the figure). Obviously, much more complex situations can occur, but this is for illustration purposes only. In Figure 6.2, the '+' symbol refers to wind-induced over-pressure $(C_w > 0)$, while the '−' sign refers to wind-induced under-pressure $(C_w < 0)$.

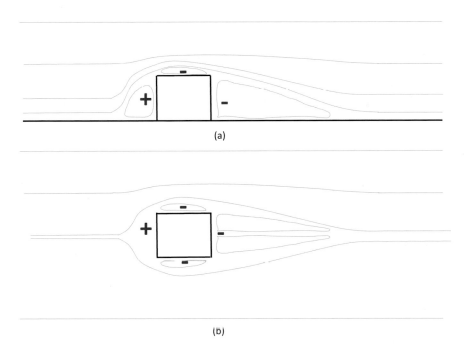

(a)

(b)

FIGURE 6.2 Illustration of wind-induced pressure differences for an isolated cube-shaped building, with wind approaching perpendicular to one side of the building (from left to right in the figure). The '+' symbol refers to wind-induced over-pressure $(C_w > 0)$, while the '−' sign refers to wind-induced under-pressure $(C_w < 0)$. (a) Side view and (b) top view.

TABLE 6.4

Wind-Induced Pressure Differences (in Pa) for a Range of Wind Coefficients and Wind Velocities (in m/s), for Ambient Temperature Equal to 20°C and Ambient Pressure Equal to 1,01,325 Pa

v_{wind} (mls)	0	5	10	15	20	25	30
C_w							
0	0	0	0	0	0	0	0
0.1	0	1.506181	6.024723	13.55563	24.09889	37.65452	54.22251
0.2	0	3.012362	12.04945	27.11125	48.19779	75.30904	108.445
0.3	0	4.518542	18.07417	40.66688	72.29668	112.9636	162.6675
0.4	0	6.024723	24.09889	54.22251	96.39557	150.6181	216.89
0.5	0	7.530904	30.12362	67.77814	120.4945	188.2726	271.1125
0.6	0	9.037085	36.14834	81.33376	144.5934	225.9271	325.3351
0.7	0	10.54327	42.17306	94.88939	168.6923	263.5816	379.5576
0.8	0	12.04945	48.19779	108.445	192.7911	301.2362	433.7801
0.9	0	13.55563	54.22251	122.0006	216.89	338.8907	488.0026
1	0	15.06181	60.24723	135.5563	240.9889	376.5452	542.2251

Only positive values for C_w are reported. Negative values are possible as well.

The over-pressure is induced in stagnation regions, i.e., at the windward side of the building. The wind coefficient has a range of possible values. The theoretical maximum value, in the case of perfect stagnation, is $C_w = 1$. While this value cannot be met in practice, C_w can get close to 1, so it is conservative to use $C_w = 1$ to calculate the possible wind-induced over-pressure, given a certain wind velocity.

The under-pressure is induced by a suction effect due to streamline curvature at the roof, the side walls and in the wake behind the leeward façade. The negative values are the highest in absolute value on the roof, where C_w can be as low as $C_w = -0.9$ [1]. At the side walls, values down to $C_w = -0.7$ are found [1]. On the leeward side, the values are smaller in absolute value, although for high-rise buildings values of $C_w = -0.5$ are reported [1].

Table 6.4 provides an overview of wind-induced pressure differences (in Pa) for a range of wind coefficients and wind velocities (in m/s), for ambient temperature equal to 20 °C and ambient pressure equal to 1,01,325 Pa. It is clear that for high wind velocities, the wind-induced pressure difference can be quite substantial, compared to the other driving forces.

Figure 6.3 illustrates the effect of wind when superposed to the stack effect, as sketched in Figure 6.1. The sketch is simplified in that the wind-induced pressure difference is assumed constant over the height of the building, which will not be the case in reality. Yet, it suffices to illustrate the basic principle: the ambient pressure line is moved to the right (increase in static pressure), while the static pressure line inside the building is assumed to be unaffected. This leads to an upward shift in the position of the neutral plane and affects the flow into the building (Figure 6.3a).

On the leeward side and the side walls, the opposite effect is observed (Figure 6.3b).

Also, on the roof, the wind induces a reduction in static pressure. In principle, this assists natural smoke extraction (Section 6.6), due to increased pressure difference over the opening for natural ventilation. It also assists in mechanical smoke extraction (Section 6.7), as can be understood from the fan characteristics (Figure 5.10). In this sense, it is conservative not to take the wind

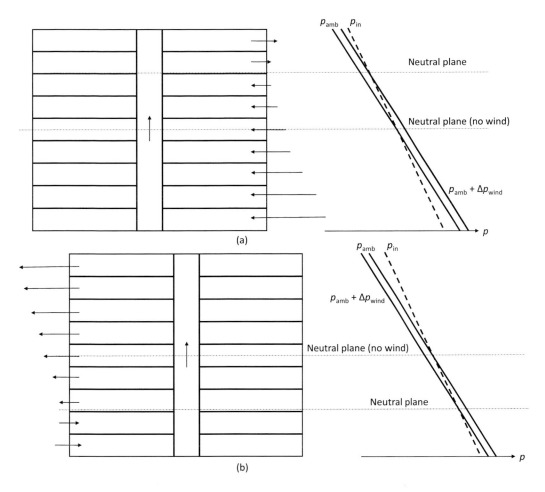

FIGURE 6.3 Illustration of the effect of wind, superposed onto the stack effect (Figure 6.1). (a) Windward side and (b) side walls or leeward side. The sketch is simplified in that the wind-induced pressure difference is assumed constant over the height of the building, which will not be true in reality.

effect into account during the design phase of the SHC system. This observation remains valid as long as the roof is mildly inclined. Indeed, if the roof inclination exceeds a certain threshold value (e.g., 25°), a stagnation zone starts to build up, similar to what is observed at the windward façade, so the roof is no longer in under-pressure. Note, however, the important discussion below, related to Figure 6.4.

Figure 6.2 also provides a sketch of the flow field. At the windward side, the flow separates from the bottom plane at some point. The position of the separation point depends on the Reynolds number of the oncoming flow, but the order of magnitude is at around $1.5H$ to H upstream of the building. The separation point is closer to the building for higher wind velocities. This implies that a region of the extent H to $1.5H$ upstream of the building is in over-pressure. This can be very important for SHC systems. For obvious reasons, it needs to be avoided for a fan to act against a wind-induced over-pressure, because this results in reduced fan efficiency (see Figure 5.10). However, another possibly dangerous situation can also be understood from Figure 6.2 in the built environment. Indeed, if another, higher, building is close to a certain building, the stagnation zone

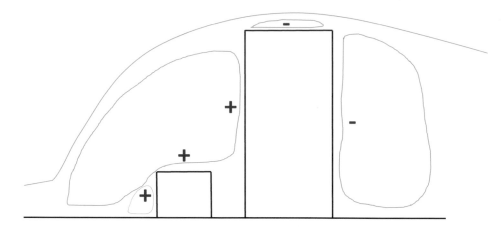

FIGURE 6.4 Illustration of the impact of a higher building on the flow field over a lower building. The roof of the lower building is no longer in under-pressure (Figure 6.2), but in over-pressure. This can have a substantial impact on the performance of the SHC system in the lower building: compare the '+' sign on the roof of the lower building to the '−' sign in Figure 6.2.

for the latter can cause the roof of the former, which is normally in under-pressure (Figure 6.2), to be in over-pressure, as sketched in Figure 6.4. Thus, the assumption of the roof being in under-pressure, assisting the smoke extraction, can become false. This can have a strong impact on the performance of the SHC system in the lower building. Note that this can be particularly problematic if the higher building is constructed at a moment when the lower building was already in place. This illustrates the possibly important complex interaction between buildings in the context of smoke and heat control.

On the leeward side, the recirculation zone is longer than on the windward side. The point of reattachment is at about $4H$ to $6H$ behind the building. The fact that this zone is more stretched also explains why the wind coefficient at the leeward side is smaller in absolute value than at the windward side. For a high-rise building, the negative wind coefficient is higher in absolute value than for a lower building, because it behaves more like a side wall: the presence of the floor is not felt as strongly and the floor is more of a flow around the building than the flow over a building. The characteristic length scale for the wake zone is then no longer the building height, but the width W of the windward façade. The wake zone depends on the Reynolds number of the wind flow but extends to about $3W$.

On the side walls, the effect is similar to that on the roof. The most negative wind coefficient values are found near the separation zone, i.e., close to the windward side.

6.5 PRESSURIZATION

Pressurization is a technique that can be applied to try and keep a certain zone inside a building smoke-free in case of fire. A typical example is a stairwell, to be kept smoke-free in case of a fire outside of the stairwell. The primary objective of a pressurization system is thus to retain tenable conditions in the protected areas (typically a stairwell), used as an escape route. In other words, life safety for evacuation is key. Pressurization can also assist firefighting and prevent smoke damage to the property, but people or property in the room of the fire are not protected by the pressurization approach. For some design calculations, the reader is referred to Refs. [1,2,4,7].

The basic idea of pressurization is simple: an over-pressure is created inside the stairwell, such that the driving force (from the fire) for the smoke to flow into the stairwell is overcome.

This over-pressure is created by one or more centrifugal fans. Care must be taken with respect to the possible effect of wind, as discussed in Section 6.4: wind-induced under-pressure at the location of the fan can drastically reduce the effectiveness of the fan.

It must not only be determined where the fresh air is supplied but also where the air and smoke must leave the building and what pathways they will follow during this process. In this manner, a global pressure gradient is created inside the building, with the highest pressure in the compartment to be protected and decreasing pressure towards areas that are farther away from escape routes.

In principle, relatively low flow rates are required in the pressurization approach. However, an important issue for the concept of pressurization concerns open doors. Indeed, as soon as a door is open, the pressure difference will cause a flow through that opening and, more importantly, the pressure in the stairwell will drop, particularly if there is a path without strong pressure loss for the flow towards ambient. For tall buildings, the risk can be reduced by using multiple injection points or compartmentation, but it cannot be eliminated. In the same spirit of trying to avoid unacceptable pressure loss, the injection point for air is often at the top of the building and particularly not at the level of the exit doors. Indeed, if the exit doors are open (e.g., for evacuation purposes), there is short-circuiting of the intake air flow directly towards the exit and the over-pressure is lost inside the stairwell. Note, however, that even when the injection point is far away from the exit doors, there will be substantial loss in over-pressure if there is an easy path for the intake air to flow towards the exit. This is a major drawback of the pressurization approach.

The above may give the impression that an as high as possible pressure difference Δp is desirable. This is not the case. Whereas Δp must be high enough $\left(\Delta p > \Delta p_{\min}\right)$ to overcome the other driving forces (e.g., stack effect, fire-induced buoyancy, wind), it must also not be too high $\left(\Delta p < \Delta p_{\max}\right)$, because too high a pressure inside the stairwell may render opening doors towards the stairwell impossible, thereby making evacuation impossible. As an order of magnitude, a force of $100\,\mathrm{N}$ can be used as the maximum force on a door, for the door to still be opened in a reasonably easy manner. The force required to open a door, with area A and width W, where the doorknob is at a distance d from the edge, closest to the doorknob, reads:

$$F = F_{\mathrm{dc}} + A\Delta p \frac{W/2}{W-d} + F_{\mathrm{fric}} \frac{W/2}{W-d}. \tag{6.10}$$

In Eq. (6.10), F_{dc} is the mechanical force exerted by the door closer (which equals zero in the absence thereof), F_{fric} is a force due to friction at the floor (which equals zero in the absence thereof) and the middle term expresses the force required to overcome the force due to the pressure difference over the door. The pressure force is assumed to act in the middle of the door, due to the integration of an assumed constant pressure difference over the door. Thus, the moment associated with this force reads $A\Delta p \frac{W}{2}$. The force exerted at the doorknob has a moment $F(W-d)$. This explains the factor $\dfrac{W/2}{W-d}$. The same reasoning holds for the friction force F_{fric}, assuming the friction shear stress is uniform and thus the friction force applies in the vertical mid-plane of the door.

Consider now a situation in absence of friction ($F_{\mathrm{fric}} = 0\,\mathrm{N}$) and in absence of a door closer ($F_{dc} = 0\,\mathrm{N}$). In such conditions, using $F = 100\,\mathrm{N}$, the maximum pressure difference can be calculated as $\Delta p_{\max} = \dfrac{200(W-d)}{AW}$. For a door of $2.2\,\mathrm{m}$ height and $0.9\,\mathrm{m}$ width, assuming $d = 0.1\,\mathrm{m}$, this yields $\Delta p_{\max} = 90\,\mathrm{Pa}$. In Ref. [4], the value $\Delta p_{\max} = 60\,\mathrm{Pa}$ is mentioned, which is more conservative. Dampers can be used to guarantee that Δp does not exceed $\Delta p_{\max}$.

Finally, it is mentioned that the system of de-pressurization is the counterpart of pressurization, in that then the pressure in the room of the fire is kept lower than the pressure in the remainder of the building. The considerations mentioned for the pressurization approach prevail.

6.6　NATURAL VENTILATION

Natural ventilation can be applied as an SHC system. It is typically assumed to be effective in the growing phase of the fire (Section 5.2). The basic formulae and fluid mechanics aspects have been discussed in Section 5.2.3. The most typical setting is the use of horizontal openings, as discussed in Section 5.2.3.1.

The basic figure is Figure 5.7, but this can also be generalized for an atrium configuration, as sketched in Figure 6.5. The basic concept is that there is a smoke layer of thickness D underneath the ceiling, in which there is a ventilation opening with area A_v. The thickness, D, and the smoke layer average temperature, T_s, can vary over time, but often calculations are made for steady-state conditions. The discussion in the present section is restricted to steady-state conditions for the sake of simplicity. It is noted that, in the case of an atrium, the fire in the adjacent room can be fully developed, with emerging flames, but the assumption is that smoke is entering the smoke layer, not flames.

The basic driving force is again buoyancy, causing an upward flow of the hot smoke gases, as explained in Section 5.2.3. The upward flow induces air entrainment into the smoke plume, such that eventually a certain mass flow rate enters the smoke layer. In order for a steady situation to occur, the mass flow rate emerging from the smoke layer through the openings for natural ventilation must equal the mass flow rate entering the smoke layer at its bottom side. As explained in Section 5.2.3, it is a natural choice to have the extraction openings at a high position (i.e., above the neutral plane) and the openings for air supply at a low position (i.e., below the neutral plane height).

The fluid mechanics aspects have been explained in Section 5.2.3. Therefore, it suffices here to explain the basic steps in the calculation procedure to determine the required area A_v for the openings for natural ventilation in steady-state conditions. The reader is referred to standards (e.g., [8,9]) for more detailed calculation formulae.

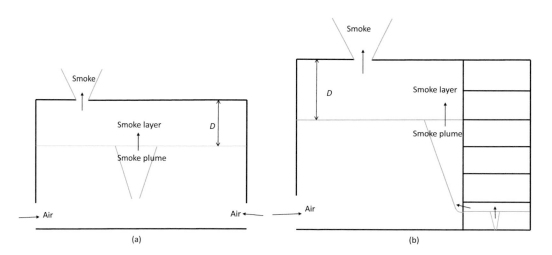

FIGURE 6.5 Schematic sketch of the concept of natural ventilation in the room of a growing fire (a) or an atrium configuration (b).

The basic equation is Eq. (5.15):

$$\dot{m}_s = \rho_s C_d A_v \sqrt{\dfrac{2gD\left(\dfrac{T_s}{T_{amb}} - 1\right)}{1 + \dfrac{T_{amb}}{T_s}\left(\dfrac{C_d A_v}{C_{d,in} A_{in}}\right)^2}} \ . \qquad (6.11)$$

From this equation, the important parameters, required for the determination of A_v are readily identified:

- The mass flow rate of smoke, $\dot{m}_s$, entering the smoke layer at its bottom side.
- The average mass density, ρ_s, and absolute temperature, T_s, of the smoke layer. These are related through the ideal gas law, Eq. (2.10), using the gas constant R for air (287 J/(kg K)) and $p_{atm} = 100$ kPa: $\rho_s = \dfrac{348}{T_s}$.
- The thickness of the smoke layer, D.
- Ambient absolute temperature, T_{amb}.
- The discharge coefficient of the opening for natural ventilation, C_d.
- The aerodynamic area, i.e., the product of the discharge coefficient and the geometric area, of the inlet opening, $C_{d,\,in} A_{in}$.

In case of multiple openings at the same height, the areas are added.

For some of these quantities, direct design choices are made. For example, ambient temperature can be assumed to be equal to 15°C $\left(T_{amb} = 288 \text{ K}\right)$. Obviously, other choices can be made, depending on typical weather conditions.

Another direct design choice concerns the thickness of the smoke layer. In other words, the value for D is prescribed in the design. It is clear from Eq. (6.11) that a larger value for D will correspond to a smaller area A_v (assuming that $\left(\dfrac{C_d A_v}{C_{d,\,n} A_{in}}\right)$ is either small or fixed, see below). This is logical: less exhaust opening area corresponds to an increased smoke layer thickness.

Another very important quantity is the mass flow rate of smoke entering the smoke layer at its bottom side $\left(\dot{m}_s\right)$. This mass flow rate must be calculated, using expressions as provided in Chapter 4 (Sections 4.2 and 4.3). The larger the mass flow rate, for a given smoke layer thickness D, the larger the required area A_v. The smoke mass flow rate, $\dot{m}_s$, is determined through the fire size (in terms of fire HRR, $\dot{Q}$, as well as floor area, A_f, and shape or perimeter, P_f) and the geometry, in particular the height of the room of the fire (Figure 6.2, left) or the atrium (Figure 6.2, right). Note that D also determines the rise height of the smoke plume until the bottom side of the smoke layer. As explained in Chapter 4, the smoke mass flow rate, $\dot{m}_s$, increases with a rise in height. Also note that, in the case of an atrium with fire in the adjacent room, the width of the spill plume entering the atrium is also an important parameter, as explained in Chapter 4.

As already mentioned, another very important design parameter concerns the size of the fire. In fact, only the convective part of the fire HRR, $\dot{Q}_{conv}$, is the driving force for the smoke and determines how much heat is transported by the smoke, as explained in Chapter 4. Assuming that there are no heat losses to the structure and that all air entrained into the smoke plume is at ambient temperature, T_{amb}, the average temperature of the smoke entering the smoke layer at its bottom side reads:

$$T_s = T_{amb} + \dfrac{\dot{Q}_{conv}}{\dot{m}_s c_p} \qquad (6.12)$$

The specific heat is assumed to be equal to $c_p = 1 \text{ kJ}/(\text{kg K})$, which is approximately the value for air at ambient temperature.

From T_s, the smoke mass density is computed as $\rho_s = \dfrac{348}{T_s}$, with the assumptions as mentioned above.

Assuming that the discharge coefficients for the openings are known, the only remaining unknown is the opening area for the air intake. Intake air is normally supplied by natural ventilation, not mechanically. The opening area for the air intake is obviously determined by the geometry. If the inlet opening area is much larger than the opening for smoke extraction, Eq. (6.11) simplifies to:

$$\dot{m}_s \approx \rho_s C_d A_v \sqrt{2gD\left(\frac{T_s}{T_{amb}} - 1\right)}, \qquad\qquad if \ C_{d,in} A_{in} \gg C_d A_v \qquad\qquad (6.13)$$

Sometimes, the ratio $\left(\dfrac{C_d A_v}{C_{d,in} A_{in}}\right)$ is fixed at the onset of the calculations. Whereas this may seem strange at first sight, it is a design assumption that can be made if, e.g., the floor area of a large compartment is subdivided into multiple (N_s) sections. Each section is assumed to have an opening for ventilation with an aerodynamic area $C_d A_v$. The fire is then assumed to be inside one of these sections and all other openings are assumed to serve as opening for the intake of fresh air. In other words, in contrast to the sketch in Figure 6.5, the intake air is then supplied from the ceiling. In such circumstances, $\dfrac{C_d A_v}{C_{d,in} A_{in}} = \dfrac{1}{N_s - 1}$. Figure 6.6 illustrates this.

It is recalled here that, as explained in Section 5.2.4, the intake of fresh air is essential for a natural ventilation system to operate properly. The phenomenon of plug-holing as discussed in Section 5.2.4 is also recalled, because this can be a reason for subdivision of $C_d A_v$ into multiple openings, such that the sum equals $C_d A_v$. The reader is referred to standards (e.g., [8,9]) to learn how this issue is dealt with in different design calculations.

An important assumption in all of the above is, as mentioned in Section 5.2.3.1, that the velocity of the smoke entering the smoke layer is low (in fact, effectively zero). This need not be the case. The possible effect hereof has been illustrated in Ref. [10], through a series of CFD

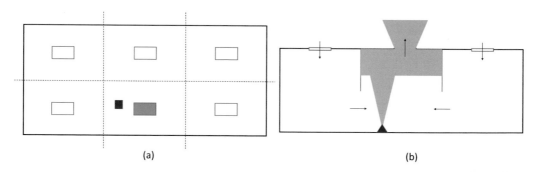

(a)   (b)

FIGURE 6.6 Illustration of the concept of using vents in the ceiling for the intake of fresh air. Left: top view; right: side view. The floor area is subdivided into N_s 'compartments' (dashed lines in left figure) and the fire is assumed to be in 1 compartment (red box in panel (a) and red cone in panel (b)). In this sketch, $N_s = 6$. The opening in blue (a) acts as an exhaust vent, while the other openings act as inlet openings for fresh air. The subdivision into 'compartments' can be done with, e.g., smoke curtains (solid thin lines in (b)). Black arrows indicate flow directions.

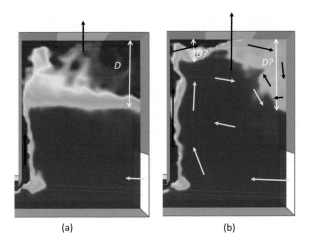

(a)   (b)

FIGURE 6.7   An example of temperature contours and flow patterns in an atrium configuration with moderate (a) and high (b) smoke extraction mass flow rates (redrawn after [9]). With high smoke extraction mass flow rate the smoke layer no longer has a horizontal bottom side and the formulae no longer prevail. [Note: In Ref. [9], the smoke extraction mass flow rate was imposed as a mechanical extraction rate, but similar arguments prevail for natural ventilation, increasing $C_d A_v$.]

(Computational Fluid Dynamics) simulations. One observation was that, as expected intuitively, an opening for smoke extraction is more effective when positioned well-aligned with the smoke plume entering the smoke layer. The upward momentum of the smoke plume assists the smoke extraction, driven by the static pressure difference over the opening. However, there is another important aspect in that extreme care must be taken if high extraction mass flow rates are strived for in the design. Indeed, if this is the case, flow patterns can emerge inside the building that destroys the concept of a steady smoke layer. This is also illustrated in Ref. [10] for an atrium configuration. Figure 6.7 shows the flow pattern in combination with temperature contours for moderate (a) and high (b) smoke extraction mass flow rates. The drastic impact on the smoke layer shape is clear. The configuration in the right figure does not meet the basic assumption of a steady smoke layer with horizontal bottom side, as sketched in Figure 6.5; so, the formulae above are no longer valid. This is an important caveat in design calculations.

6.7   MECHANICAL VENTILATION

6.7.1   Vertical Ventilation

For vertical mechanical ventilation, the basic discussion is essentially very similar to Section 6.6. The determination of the aerodynamic area, $C_d A_v$, required to extract the desired mass flow rate of smoke, is replaced by the determination of one or multiple fans to guarantee the desired smoke extraction mass flow rate. Similar to the natural ventilation approach (Section 6.6), it is a natural choice to have the extraction fans at a high position (i.e., above the neutral plane) and the openings for air supply at a low position (i.e., below the neutral plane height). The discussion in the present section is restricted to steady-state conditions. The smoke layer is assumed to have a horizontal bottom side again, as in Section 6.6. The comments made, based on Figure 6.7, prevail for mechanical ventilation.

As explained in Section 6.6, the mass flow rate, $\dot{m}_s$, entering the smoke layer at its bottom side is determined by the choice of smoke layer thickness, D, the geometry of the building and the design fire.

The average smoke temperature is calculated from Eq. (6.12). From the smoke density and the smoke mass flow rate, the volume flow rate for mechanical smoke extraction can be calculated as follows:

$$\dot{V}_s = \frac{\dot{m}_s}{\rho_s}. \tag{6.14}$$

From the total volume flow rate and the required number of extraction points to avoid plug-holing (see Section 5.2.4), fans can be chosen to meet the required volume flow rate per fan.

Also, in the case of mechanical ventilation, the aerodynamic area of the openings for the intake of fresh air is very important. Normally, the fresh air is supplied by means of natural ventilation, just as is the case for natural smoke extraction. In cases where both air supply and smoke extraction are done mechanically, special care must be taken. Indeed, if the mass flow rates do not match, strong under-pressure (if the extraction mass flow rate exceeds the air supply rate) or over-pressure (if the opposite is true) can build up in time. Also, pressure fluctuations can occur, as briefly discussed below. Assuming now that the air supply is through natural ventilation, the aerodynamic area of the openings of the intake of fresh air is very important, because this will determine the pressure drop for the flow at the inlet openings. The larger this pressure drop, the larger the static pressure difference over the fan and thus the lower the volume flow rate through that fan (see Figure 5.10).

An additional reduction in the effectiveness of the extraction fans can be caused by ductwork or shafts, through which the smoke must flow. Indeed, Figure 6.5 sketches the situation for natural ventilation, where the smoke flows directly into the open atmosphere. The same situation is possible for mechanical extraction, but another situation is that smoke is collected in ductwork and shafts before it is released into the open atmosphere. Special care must be taken concerning pressure losses due to the flow through this ductwork and shafts because these imply pressure differences to be overcome by the extraction fans and thus lead to lower volume flow rates (Figure 5.10). This is discussed further in Section 6.8.

As briefly mentioned above, fluctuating pressures can occur, particularly in situations with mechanical air supply and smoke extraction, interacting with a fire source. Whether or not this will occur, depends on the flow rates, the fire size and the size of the compartment. It is related to the Global Equivalence Ratio (GER) inside the compartment. The basic phenomenon is as follows:

- Initially, a small fire grows. There is no lack of oxygen (low GER value).
- As the temperature inside the compartment increases, the pressure increases as well, for given supply and extraction mass flow rates, due to the ideal gas law. Note that the extraction mass flow rate decreases for a given volume flow rate as the density decreases (due to the temperature increase) so that the pressure inside increases even more.
- The rise in pressure reduces the supply flow rate of fresh air, due to the stronger pressure difference the fan has to overcome (Figure 5.10).
- Consequently, the GER inside the compartment increases. If this increase is such that the fire becomes strongly under-ventilated, the HRR inside the compartment decreases and the temperature decreases. This causes a drop in pressure inside the compartment.
- This drop in pressure results in an increase in the supply rate of fresh air, which leads to an increase in fire HRR and an increase in temperature inside the compartment again, so that a period cycle can occur (until burn-out of the fuel).

Such phenomena have been reported in, e.g., Ref. [11], as mentioned in Section 5.5.

As a final comment in this section, it is mentioned that in the context of tunnel ventilation, the term 'transverse ventilation' is often used instead of 'vertical ventilation'. In such configurations, the smoke is typically collected in exhaust ducts, rather than issued directly into the ambient, but this does not strongly affect the flow phenomena underneath the fan. This is also called 'smoke extraction', which is briefly discussed in Section 6.5.

6.7.2 Horizontal Ventilation

Horizontal ventilation can be applied in tunnels and underground structures, such as car parks. Since there are quite substantial differences between these configurations, they are discussed separately. Regardless of the geometry, the horizontal momentum is essential. Indeed, it is the momentum, i.e., the product of mass flow rate and velocity, induced by the ventilation system, that competes with the fire-induced smoke momentum. It is important to stress this because in tunnel configurations, often the term 'critical ventilation velocity' is used, while 'critical momentum' would be a more logical and less misleading term.

6.7.2.1 Tunnels

In tunnel configurations, forced horizontal ventilation is also called 'longitudinal ventilation' [12,13]. Usually, the longitudinal ventilation is imposed by a mechanical system, but in principle in inclined tunnels (so that the flow is no longer 'horizontal') also the fire-induced buoyancy force or stack effect can be relied upon (see Section 6.1). Yet, it is clear that mechanical systems, which are typically also in place for daily ventilation (e.g., to remove or dilute exhaust gases in the tunnel), are more effective, particularly if they operate in the same sense as the natural fire-induced buoyancy force.

As mentioned above, it is common practice to discuss longitudinal ventilation in tunnels in terms of velocities. This is to a certain extent misleading and confusing because eventually the flow field is determined by momentum, which is the product of velocity and mass flow rate. Yet, the commonly used terminology is adopted here. The airflow inside the tunnel, induced by the ventilation system, is essentially one-dimensional, given the relatively small height and width of the tunnel, compared to its length. As such, discussing ventilation velocities v_{vent} or momentum is equivalent per se. Indeed, the momentum equals $\rho_{\text{amb}}WHv_{\text{vent}}^2$, where W is the tunnel width and H is the tunnel height. Air is assumed to be at ambient conditions.

Figure 6.8 illustrates the concept of the critical ventilation velocity (a) and smoke back-layering (b). The fire-induced buoyancy force pushes the hot gases upwards, as explained earlier. The

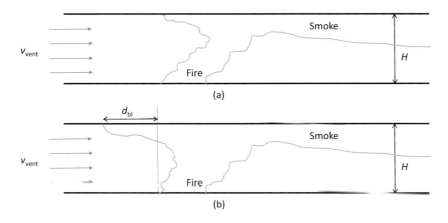

FIGURE 6.8 Sketch of the concept of longitudinal ventilation at 'critical ventilation velocity' $\left(v_{\text{vent}} = v_{\text{cr}}\right)$ (a) and smoke back-layering $\left(v_{\text{vent}} < v_{\text{cr}}\right)$ (b). The flames from the fire might reach the ceiling, depending on the fire HRR and area and on the tunnel height. Downstream of the fire, stratification may be lost.

ventilation system creates momentum in the airflow to compete with this fire-induced momentum of the smoke. As explained in Chapter 4, it is the convective part of the total HRR that induces the momentum and this momentum is different in the continuous flame region, the intermittent flame region and the smoke plume region. The regime met near the tunnel ceiling depends on the size of the fire (in terms of HRR and geometry), on the one hand, and on the tunnel height, on the other hand. If a continuous flame regime exists at the ceiling level, the unburnt fuel reaching the ceiling will burn underneath the ceiling as it mixes with oxygen. The gases expand but do not generate as much momentum inside the smoke as they would have had they burnt in the fire plume (causing additional momentum through air entrainment).

If the ventilation momentum is such that the oncoming velocity, v_{vent}, exceeds a critical value, v_{cr}, all the smoke is pushed backward from the fire source. This makes the region upstream of the fire essentially smoke-free, so that evacuation and firefighting can readily be done through that upstream region. Obviously, the downstream zone is not protected at all, and smoke stratification may be lost. Moreover, care must be taken with respect to the HRR. Indeed, in case of a large potential fire source, the fire can become under-ventilated due to lack of available oxygen, given the relatively small cross-sectional area of many tunnels. If longitudinal ventilation is applied, fresh oxygen is supplied to the fuel source and the fire HRR can increase substantially. The reader is referred to [12,13] for a more extensive discussion on this topic.

In line with the reasoning above, it is found that the required ventilation momentum, $\rho_{amb}WHv_{vent}^2$, becomes independent of (the convective part of) the fire HRR in situations where there is a continuous flame region underneath the ceiling near the fire source, whereas it scales as $\dot{Q}_{conv}^{2/3}$ in the smoke plume regime, respectively. Indeed, the asymptotic values for small and high HRRs are recovered in the theoretical paper [14] and the Wu-Bakar correlation [15]. In Ref. [13], it is recommended to use the correlation presented in Ref. [16], which also confirms this. In that correlation, use is made of the non-dimensional (total) HRR:

$$\dot{Q}^* = \frac{\dot{Q}}{\rho_{amb}c_p T_{amb} g^{1/2} H^{5/2}}. \tag{6.15}$$

The critical ventilation velocity is also expressed in a non-dimensional manner:

$$v_{cr}^* = \frac{v_{cr}}{\sqrt{gH}}. \tag{6.16}$$

The correlation in Ref. [16] then reads:

$$v_{cr}^* = 0.43 \text{ if } \dot{Q}^* > 0.15; \quad v_{cr}^* = 0.81\dot{Q}^{*1/3} \text{ if } \dot{Q}^* \leq 0.15. \tag{6.17}$$

Note that Eq. (6.17) indeed confirms the exponents as mentioned. The continuous flame regime corresponds to high values of Eq. (6.15). In that case, Eq. (6.17) shows that the required ventilation momentum becomes independent of the fire HRR. The critical ventilation velocity is proportional to $v_{cr} \propto \sqrt{H}$ (Eqs. 6.16 and 6.17), in line with Eq. (17b) in the theoretical paper [14]. For low values of HRR (smoke plume regime underneath the ceiling), the critical velocity scales as $v_{cr} \propto \dot{Q}^{1/3} H^{-1/3}$, in line with Eq. (17a) in Ref. [14]. This implies that the ventilation momentum indeed scales proportionally to $\dot{Q}_{conv}^{2/3}$ in this regime.

In Ref. [13], it is also explicitly stated not to use the Kennedy correlation from Ref. [17], which is built upon Ref. [18]. Indeed, from the above, it can be understood that the Kennedy correlation, which scales the critical ventilation velocity to $\dot{Q}_{conv}^{1/3} W^{-1/3}$, regardless of the regime, is in fact not valid. It only scales properly with the HRR for low HRR values, but even then the scaling with tunnel height is not correct. In order to avoid confusion, the correlation is not presented here.

A second important notion concerns smoke back-layering. Indeed, if $v_{vent} < v_{cr}$, the ventilation momentum is not sufficient to push all the smoke backwards, downstream of the fire source. Then the smoke can travel upstream as well. This is called 'back-layering'. The back-layering distance, d_{bl}, increases as the ventilation velocity deviates more from its critical value. In Ref. [19], an expression is formulated, based on the difference, $v_{cr} - v_{vent}$, derived from CFD simulations in a car park (which resembles a very wide tunnel in Ref. [19]). In Ref. [16], a correlation is provided, based on experiments, expressing the back-layering distance as a function of the ratio of the two velocities:

$$\frac{d_{bl}}{H} = 18.5 \ln\left(\frac{v_{cr}}{v_{vent}}\right). \tag{6.18}$$

More research would be valuable to confirm the general validity of expression (6.18).

Finally, it is mentioned that in the expressions above, H is the tunnel height. This is clearly the characteristic height when the fire source is set flush with the floor. In Ref. [20], it is illustrated that Eq. (6.17) still prevails when H is defined as the distance between the vertical position of the fire source and the tunnel ceiling.

6.7.2.2   Other Underground Structures

The specific feature of tunnels is the fact that the flows are essentially one-dimensional (except near the fire source). This need not be the case in general underground structures, such as subway stations or car parks. In this section, some key features are discussed in the context of smoke and heat control in large closed car parks.

It is essential to appreciate that the flow field pattern can now be quite complex, because particularly the width, W, can now be much larger than in tunnels (whereas the height, H, can be quite similar as in tunnels). As mentioned in Section 6.7.2.1, the flow field is determined by momentum. In combination with the observation that the flow pattern can now be complex, it is no longer meaningful to build arguments on the basis of velocities. Indeed, the combination of velocity and mass flow rate is required to push the fire-induced smoke into a certain direction (or, formulated in another manner, to extract the smoke in a certain direction).

If forced mechanical ventilation is used in such circumstances, it is common practice to extract the smoke mechanically by means of large centrifugal fans and to supply the intake air naturally. If both the air supply and the smoke extraction are mechanical, special care must be taken in terms of pressure build-up (over-pressure or under-pressure), as explained in Section 6.6. In case of a car park underneath a building, it is common practice to put the building in over-pressure, relative to the car park, very much as explained in Section 6.5. Yet, this must not be confused with possible under-pressure, caused by a smoke extraction mass flow rate exceeding the supply mass flow rate of replacement air. Indeed, such an under-pressure can cause problems in opening doors, as explained in Section 6.5.

A few statements, obtained from reviewing recent research in the context of smoke and heat control in large car parks, have been formulated in Ref. [19]:

*   There are substantial differences in ventilation requirements among various existing guidelines. These differences relate to the accordingly envisaged fire safety strategy. Much higher smoke extraction rates are required, e.g., when the design objective is to assist a fire service intervention, as opposed to the situation where the only objective is smoke clearance after the fire.
*   In contrast to tunnel fires, fires in large car parks are unlikely to be under-ventilated. Consequently, the fire size (in terms of HRR and geometry) is unlikely to increase substantially due to the forced ventilation. The primary effect of forced ventilation in large

car parks is thus a reduction of the temperatures, which in turn can lead to a slower fire spread to neighbouring cars. Note that this may not be the case for small car parks.

- As the flow follows the path of least resistance, the ventilation air tends to bypass the fire source if possible. Indeed, the fire causes a pressure rise, so if the airflow finds a way to bypass this, it will do so.
- Complex flow patterns can occur in car parks, including recirculation and stagnation zones. This results in the complex transport of the smoke.
- The possible presence of beams underneath the ceiling in different directions can strongly affect the performance of the SHC system, in the sense that they can change the required momentum of the airflow substantially. Indeed, if the fire-induced smoke momentum is channelled in between the beams, the smoke momentum is no longer omnidirectional. It becomes stronger aligned with the beams and much weaker in the perpendicular direction.

In a nutshell, for the SHC system to be effective in assisting firefighting in large car parks, two main conditions must be fulfilled:

- The airflow momentum must be sufficient to overcome the flow resistance caused by the fire-induced smoke flow. This is similar to the discussion in Section 6.7.2.1 for tunnels.
- The airflow must be guided to reach the fire source. If the fire source is positioned in a stagnation or recirculation zone, the airflow will essentially bypass it and the effect of the ventilation will be very limited. Also, fires near a wall cause a more challenging situation for the SHC system: not only is the fire development faster, but also the fire-induced flows are stronger. The possible complexity in the flow pattern is a fundamental difference from Section 6.7.2.1.

A few aspects have been illustrated in Ref. [21], presenting CFD simulations results, backed up by full-scale experimental data [22]. The car park was a rectangular box of ~28.6 m × 30 m × 2.7 m. Figure 6.9 shows a few simulation results for a fire source of ~500 kW (which is small, compared to a typical HRR of a modern car, which is beyond 5 MW [19]) and a smoke extraction flow rate of 2,00,000 m³/h. The fire source is positioned on the floor in the middle of

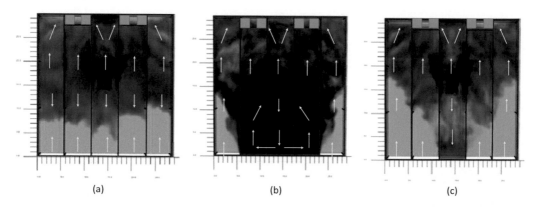

(a)   (b)   (c)

FIGURE 6.9 Illustration of the impact of the inlet configuration on the smoke distribution in the car park of Refs. [19,20]. The fire HRR is ~500 kW. The fire source is in the middle of the car park. The extraction rate is 2,00,000 m³/h (four extraction fans of 50,000 m³/h each, indicated as blue squares). The arrows indicate the flow pattern. (a) Inlet completely open; (b) middle part (3/5) of the inlet blocked; (c) middle part (1/5) of the inlet blocked.

the car park. The impact on the smoke distribution due to differences in configuration at the inlet for air supply is illustrated.

As long as the flow is unidirectional, the observations are in line with the discussion presented for tunnels. In other words, the car park essentially behaves like a wide tunnel. However, substantial changes in the smoke pattern are observed when the inlet opening is partially blocked. As the arrows indicate, recirculation zones occur behind the blocked parts of the inlet (best seen in the middle sketch of Figure 6.9). If smoke enters this recirculation zone, it is trapped. Increasing the extraction flow rate does not remedy this at all [21]. The right sketch illustrates this can even happen when a relatively small fraction of the inlet area is blocked. Note that this is observed for a small fire source. For larger fire sources, this prevails a fortiori, since higher extraction flow rates are in principle required to extract the smoke. This increases the probability of recirculation zones.

The key result of Figure 6.9 is not the observation for this specific set-up. Rather, it is the illustration that car parks are far more complex in terms of flow patterns than tunnels and, consequently, that recirculation and stagnation zones need to be avoided. In other words, the momentum of the ventilation air flow must be guided over the fire source. Indeed, the middle and right sketches in Figure 6.9 reveal that the air bypassing the fire source is possible. This guidance can be done by means of walls or smoke screens. However, fixed walls are often undesired for security reasons and smoke screens are not always a viable option, as they must not be blocked by cars.

Finally, it is mentioned here that jet fans can be useful in reducing temperatures locally and in clearing recirculation or stagnation zones, but their impact is only very local if the global extraction flow rate is high [21]. They do not have a substantial impact on the global smoke pattern [21].

6.8 SMOKE EXTRACTION

The term 'smoke extraction' indicates that fire-induced smoke is extracted through openings in the neighbourhood of the fire. In a tunnel or a car park, e.g., the smoke is then often extracted by mechanical ventilation and is collected in exhaust ductwork, to be guided towards ambient. Natural ventilation is also possible for tunnels, relying on the stack effect of vertical shafts. The effectiveness thereof requires further systematic research.

Figure 6.10 illustrates the principle and the effect on the pressure lines and corresponding flow. The opening (e.g., a door or a window) connects the fire compartment to an adjacent compartment or ambient. Therefore, the pressure line there is given the symbol p_a. This line is essentially unaffected by the extraction system. Note that the extraction can be combined with, e.g., pressurization of a stairwell, which can be the 'adjacent' compartment. In that case, the static pressure p_a increases, i.e., moves to the right. This strengthens the effect of the smoke extraction system. Indeed, the effect of the smoke extraction is a drop in pressure inside the fire compartment, p_i. As a consequence, the position of the neutral plane (Section 5.2.3), z_N, moves to a higher position. In Figure 6.10, the situation becomes such that no smoke leaves the fire compartment through the opening. In other words, all the smoke is collected by the smoke extraction system, e.g., into the ductwork. Thus, while the fire compartment is not protected, the adjacent compartments are protected against the smoke. Obviously, this change in the situation leads to an increase in the mass flow rate of fresh air into the fire compartment. This illustrates a potential risk of the smoke extraction approach in case of originally under-ventilated fires. Indeed, the increase in the supply of fresh oxygen will then reduce the global equivalence ratio and can lead to more rapid fire growth, as explained in Chapter 5.

In a tunnel, smoke extraction is a type of transverse ventilation and the basic concept is such that the incoming air flows, induced by the smoke extraction, are supplied with sufficient

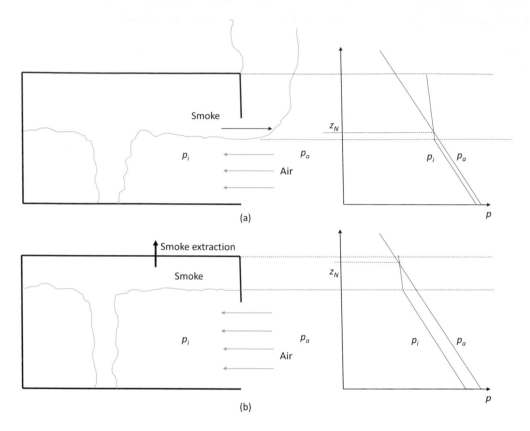

FIGURE 6.10   Illustration of the effect of a smoke extraction system. The static pressure inside the fire compartment drops so that the neutral plane position becomes higher and the flow into the compartment increases. The pressure in the adjacent compartment or ambient is assumed fixed. Top: no smoke extraction; bottom: smoke extraction system activated.

momentum (or velocity, see Section 6.7.2.1) from both sides, so smoke spreading is prevented [13]. In contrast to the longitudinal ventilation, discussed in Section 6.7.2.1, this strategy can maintain the smoke stratification and the smoke is not pushed into a certain direction (in which case a portion of the tunnel is not protected at all). If, however, the extraction capacity (and thus the cost involved) is reduced, smoke can spread to a much larger region, essentially as discussed in Section 6.7.2.1.

In any case, it is clear that through the strategy of smoke extraction, also heat is extracted, as the smoke removed is hot. As such, this is beneficial in that the fire spread is slowed down. This feature is not specific to tunnels. It also applies to, e.g., car parks or any type of compartment. As for all types of ventilation, it is important that fresh air is supplied to replace the extracted smoke. If this is not foreseen, the pressure in the fire compartment will drop and the smoke extraction will turn ineffective due to excessive pressure differences to be overcome by the extraction fan, as explained in Section 5.2.4 (Figure 5.10).

It is common practice that multiple ducts, with multiple openings each, are connected to a limited number of large fans. Smoke dampers in the network of ducts prevent smoke from flowing into certain parts of the ductwork. However, an important aspect is then the pressure losses in the ducts. Correlations are available to estimate the pressure losses. This is considered beyond the scope of this book.

6.9 POSITIVE PRESSURE VENTILATION

Positive Pressure Ventilation (PPV) refers to mobile systems, intended to create an over-pressure. Figure 6.11 sketches the working principle, which is easy to understand: the PPV fans are positioned in front of the door of a building and generate a cone-shaped air stream, into which air from the surroundings is entrained (see Section 2.11). This creates an over-pressure outside the door opening. Once an exhaust opening is created, e.g., by breaking or opening one or more windows, a flow path is created for the air and smoke through one or more pathways inside the building at hand. This technique can be used for smoke clearance after the extinguishment of the fire, but it is also possible to create ventilation during the fire service intervention, during or even before fire extinction. It is then also called 'Positive Pressure Attack' (PPA). For obvious reasons, it is important that no persons are present along the pathway for the smoke through the building.

When used during fire service interventions, PPV systems can make (or keep) a staircase smoke-free and can remove smoke from the fire rooms. As such, PPV can reduce temperatures and toxic gas levels and can improve visibility for firefighters. Obviously, this is not necessarily true if the fire is under-ventilated at the moment of activation of the PPV system. Indeed, in such circumstances, when the GER is initially too high for the fire to develop rapidly, the supply of fresh oxygen with the air by the PPV system can lead to very rapid fire growth. If the fire is fuel-controlled at the moment of activation of the PPV system, the primary effect is the removal of smoke and heat, which is beneficial for the firefighting operation and which slows down the development of the fire. Another issue is that care must be taken that the PPV system does not blow hot gases (smoke or flames) back through the upper part of the ventilation opening, through which the firefighters plan to enter. This is particularly dangerous for a fire at floor level. In general, due to the complexity, it is important to be very cautious when applying the technique.

From the experimental study of Ref. [23], it is concluded that effective use of PPV is only applicable to a limited number of storeys, due to friction losses for the flow, due to which the generated over-pressure decreases with distance from the inlet opening.

In Ref. [24], a full-scale experimental study is presented on the dependence of the effectiveness of PPV with a single and multiple fans on the positioning of the fans, in particular the distance from the door opening and the angle of the fan axis in relation to the ventilation opening and the floor. An interesting observation in Ref. [24] was that the PPV with a single fan became more

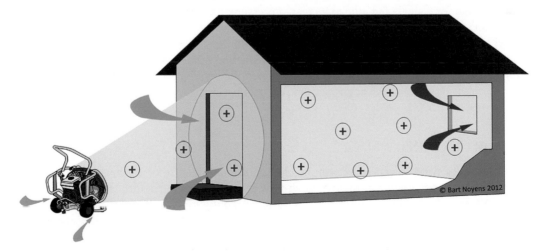

FIGURE 6.11 Sketch of the principle of positive pressure ventilation. (Adapted from Ref. [22]; thanks to Bart Noyens)

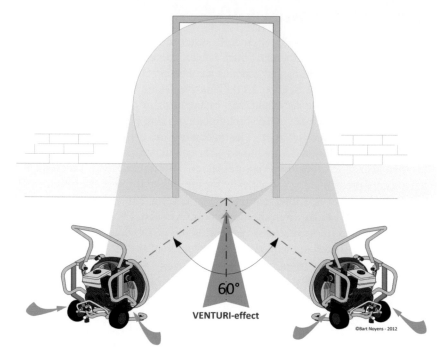

FIGURE 6.12 Positioning of multiple fans in V-shape. An angle between the fan axes of 60° is reported as optimum in Ref. [22]. (Adapted from Ref. [22]; thanks to Bart Noyens)

effective as the fan was positioned closer to the door opening (which served as a ventilation opening). Considering firefighting operational issues as well, a distance of 1.6 m from the door opening is suggested in Ref. [24] when a single fan is applied.

The use of multiple fans is also discussed in Ref. [24]. With two fans, a V-shaped set-up with an inner angle of 60° between the fan axes, as sketched in Figure 6.12, is found to be more effective than a set-up in series or in parallel. The reduction in PPV effectiveness on higher floors is more substantial than for a single fan, due to increased pressure (and mass) losses per floor level, but if multiple fans are at hand, it is still better to use them, because all mass flow rates measured were reported to be higher than with a single fan in Ref. [24].

Another interesting observation is the dramatic reduction in PPV effectiveness if the entrainment of air into the fan and its generated cone-shaped flow is hindered. In such circumstances, even lower mass flow rates have been measured than with a single fan that is perpendicular to the door opening.

6.10   AIR CURTAINS

The dispersion of fire-induced smoke can be blocked through the use of air curtains. The basic concept concerns aerodynamic sealing: high-velocity air is provided through narrow slots. Figure 6.13 provides a side view sketch for a tunnel configuration. The fire-induced smoke approaches the air curtain. In Figure 6.13, the air curtain is already activated and the smoke has just reached the air curtain.

The principle is that the strong vertical downward momentum of the air curtain is so strong that it can block the horizontal momentum of the oncoming fire-induced smoke. Whether or not

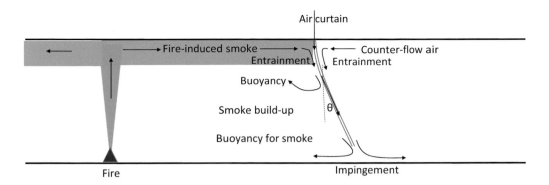

FIGURE 6.13 Sketch of the effect of an air curtain for a tunnel configuration. The air curtain flow is issued from a narrow slit, covering practically the entire tunnel width. The sketch shows a side view. Arrows indicate flows. Sketch of the situation before the smoke reaches the (already activated) air curtain.

this is possible will obviously depend on both momentums. Note that this concerns smoke blocking; radiation from the fire and the smoke is obviously not blocked by the air curtain, as air is transparent for radiation.

If the smoke is effectively blocked by the air curtain, the region upstream of the air curtain is not kept smoke-free. On the contrary, the hot smoke layer beneath the ceiling is disturbed by the flow induced by the air curtain, primarily due to entrainment into the air curtain flow. This entrainment also causes an airflow from the side which is kept smoke-free (the right side in Figure 6.13), thus acting as counter-flow air versus the smoke. This assists in smoke blocking. At the left-hand side, the smoke is pushed upward by the buoyancy, causing a locally thickened smoke layer. Part of the smoke, however, will also be dragged down by the air curtain and impinges onto the floor, to be pushed back towards the fire. This smoke is cooled by the cold air, but nevertheless remains at a higher temperature than the air, so it will also move upward due to buoyancy. It is clear that the region between the fire and the air curtain will be filled with smoke quite rapidly. However, the zone behind the air curtain can remain smoke-free.

If the air curtain momentum is insufficient to block the smoke, the evolution in the zone between the fire and the air curtain remains similar to what has just been described. However, part of the smoke will now also cross the air curtain. This smoke is to a large extent dragged down by the air curtain and is also cooled while flowing through the air curtain flow. As a consequence, there is no stratified situation, with a clear hot smoke layer beneath the ceiling, downstream of the air curtain. This is clearly not a desirable configuration.

It is noted that, inevitably, some deflection of the vertically downward air curtain flow will occur if the flow does not contain a horizontal momentum (velocity) component. The deflection angle, θ, i.e., the angle between the vertical direction and the actual flow direction, will depend on both the fire-induced smoke momentum and the air curtain flow momentum. If there is additional longitudinal ventilation (Section 6.7.2.1), this will obviously also result in a deflection of the air curtain.

More research is required to quantify the above.

As the final comment to this section, it is mentioned that in Ref. [25], a similar principle is described in terms of 'door vents', in the context of keeping a staircase smoke-free in case of a fire in an adjacent compartment. Indeed, high momentum air is blown into the compartment of the fire, thus keeping the fire-induced smoke inside. This approach could provide an alternative to a pressurization system, although further development and quantification are still required.

6.11 EXERCISES

1. Consider a tunnel with horizontal forced mechanical longitudinal ventilation (Section 6.4.2.1).
    a. Assuming that the ambient temperature is 15°C, determine the value of the heat release rate for which $\dot{Q}^* = 0.15$ if the tunnel height equals 6 m. (Answer: 14.6 MW)
    b. Repeat exercise a, but for a tunnel height equal to 12 m (i.e., twice as high). (Answer: 82.7 MW)
    c. Calculate the critical ventilation velocity from Eq. (6.18) for the tunnel of question a for a heat release rate equal to 10 MW. (Answer: 2.91 m/s)
    d. Repeat exercise c for a heat release rate equal to 20 MW. (Answer: 3.3 m/s)
    e. If the ventilation velocity is set to 2.5 m/s, calculate the smoke back-layering distance from Eq. (6.19) for exercise c. (Answer: 30.8 m)
    f. Repeat exercise e for the settings in exercise d. (Answer: 42.5 m)
2. Consider a 50-storey high-rise building. Each storey is 2.7 m high. The openings are distributed uniformly over the building.
    a. Before the onset of fire, the temperature inside the building is 21 °C and the ambient temperature is 10°C. Calculate the pressure difference between the inside and the outside of a window on the top floor of the building. (Answer: 30.2 Pa)
    b. Repeat exercise a, with ambient temperature equal to −10°C. (Answer: 91.7 Pa)
    c. What is the value of the wind velocity, yielding a pressure difference equal to the pressure difference by the stack effect as described in question a, if the wind coefficient equals 0.8? (Answer: 7.8 m/s)
    d. Repeat exercise c for the settings of exercise b. (Answer: 13.1 m/s)
3. Consider an office in which an SHC system is installed as sketched in the figure. The dimensions of the compartment are: $L = 8$ m, $W = 4$ m, $H = 4$ m. A steady-state heptane square fire source (2 m², 500 kW) is positioned in the centre, flush with the floor. The ambient temperature and pressure are $T_{amb} = 20$ °C and $p_{amb} = 1,01,325$ Pa, respectively.

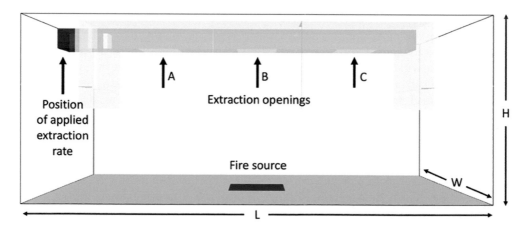

Calculate the following, in the context of a smoke extraction system:
    a. The required volumetric smoke extraction rate to be provided by the SHC system in order to prevent smoke from descending below a height of 2.5 m. (Answer: $\dot{V}_{extr} = 7.2$ m³/s)
    b. The required width and height of the duct if the maximum velocity in the duct is limited to 10 m/s. Consider a height-to-width ratio of 1:2 for the duct. (Answer: $W_{duct} = 1.2$ m, $H_{duct} = 0.6$ m)

c. Using a Moody diagram, the estimated pressure loss over a 10 m long straight duct, assuming fully developed flow and a smooth inner surface of the duct. (Answer: $\Delta p = 7.8$ Pa)

REFERENCES

1. J.H. Klote and J.A. Milke (2002) *Principles of Smoke Management*, ASHRAE.
2. J.H. Klotee (2002) "Smoke control", in M.J. Hurley et al. (Eds.), *SFPE Handbook of Fire Protection Engineering*, 3rd Edition, Chapter 4-12. Springer: New York.
3. H.P. Morgan, B.K. Ghosh, G. Garrad, R. Pamlitschka, J.-C. De Smedt and L.R. Schoonbaert (1999) *Design Methodologies for Smoke and Heat Exhaust Ventilation*. BR 368, BRE Publications: Garston, Hertfordshire.
4. CIBSE Guide E (1997) *Fire Engineering*. The Chartered Institution of Building Services Engineers: London, UK.
5. A. Stec and R. Hull (2010) *Fire Toxicity*. Woodhead Publishing: Sawston.
6. B. Blocken (2014) "50 years of Computational Wind Engineering: Past, present and future", *Journal of Wind Engineering and Industrial Aerodynamics,* Vol. 129, pp. 69–102.
7. CEN EN 12101-6 (2005) Smoke and heat control systems – Part 6: Specification for pressure differential systems.
8. CEN TR 12101-5 (2005) Smoke and heat control systems – Part 5: Guidelines on functional recommendations and calculation methods for smoke and heat exhaust ventilation systems.
9. NFPA 92B (2009) Smoke management systems in Malls, Atria and Large Spaces.
10. N. Tilley and B. Merci (2013) "Numerical study of smoke extraction for adhered spill plumes in atria: impact of extraction rate and geometrical parameters", *Fire Safety Journal*, Vol. 55, pp. 106–115.
11. H. Prétrel, W. Le Saux, L. Audouin (2012) "Pressure variations induced by a pool fire in a well-confined and force-ventilated compartment", *Fire Safety Journal*, Vol. 52, pp. 11–24.
12. A. Beard and R. Carvel (2005) *The Handbook of Tunnel Fire Safety*. Thomas Telford Publishing: London.
13. H. Ingason, Y.Z. Li and A. Lönnermark (2015) *Tunnel Fire Dynamics*. Springer: New York.
14. J.P. Kunsch (2002) "Simple model for control of fire gases in a ventilated tunnel", *Fire Safety Journal*, Vol. 37(1), pp. 67–81.
15. Y. Wu and M.Z.A. Bakar (2000) "Control of smoke flow in tunnel fires using longitudinal ventilation systems: A study of the critical velocity", *Fire Safety Journal*, Vol. 35 (4), pp. 363–390.
16. Y.Z. Li, B. Lei and H. Ingason (2010) "Critical velocity and backlayering distance in tunnel fires with longitudinal ventilation taking thermal properties of wall materials into consideration", *Fire Safety Journal*, Vol. 45, pp. 361–370.
17. W.D. Kennedy (1996) Critical velocity: Past, present and future, in: *Seminar of Smoke and Critical Velocity in Tunnels*. JFL Lowndes: London, pp. 305–322.
18. P.H. Thomas (1958) The movement of buoyant fluid against a stream and the venting of under-ground fires, Fire Research Note No. 351, Fire Research Station.
19. B. Merci and M. Shipp (2013) "Smoke and heat control for fires in large car parks: lessons learnt from research?" *Fire Safety Journal*, Vol. 57, pp. 3–10.
20. Y. Liu, Z. Fang, Z. Tang, T. Beji and B. Merci (2020) "Analysis of experimental data on the effect of fire source elevation on fire and smoke dynamics and the critical velocity in a tunnel with longitudinal ventilation", *Fire Safety Journal,* Vol. 114, 103002.
21. X. Deckers, S. Haga, N. Tilley and B. Merci (2013) "Smoke control in case of fire in a large car park: CFD simulations of full-scale configurations", *Fire Safety Journal*, Vol. 57, pp. 22–34.
22. X. Deckers, S. Haga, B. Sette and B. Merci (2013) "Smoke control in case of fire in a large car park: Full-scale experiments", *Fire Safety Journal*, Vol. 57, pp. 11–21.
23. S. Kerber, D. Madrzykowski and D. Stroup (2007) Evaluating positive pressure ventilation in large structures: high-rise pressure experiments. NISTIR 7412, Gaithersburg, MD, USA National Institute of Standards and Technology.
24. K. Lambert and B. Merci (2014) "Experimental study on the use of positive pressure ventilation for fire service interventions in buildings with staircases", *Fire Technology*, Vol. 50 (6), pp. 1517–1534.
25. A. Bittern (2015) Novel smoke control for tall buildings, PhD Thesis at The University of Edinburgh (UK).

7 Impact of Water on Fire and Smoke Dynamics

7.1 INDIVIDUAL WATER DROPLET

7.1.1 DROPLET DYNAMICS

A droplet exchanges momentum with the surrounding gas phase as it flows. The Lagrangian transport equation reads [1]:

$$m_d \frac{d\vec{u}_d}{dt} = \vec{F}_g + \vec{F}_D,$$ (7.1)

where m_d and $\vec{u}_d$ are the droplet mass and velocity vector, respectively; t is the time; and $\vec{F}_g$ and $\vec{F}_D$ are the gravitational and the drag force vectors, respectively. Other forces like the lift force and the Basset force can generally be ignored for the application of interest.

The gravitational force is expressed as follows:

$$\vec{F}_g = m_d\, \vec{g}.$$ (7.2)

The drag force is generally expressed as follows:

$$\vec{F}_D = -\frac{1}{2}\rho C_D A_{d,c}\left(\vec{u}_d - \vec{u}\right)\left|\vec{u}_d - \vec{u}\right|,$$ (7.3)

where ρ is the gas density, C_D is the drag coefficient, $A_{d,c}$ is the cross-sectional area of the droplet and u is the gas velocity.

In the remainder of the chapter, the droplet is assumed to be spherical and thus:

$$m_d = \rho_l \frac{\pi}{6} d_d^3$$ (7.4)

and

$$A_{d,c} = \frac{\pi}{4} d_d^2,$$ (7.5)

where ρ_l is the liquid water density and d_d is the droplet diameter.

Several correlations exist for the drag coefficient, C_D. The reader is referred to specialized literature. As an example, in McGrattan et al. [2], the following expression is mentioned:

$$C_D = \begin{cases} \dfrac{24}{\mathrm{Re}_d}, & \mathrm{Re}_d \leq 1 \\[2mm] 24\left(0.85\,\mathrm{Re}_d^{-1} + 0.15\,\mathrm{Re}_d^{-0.313}\right), & 1 < \mathrm{Re}_d \leq 1{,}000, \\[2mm] 0.44, & \mathrm{Re}_d > 1{,}000 \end{cases}$$ (7.6)

DOI: 10.1201/9781003204374-7

where Re_d is the droplet Reynolds number. The latter is defined as follows:

$$Re_d = \frac{\rho\, d_d \left| \vec{u}_d - \vec{u} \right|}{\mu}, \tag{7.7}$$

where the density, ρ, is typically taken as the ambient density. However, the dynamic viscosity is calculated using the '1/3' rule [1]:

$$\mu = \mu_{d,s} + \frac{1}{3}\left(\mu_{amb} - \mu_{d,s}\right), \tag{7.8}$$

where the subscripts 'd,s' and 'amb' refer to the droplet surface and the surrounding ambient environment, respectively. Equation (7.8), which is particularly important for an evaporating droplet (see Sections 7.1.2 and 7.1.3), requires the knowledge of the vapour mass fraction, Y, and the temperature, T, which are defined in a similar way as follows:

$$Y = Y_{d,s} + \frac{1}{3}\left(Y_{amb} - Y_{d,s}\right) \quad \text{and} \quad T = T_{d,s} + \frac{1}{3}\left(T_{amb} - T_{d,s}\right). \tag{7.9}$$

It is interesting to examine how quickly a droplet adjusts its velocity vector to the local gas flow field. This is carried out by estimating the relaxation time, which can be modelled by the Schiller–Neumann correlation [1]:

$$\tau_d = \frac{\rho_l\, d_d^2}{18\mu} f_1\left(Re_d\right) \tag{7.10}$$

with

$$f_1 = \begin{cases} \left(1 + 0.15\, Re_d^{0.687}\right)^{-1}, & Re_d \leq 1{,}000 \\[2mm] 54.55\, Re_d^{-1}, & Re_d > 1{,}000 \end{cases}. \tag{7.11}$$

For sufficiently high values of Re_d ($>1{,}000$):

$$\tau_d \simeq 3\frac{\rho_l}{\rho} \frac{d_d}{\left|\vec{u}_d - \vec{u}\right|}, Re_d > 1{,}000. \tag{7.12}$$

The relaxation time increases with the droplet diameter and mass density, and decreases as the gas phase is more dense and as the relative droplet velocity increases (i.e., as the gas phase exerts more momentum onto the droplet).

The ratio of the droplet relaxation time to a flow-related time scale (e.g., the turbulent integral time scale or the Kolmogorov scale) is called the Stokes number:

$$St = \frac{\tau_d}{\tau_{flow}}. \tag{7.13}$$

If the Stokes number is large (i.e., $St \gg 1$), the droplet behaves in a ballistic manner, i.e., it does not follow turbulent flow motions or streamlines and flies through vortices. If the Stokes number is small (i.e., $St \ll 1$), the droplet motion follows the gas phase very accurately. For example, in flow visualization, seeding particles are chosen such that the Stokes number is sufficiently low ($St < 0.1$).

It is instructive also to determine the equilibrium (or 'terminal') value for the vertical drop velocity. This corresponds to Eq. (7.1), with the left-hand side equal to zero (i.e., there is no more acceleration or deceleration):

$$\left|\vec{u}_d - \vec{u}\right|_{eq} = \sqrt{\frac{4}{3}\frac{\rho_l}{\rho}\frac{d_d}{C_D}g}. \tag{7.14}$$

Equation (7.14) reveals that for a given situation, the equilibrium value for the vertical drop velocity increases, proportional to the square root of the droplet diameter (if the drag coefficient C_D is independent of the Reynolds number).

The transport equation (7.1) can be solved numerically, for each of the three coordinate directions, to determine droplet trajectories for given initial and boundary conditions. Alternatively, in a more simplified approach, two equations for the tangential and vertical velocity components are solved; e.g., in Tang et al. [3], an extensive analysis of droplet trajectories under variable circumstances in still surroundings is carried out. In an even more simplified approach [4], considering constant drag and decoupled velocity component equations, analytical expressions are available for the droplet velocity and position as a function of time.

In order to illustrate some of the aspects discussed above, let us consider the example of a 0.5 mm – diameter droplet injected at two different elevation angles (angle with the vertical axis), $\phi = 0°$ and $\phi = 60°$, and with an initial velocity $u_{d,0} = 10$ m/s. The initial velocities of each component (radial and then vertical) are calculated as: $u_{d,r,0} = u_{d,0} \sin \phi$ and $u_{d,z,0} = u_{d,0} \cos \phi$. Therefore, for the droplet injected vertically, $u_{d,r,0} = 0$ m/s and $u_{d,z,0} = 10$ m/s, and for the droplet injected sidewards, $u_{d,r,0} = 8.66$ m/s and $u_{d,z,0} = 5$ m/s. The calculations are carried out with the Fire Dynamics Simulator (FDS) [2].

The trajectories of the two droplets are displayed in Figure 7.1a. The velocity magnitudes, i.e., $u_d = \sqrt{u_{d,r}^2 + u_{d,z}^2}$, displayed in Figure 7.1b show that an equilibrium (terminal) velocity value of about 4 m/s is reached when the droplet is injected sidewards. Obviously, the droplet injected vertically downwards reaches the distance of 5 m below the injection in a shorter time (i.e., about 0.90 s) in comparison to the droplet injected sidewards (i.e., 1.25 s) (see the 'end' of the curve in Figure 7.1b). Thus, less time is 'available' for the equilibrium velocity to be reached. Nevertheless, the asymptotic behaviour indicates a similar value of about 4 m/s.

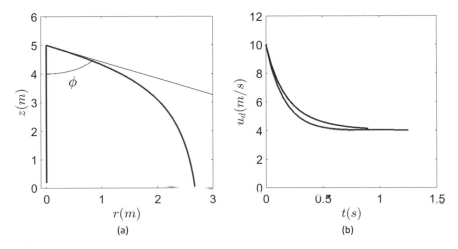

FIGURE 7.1 Trajectory (a) and velocity profile (b) of a 0.5 mm diameter droplet injected with an initial velocity of 10 m/s downwards (red curves) and with an elevation angle $\phi = 60°$ (blue curves).

7.1.2 MASS TRANSFER

One of the pioneering analyses of heat and mass transfer of isolated evaporating droplets is Spalding [5], and it is revisited in Chrigui et al. [6]. It is assumed that the droplet evaporates in an infinite environment, where the conditions at infinity are not affected by the presence of the droplet. Consider now a water droplet, evaporating in air, where the only flow is a radial expansion flow, characterized by a radial velocity u_r (see Figure 7.2). Air diffuses towards the water droplet, and water vapour diffuses into the environment. The droplet temperature T_d is not necessarily constant (i.e., T_d can vary with time). It is also not necessarily uniform. If it is assumed uniform, this corresponds to the assumption of infinite conductivity inside the droplet. This point is addressed in Section 7.1.3.1.

First consider conservation of total mass. Figure 7.2 provides the basic sketch of the configuration of an isolated evaporating droplet with (decreasing) radius r_d ($= d_d/2$), releasing vapour. As mentioned above, the vapour flows radially outwards with radial velocity u_r. The conservation of mass can be expressed, using the dashed circle in Figure 7.2 as control volume. Conservation of mass within the control volume with radius r over a time step Δt, during which the droplet radius decreases from r to r_d, reads:

$$\rho_l \frac{4}{3}\pi r_d^3 + \rho \frac{4}{3}\pi \left(r^3 - r_d^3\right) + \rho 4\pi r^2 u_r \Delta t = \rho_l \frac{4}{3}\pi r^3. \tag{7.15}$$

The difference between the first term on the left-hand side and the right-hand side reads:

$$\Delta m_d = \rho_l \frac{4}{3}\pi r_d^3 - \rho_l \frac{4}{3}\pi r^3. \tag{7.16}$$

Using the assumption that $\rho \ll \rho_l$, so that the second term on the left-hand side of Eq. (7.15) can be ignored, and taking the limit $\Delta t \to 0$, Eq. (7.15) becomes:

$$\frac{dm_d}{dt} = \dot{m}_d = -\rho 4\pi r^2 u_r. \tag{7.17}$$

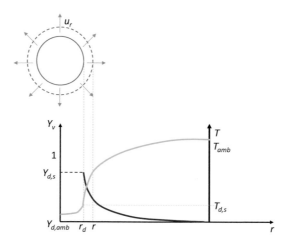

FIGURE 7.2 Sketch of an isolated evaporating droplet. r denotes the radial direction. The dashed circle denotes a control volume for expression (7.15). The green graph denotes temperature. The red line is mass fraction of droplet vapour (which is not defined inside the liquid phase, hence the dashed line). The subscript 's' refers to the droplet surface, while 'amb' refers to ambient condition. 'd' refers to 'droplet'.

Expressed in words, Eq. (7.17) states that the loss in liquid mass per unit time flows instanta-neously radially outwards from the droplet into the gas phase. Note that ρ is the local gas density, i.e., the density of the mixture of surrounding gas (air) and vapour (water vapour). Eq. (7.17) expresses the conservation of mass.

The mass fraction of vapour in the gas is denoted as Y_d. This non-dimensional quantity expresses the ratio of the local amount (in kg) of vapour, stemming from the droplet, to the total local amount (in kg) of gas. The subscript 'd' refers to the droplet, i.e., the droplet is supposed to contain only one single component. Multi-component evaporation is not discussed here. The reader is referred to specialized literature on that topic (e.g., Jenny et al. [1]; the references therein). In case of a water droplet evaporating in air, the composition of the environment (air) differs from the composition of the evaporating droplet (water). As a consequence, the water vapour mass fraction at the droplet surface, $Y_{d,s}$, is lower than 1 (see Figure 7.2). The value can be determined for the Clausius–Clapeyron equation, Eq. (2.138) (although the equation may not be very accurate for small droplets [1]). If the environment were of the same composition as the evaporating droplet (e.g., a water droplet evaporating in pure steam), the vapour mass fraction Y_d equals 1 everywhere. The situation $Y_{d,s} < 1$ is typical for fire situations. As a consequence of the gradient in mass fraction Y_d, there is, in addition to the convective flux (due to the radial expansion velocity u_r) of vapour away from the droplet surface, also a diffusion flux. At the same time, as mentioned, there is diffusion of environment vapour (e.g., air) towards the droplet surface.

From these ingredients, the transport equation for the vapour mass fraction reads:

$$-\dot{m}_d = 4\pi r^2 \rho Y_d u_r - 4\pi r^2 \rho D \frac{\partial Y_d}{\partial r}, \quad r \geq r_d. \tag{7.18}$$

The left-hand-side term in Eq. (7.18) is a source term at the droplet surface, releasing a mass of vapour per unit time. The first term on the right-hand side in Eq. (7.18) is a convective mass flux out of the control volume by the radially outward expansion velocity (recall indeed that ρ is the local gas density, i.e., the density of the mixture of surrounding gas (air) and vapour (water vapour), so that multiplication with Y_d is necessary to obtain an amount of kg vapour per second). The second term on the right-hand side in Eq. (7.18) is a diffusion flux out of the control volume (if $Y_{d,s} > Y_{d,\text{amb}}$, which is the typical situation for fire) where D denotes the mass diffusivity (in m^2/s) of the vapour into the gas.

Inserting Eq. (7.17) into Eq. (7.18) gives:

$$-\dot{m}_d Y_d - 4\pi r^2 \rho D \frac{\partial Y_d}{\partial r} = -\dot{m}_d. \tag{7.19}$$

This is a first-order differential equation which must simultaneously satisfy two boundary conditions:

$$Y_d = Y_{d,s} \text{ at } r = r_d \quad \text{and} \quad Y_d = Y_{d,\text{amb}} \text{ at } r = r_\infty \tag{7.20}$$

The integration of Eq. (7.19) gives:

$$Y_d = 1 - \left(1 - Y_{d,\text{amb}}\right) \exp\left[\left(\frac{r_\infty - r}{r \, r_\infty}\right)\left(\frac{\dot{m}_d}{4\pi \rho D}\right)\right]. \tag{7.21}$$

Equation (7.21) can be rearranged as follows:

$$\dot{m}_d = -4\pi\,\rho D\,r\left(\frac{r_\infty}{r_\infty - r}\right)\ln\left(\frac{1 - Y_{d,\text{amb}}}{1 - Y_d}\right). \tag{7.22}$$

Finally, by considering the boundary condition at $r = r_d = d_d/2$ and taking:

$$B_M = \frac{Y_{d,s} - Y_{d,\text{amb}}}{1 - Y_{d,s}} \tag{7.23}$$

and

$$\text{Sh} = \left(\frac{2r_\infty}{r_\infty - r_d}\right), \tag{7.24}$$

Equation (7.22) becomes:

$$\dot{m}_d = -\pi\,\rho D\,d_d\,\text{Sh}\,\ln(1 + B_M). \tag{7.25}$$

The variable B_M denotes the Spalding number, which expresses the rate of increase in the evaporation rate.

The variable Sh denotes the Sherwood number, which can be rewritten as follows [7]:

$$\text{Sh} = 2 + \frac{d_d}{\delta_M}, \tag{7.26}$$

where $\delta_M = r_\infty - r_d$ is referred to as the 'film' thickness.

The most classical expression for the Sh number is the expression of Ranz and Marshall [8] in forced convective environment without taking into account Stefan's flux.

$$\text{Sh} = \text{Sh}_0 = 2 + 0.6\,\text{Re}_d^{1/2}\,\text{Sc}^{1/3}, \tag{7.27}$$

where Sc is the Schmidt number.

The 'film' thickness in this case is expressed as follows:

$$\delta_{M,0} = \frac{d_d}{\text{Sh}_0 - 2}, \tag{7.28}$$

Note that if the droplet is not in motion, i.e., $\text{Re}_d = 0$, then $\text{Sh} = 2$ (infinite thickness of the film). This is called a purely diffusive transport and the evaporation rate is expressed as follows:

$$\dot{m}_d = -4\pi\,r_d\,\rho D\ln(1 + B_M). \tag{7.29}$$

Furthermore, if $Y_{d,s}$ is very low, Eq. (7.29) can be further simplified to:

$$\dot{m}_d = -4\pi\,r_d\,\rho D\left(Y_{d,s} - Y_{d,\text{amb}}\right). \tag{7.30}$$

In fact, expression (7.27) is valid for low B_M. For high B_M, and taking into account Stefan's flux, the Sh is typically expressed as follows [7]:

$$Sh = 2 + \frac{Sh_0 - 2}{F_M}, \tag{7.31}$$

where F_M is a correction factor (for which correlations are available in, e.g., Sirignano [7]) which is always higher than 1 and accounts for the increase in the film thickness due to the blowing effect (high B_M):

$$F_M = \frac{\delta_M}{\delta_{M,0}}. \tag{7.32}$$

A key quantity in the calculation of the droplet evaporation rate, $\dot{m}_d$, discussed in this section, is the mass fraction of vapour at the droplet surface, $Y_{d,s}$, which is linked (through the Clausius–Clapeyron equation) to the droplet surface temperature, $T_{d,s}$. The latter is calculated using the heat transfer analysis discussed in the following section.

7.1.3 HEAT TRANSFER

7.1.3.1 Conductivity within the Droplet

Heat transfer within the droplet occurs primarily by conduction. In order to estimate whether the droplet can be assumed to be thermally thin, it is important to check the following criterion based on the Biot number:

$$Bi = \frac{h \, d_d}{2 \, k_l} < 0.1, \tag{7.33}$$

where k_l is the liquid thermal conductivity and h is the convective heat transfer coefficient. Note that the radius of the droplet ($d_d/2$) is taken as characteristic length scale here. If the droplet is thermally thin, it is then characterized by a single temperature, i.e., $T_{d,s} = T_d$, which greatly simplifies the heat transfer analysis. This is referred to in the literature as the infinite conductivity concept and is very often used as a 'standard' approach in simulations of water-based fire suppression as well as spray combustion applications.

A more advanced approach consists of solving Fourier's equation for conduction in order to obtain the temperature profile within the droplet (i.e., temperature as a function of the distance from the surface). However, this approach does not consider convective currents that may take place within the droplet due to friction at the liquid surface. This can be accounted for indirectly though by using the 'effective conductivity concept' where a higher thermal conductivity is used in order to account for improved heat transfer due to convection.

The infinite conductivity concept is used in the remainder of this chapter.

7.1.3.2 Convective Heat Transfer

The general expression for convective heat transfer is:

$$\dot{q}_{conv} = h \, A_d \left(T_g - T_d \right), \tag{7.34}$$

where A_d (m^2) is the droplet surface area (available for convective heat exchange) and T_g is the 'local' gas temperature (in K or °C) with which the droplet or film exchanges heat. The convection coefficient h (W/(m^2 K)) is defined as follows:

$$h = \frac{k \, Nu}{d_d}, \tag{7.35}$$

where k (W/(m K)) is the local conduction coefficient of the gas and Nu is the Nusselt number (which will be discussed in Section 7.1.3.4).

7.1.3.3 Radiative Heat Transfer

The rate of radiative heating of a droplet is expressed as follows [2]:

$$\dot{q}_{\text{rad}} = A_d\,\varepsilon\left(\frac{U}{4} - \sigma\,T_d^4\right),\tag{7.36}$$

where ε is the emissivity, σ is the Stefan–Boltzmann constant and U is the integrated radiant intensity.

7.1.3.4 Energy Balance

Similarly to Eq. (7.19) for mass transfer, and simplifying to the case of steady state for the droplet surface temperature, the energy balance for heat transfer (without considering the source terms for convection and radiation) reads:

$$4\pi r^2 k\,\frac{\partial T}{\partial r} + \dot{m}_d\left[c_{pv}\left(T - T_{d,s}\right) + L_v\right] = 0.\tag{7.37}$$

This is a first-order differential equation which must simultaneously satisfy two boundary conditions:

$$T = T_{d,s}\ \text{ at }\ r = r_d\ \text{ and }\ T = T_{\text{amb}}\ \text{ at }\ r = r_\infty.\tag{7.38}$$

The integration of Eq. (7.37), using the first boundary condition in Eq. (7.38), gives:

$$T = T_{d,s} + \frac{L_v}{c_{pv}}\left(-1 + \exp\left[\frac{\dot{m}_d\,c_{pv}}{2\pi\,k\,r_d}\right]\left[\frac{(r - r_d)}{2r}\right]\right).\tag{7.39}$$

Equation (7.39) can be rearranged as follows:

$$\dot{m}_d = -4\pi\,\frac{k}{c_{pv}}\,r_d\left(\frac{r}{r - r_d}\right)\ln\left(\frac{c_{pv}\left(T - T_{d,s}\right) + L_v}{L_v}\right).\tag{7.40}$$

Finally, by considering the boundary condition at $r = r_\infty$ and taking:

$$B_T = \frac{c_{pv}\left(T_{\text{amb}} - T_{d,s}\right)}{L_v}\tag{7.41}$$

and

$$\text{Nu} = \left(\frac{2r_\infty}{r_\infty - r_d}\right).\tag{7.42}$$

Equation (7.40) becomes:

$$\dot{m}_d = -\pi\,\frac{k}{c_{pv}}\,d_d\,\text{Nu}\,\ln\left(1 + B_T\right).\tag{7.43}$$

Similarly to B_M, the variable B_T denotes the Spalding number, which expresses the rate of increase of the evaporation rate. Similarly to Sh (see Eq. (7.27)), the Nusselt number can be estimated as follows:

$$Nu = 2 + 0.6 \, Re_d^{1/2} \, Pr^{1/3}, \tag{7.44}$$

where Pr is the local Prandtl number in the gas phase.

The analogy can be further extended to the 'film' thickness δ_H and the correction factor F_H, which are the counterparts of δ_M and F_M for mass transfer [7].

It is of interest to note that:

$$\dot{m}_d = \frac{\pi}{6} \rho_l \frac{d\left(d_d^3\right)}{dt}. \tag{7.45}$$

Using the chain rule, the derivative in Eq. (7.45) can be expanded as follows:

$$\frac{d\left(d_d^3\right)}{dt} = \frac{d\left(d_d^3\right)}{d\left(d_d^2\right)} \frac{d\left(d_d^2\right)}{dt}. \tag{7.46}$$

The last term on the right-hand side of Eq. (7.46) can be further developed as follows:

$$\frac{d\left(d_d^3\right)}{d\left(d_d^2\right)} = \frac{d\left[\left(d_d^2\right)\right]^{3/2}}{d\left(d_d^2\right)} = \frac{3}{2}\left(d_d^2\right)^{1/2} = \frac{3}{2}d_d. \tag{7.47}$$

Combining Eqs. (7.45), (7.46) and (7.47) gives:

$$\dot{m}_d = \frac{\pi}{4} \rho_l \, d_d \frac{d\left(d_d^2\right)}{dt}. \tag{7.48}$$

Inserting Eq. (7.48) in Eq. (7.43) gives:

$$\frac{d\left(d_d^2\right)}{dt} = -\frac{4k}{\rho_l c_{pv}} \, Nu \, \ln\left[1 + \frac{c_{p,v}\left(T_{amb} - T_{d,s}\right)}{L_v}\right]. \tag{7.49}$$

In case the droplet reaches a saturation (wet-bulb) temperature, $T_{d,sat}$, and $Nu = 2$, Eq. (7.49) becomes:

$$\frac{d\left(d_d^2\right)}{dt} = -K, \tag{7.50}$$

where the parameter, K, can be considered as constant (it is called the evaporation constant) and is expressed as follows [4]:

$$K = \frac{8k}{\rho_l c_{pv}} \, \ln\left[1 + \frac{c_{p,v}\left(T_{amb} - T_{d,sat}\right)}{L_v}\right]. \tag{7.51}$$

The integration of Eq. (7.50) gives:

$$d_d^2(t) = d_{d,0}^2 - K t. \tag{7.52}$$

where $d_{d,0}$ is the initial droplet diameter prior to the evaporation process at a constant saturation temperature. Equation (7.52) is known as the 'd^2 – law'.

So far we considered a constant droplet temperature. In the following, we will consider the variation of T_d with time. The energy equation for the droplet reads:

$$\frac{d\left[m_d\, h_l(T)\right]}{dt} = -4\pi\, r^2 \rho\, u_r\, h_v(T) + \dot{q}, \tag{7.53}$$

where $h_l(T)$ is the sensible enthalpy per unit mass of the liquid and $h_v(T)$ is the sensible enthalpy per unit mass of the vapour. The first term on the right-hand side represents the flux of energy per unit time flowing out of the control volume (in Figure 7.2) by the radially outward expansion velocity. The parameter $\dot{q}$ is a heat source term from the gas to the droplet, which can be expressed as follows:

$$\dot{q} = \dot{q}_{conv} + \dot{q}_{rad}. \tag{7.54}$$

The sensible enthalpy per unit mass of the liquid is expressed as follows:

$$h_l(T) = \int_{T_{ref}}^{T} c_{p,l}(T')dT', \quad \text{if} \quad T \leq T_{evap}. \tag{7.55}$$

where T_{ref} is a reference temperature and T_{evap} is an evaporation temperature.

The sensible enthalpy per unit mass of the vapour is expressed as follows:

$$h_v(T) = \left(\int_{T_{ref}}^{T} c_{p,l}(T')dT' \right) + L_v + \left(\int_{T_{evap}}^{T} c_{p,v}(T')dT' \right), \quad \text{if } T \geq T_{evap}, \tag{7.56}$$

where T_{ref} is a reference temperature and T_{evap} is an evaporation temperature.

Assuming that the vapour leaves the control volume at $T = T_{evap}$, Eq. (7.56) becomes:

$$h_v(T) = \left(\int_{T_{ref}}^{T_{evap}} c_{p,l}(T)dT \right) + L_v = h_l(T) + L_v. \tag{7.57}$$

Inserting Eqs. (7.57) and (7.17) into Eq. (7.53) gives:

$$\frac{d\left[m_d\, h_l(T)\right]}{dt} = \dot{m}_d\, h_l + \dot{m}_d\, L_v + \dot{q}. \tag{7.58}$$

By assuming that the liquid-specific heat does not vary significantly with time and that $T_{evap} = T_d$, we have by definition:

$$\frac{d\left[m_d\, h_l(T)\right]}{dt} = m_d\, c_{p,l}\, \frac{dT_d}{dt} + \dot{m}_d\, h_l. \tag{7.59}$$

Equating (7.58) and (7.59) gives:

$$m_d\, c_{p,l}\, \frac{dT_d}{dt} = \dot{m}_d\, L_v + \dot{q}. \tag{7.60}$$

The temperature equation for the gas can be expressed as follows:

$$m_g\, c_g\, \frac{dT_g}{dt} = -\dot{m}_d\left(h_l - h_v\right) = \dot{m}_d\, L_v, \tag{7.61}$$

where m_g and c_g are the mass and the specific heat of the gas, respectively.

In order to illustrate some of the aspects discussed above for heat and mass transfer of a single droplet, let us consider the example of a static (i.e., not moving) 0.5 mm diameter droplet in a hot environment with an ambient temperature $T_{\mathrm{amb}} = 200°C$. The calculations are carried out with the Fire Dynamics Simulator (FDS) [2].

Figure 7.3a illustrates the 'd^2 – law' by showing the linear dependence between $(d_d/d_{d,0})^2$ and the time t. The slope corresponds to $K/d_{d,0}^2$. It is important to note that evaporation does not occur instantaneously but after a certain period of time during which the droplet diameter remains unchanged. Looking at Figure 7.3b, this period of time corresponds approximatively to a heat-up time during which the droplet (surface) temperature increases until reaching 'saturation'.

An in-depth analysis of the theory of an individual evaporating spherical droplet is provided in Sirignano [7], but also in combustion literature (e.g., Jenny et al. [1]), very useful information can be found.

7.1.4 EXERCISES

1. Calculate the terminal value of the vertical drop velocity of a spherical water droplet with diameter equal to 0.5 mm in air at 300 K at atmospheric pressure. The water droplet temperature is assumed constant, equal to 300 K. (Answer: 2.09 m/s)
2. Repeat exercise 1 for droplet diameters equal to 0.05, 0.1 and 1 mm. Explain the observations. (Answer: 0.074 m/s; 0.28 m/s; 3.93 m/s; the terminal velocity increases with diameter because the mass of the droplet increases with the third power of the radius, while

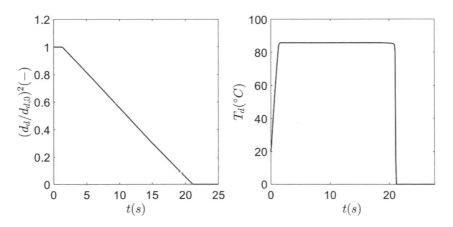

FIGURE 7.3 Diameter (a) and temperature (b) time evolution profile of a static 0.5 mm diameter droplet in a hot environment with an ambient temperature $T_{\mathrm{amb}} = 200°C$.

the drag force essentially increases with the square of the droplet diameter; the terminal velocity is not really linearly proportional to the diameter because C_D is not exactly inversely proportional to the droplet diameter (Eq. (7.6), with $1 < \mathrm{Re}_d \leq 1,000$).)

3. Repeat exercise 1 for air temperatures equal to 400, 500 and 600 K (which are representative values for smoke temperatures). The water droplet temperature is assumed constant, equal to 300 K (i.e., heat transfer is ignored). Explain the observations. (Answer: 2.25 m/s; 2.36 m/s; 2.46 m/s; the terminal velocity increases primarily due to the decrease in ρ in Eq. (7.14).)

7.2 SPRAYS OF WATER DROPLETS

In the context of fire, water is generally supplied in the form of sprays, i.e., a collection of many droplets. In this section, spray features are described in terms of shapes and droplet size distribution, the latter being strongly influenced by the nozzle type and operating conditions. Furthermore, the droplet size determines the dominant mechanisms in the interaction of water sprays with fire and smoke.

7.2.1 CHARACTERIZATION OF THE INJECTION

For obvious reasons, the water flow rate is an important characteristic of the spray. The water flow rate through an orifice is proportional to the square root of the pressure difference over the orifice (see Section 2.4). The area and discharge coefficient also appear in expression (2.79). It is common practice to quantify the flow rate as follows:

$$\dot{V}_w = K\sqrt{\Delta p}. \tag{7.62}$$

Typical units are: $\dot{V}_w$ expressed in lpm (litre per minute), Δp in bar and thus K in lpm/bar$^{0.5}$. Special care must be taken if other units are used to define K, such as gpm/psi$^{0.5}$ (gpm: gallons per minute).

In addition to the water flow rate, the water jet velocity through the nozzle can be calculated as follows:

$$u_j = C\sqrt{\frac{2\Delta p}{\rho_l}}, \tag{7.63}$$

where the constant C is a factor that accounts for friction losses in the nozzle.

Alternatively, the water jet velocity can be calculated as follows:

$$u_j = \frac{\dot{V}_w}{\pi(d_n^2/2)}, \tag{7.64}$$

where d_n is the nozzle diameter.

Equation (7.63) or (7.64) can be used to estimate the initial droplet velocity when solving Eq. (7.1) for the droplet motion.

7.2.2 SPRAY ANGLES AND SHAPES

Sprays typically cover the entire azimuthal angle range of $0 \leq \theta \leq 2\pi$ (exceptions are water curtains, Sections 7.3.1.3 and 7.3.2.4). Whereas there can be quite strong dependence of the droplet distribution and their radial velocity on θ, e.g., due to the difference between tines and slots or spaces, this is not always taken into account in numerical simulations.

However, sprays do not always cover the entire range of possible 'elevation' angles $0 \leq \phi \leq \pi$, where $\phi = 0$ refers to the vertically downward angle, $\phi = \pi/2$ refers to the horizontal direction and $\phi = \pi$ refers to the vertically upward direction. Figure 7.4 depicts a possible 'hollow-cone' spray, where $\phi_{min} \leq \phi \leq \phi_{max}$.

7.2.3 ATOMIZATION AND BREAK-UP

In Section 7.1, individual droplets were discussed. However, water supply does not occur as a collection of individual droplets. Water is supplied as a continuous flow through a piping system. The water flow then reaches the nozzle and undergoes atomization. This process strongly depends on the operating conditions, but obviously also on the nozzle type.

It is interesting to analyse the effect of the operating conditions by examining the Weber number defined as follows:

$$\text{We} = \frac{\rho_l u_j^2 d_n}{\sigma}, \tag{7.65}$$

where σ is the surface tension of water in air ($\sigma = 0.0728\,\text{N/m}$ at 20°C). Indeed, the Weber number is the ratio of flow-induced force to the surface tension. The latter refers to the tendency for liquids to try and acquire the least surface area possible and tries to keep the fluid together. As such, it is a stabilizing factor, while the flow-induced force tries to destabilize the fluid. Thus, the higher We, the more unstable the water jet, which leads to a more pronounced break-up. Additionally, the higher We, the shorter the break-up distance, i.e., distance from the nozzle at which droplets are formed (and the so-called atomization process is completed). This is shown in Zhou and Yu [9]:

$$\frac{2R_{bu}}{d_n} = 370\,\text{We}^{-1/4}, \tag{7.66}$$

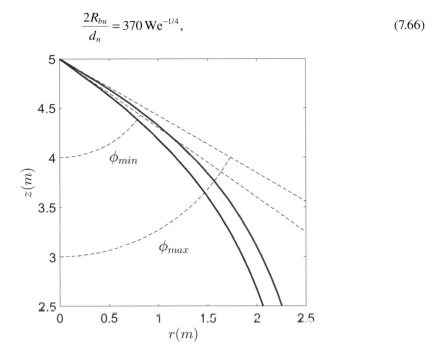

FIGURE 7.4 Sketch of a hollow-cone spray where the elevation angle is limited in the range of $\phi_{min} \leq \phi \leq \phi_{max}$. The sketch is based on the trajectory of a droplet shown in Figure 7.1 with $\phi_{min} = 55°$ and $\phi_{min} = 60°$.

where R_{bu} is the radial (break-up) distance from the nozzle.

Note that in addition to the water jet at the injection, droplets may also undergo break-up (secondary atomization), which is described using the 'droplet' Weber number defined as follows:

$$\mathrm{We}_d = \frac{\rho_l |\vec{u}_d - \vec{u}|^2 d_d}{\sigma}, \tag{7.67}$$

In other words, the 'relative droplet velocity' (see Section 7.1.1) is used to characterize the flow-induced forces and the droplet diameter is used as characteristic length scale. This number can also be used to determine to what extent the droplet shape deviates from a sphere. For $\mathrm{We}_d < 5$, the deformation from a spherical shape is negligible [7]. In the range of $10 < \mathrm{We}_d < 20$, there is continuous deformation [7] and bag break-up can occur in the range of $11 < \mathrm{We}_d < 80$ [1]. Still higher values ($80 < \mathrm{We}_d < 350$) lead to 'stripping break-up' and still higher values ($\mathrm{We}_d > 850$) to 'catastrophic break-up' [1]. These phenomena refer to 'secondary atomization' or 'secondary break-up' [1,10].

A distinction can already be made at this level (i.e., operating conditions) between sprinklers and water mist systems, as shown in Table 7.1.

By considering a nozzle diameter in the order of 10 mm for sprinklers and 1 mm for water mist and a water jet velocity in the order of 10 m/s for sprinklers and 100 m/s for water mist, Eq. (7.65) gives We ~ 10,000 for the former and We ~ 1,00,000 for the latter. As a consequence, more pronounced hydrodynamic instabilities occur near the injection of a water mist system, leading to much smaller droplets (in comparison to 'conventional' sprinkler systems). In fact, an 'average' droplet size can be estimated from Refs. [11-14]:

$$d_{v,50} = C d_n \mathrm{We}^{-1/3}, \tag{7.68}$$

where $d_{v,50}$ is the volume-median diameter (that will be explained in Section 7.2.5) and C is a constant that depends on the sprinkler (values range from 1.74 to 3.21 [12]).

Using Eqs. (7.63), (7.65) and (7.68) gives the following dependence of the $d_{v,50}$ on the pressure:

$$d_{v,50} \sim \left(\Delta p\right)^{-1/3}. \tag{7.69}$$

TABLE 7.1
Examples of Different Operating Conditions between Sprinkler and Water Mist Systems

	Sprinkler	Mist
d_n (mm)	13[*]	0.49[a]
		1.14[a]
$\dot{V}_w$ (l/s)	1.8–7.6[b]	-
ΔP (bar)	up to 50	<2
K (lpm/bar$^{1/2}$)	80–201.5[c]	0.12[a]
		0.42[a]
u_j (m/s)	-	85[a]

[a] Santangelo [10].

[b] Sheppard [11].

[c] Ren et al. [12]: values for an ESFR ('early suppression fast response') sprinkler.

Furthermore, the droplet Weber number, We_d, is much higher in water mist systems, which causes (as mentioned above) secondary atomization (or break-up) leading to even smaller droplets. Droplet collision and coalescence may occur in a water mist system, but apparently the secondary atomization overtakes these processes very close to the nozzle.

As mentioned above, the atomization process depends not only on the operating conditions, but obviously also on the nozzle type. For example, a (pendant) sprinkler nozzle typically consists of an orifice, from which the water emerges as a jet onto a deflector. This deflector contains tines and slots in between the tines. The water flow is thus redirected from the vertical axially downward direction to a radially outward direction and forms thin sheets on the tines and through the slots. These sheets subsequently break up to form drops with different sizes and velocities. This is described in, e.g., Zhou and Yu [9]. In Ren et al. [12], the impact of the tines and the slots (called 'spaces' in Ren et al. [12]) in the near-field region is shown to be significant. In Santangelo [10], a water mist system with pressure-swirl atomizer is analysed. A hollow-cone spray is generated by the nozzles. There is a central stream of high momentum, where the particle motion is essentially driven by the conversion of the static pressure in the nozzle into dynamic load in the injector. Air is entrained. Around the spray cone, droplets are 'floating' in the surrounding air and these represent the actual mist: floating and evaporating, rather than moving downwards.

At the end of the atomization (break-up stage), the droplets will follow a statistical distribution as a function of the elevation angle, ϕ ($\phi_{min} < \phi < \phi_{max}$), the azimuthal angle, θ ($0 < \theta < 2\pi$), and the droplet size, d_d. Assuming statistical independence between these three parameters, the probability density function for the water volume flux reads [15]:

$$f_v\left(\theta, \phi, d_d\right) = f_v\left(\theta\right) f_v\left(\phi\right) f_v\left(d_d\right), \tag{7.70}$$

where the first two functions in the right-hand side refer to the water volume flux angular distribution (see Section 7.2.4) and the third function refers to the droplet size distribution (see Section 7.2.5). In addition to a distribution in droplet size, a distribution in velocities can also be considered.

7.2.4 ANGULAR DISTRIBUTION OF WATER

Ignoring possible differences between tines and spaces, although they are clearly observed (e.g., Ren et al. [12]), a uniform random distribution can be used for the azimuthal range ($0 < \theta < 2\pi$), as is done in McGrattan et al. [2]:

$$f_v\left(\theta\right) = \frac{1}{2\pi}. \tag{7.71}$$

The volume flux angular probability distribution with respect to the elevation angle can be expressed as follows [2]:

$$f_v\left(\phi\right) = \exp\left[-\beta\left(\frac{\phi - \mu}{\phi_{max} - \phi_{min}}\right)^2\right]. \tag{7.72}$$

The spread parameter, β, indicates how narrow the water distribution is around the peak value; the higher β, the narrower the distribution. The default value in McGrattan et al. [2] is $\beta = 5$. A value of $\beta = 0$ corresponds to a uniform distribution, i.e., $f_v\left(\phi\right) = 1$. The parameter μ 'controls the location of the peak and may be used to approximate a hollow cone nozzle' [2]. Figure 7.5 is an illustration of Eq. (7.72) for $\phi_{min} = 0$ and $\phi_{max} = \pi/2$ using several values of β and μ.

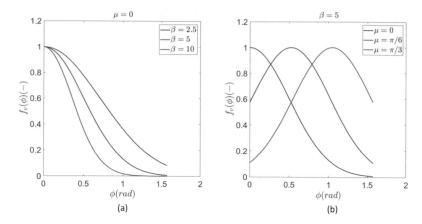

FIGURE 7.5   Illustration of Eq. (7.72) for $\phi_{\min} = 0$ and $\phi_{\max} = \pi/2$. (a) Impact of β and (b) impact of μ.

7.2.5 DROPLET SIZE DISTRIBUTION

The droplet size distribution of the water spray is expressed in terms of its Cumulative Volume Fraction (CVF), which is often assumed to fit the Rosin–Rammler (RR) distribution [16]:

$$F_v\left(d_d\right) = 1 - \exp\left[-\ln(2)\left(\frac{d_d}{d_{v,50}}\right)^{\gamma}\right],\tag{7.73}$$

where $d_{v,50}$ is the volume-median diameter (VMD) (defined such that half of the volume of the water is contained in droplets with diameter below $d_{v,50}$ and thus half of the water is contained in droplets with diameter larger than $d_{v,50}$) and γ is a distribution width parameter.

The probability density function (PDF) of the volume fraction (i.e., volume fraction distribution (VFD)) is calculated as the first derivative of the function $F_v\left(d_d\right)$:

$$f_v(d_d) = \frac{dF_v\left(d_d\right)}{d\,d_d} = \frac{\ln(2)\gamma\,d_d^{\gamma-1}}{d_{v,50}^{\gamma}}\exp\left[-\ln(2)\left(\frac{d_d}{d_{v,50}}\right)^{\gamma}\right].\tag{7.74}$$

The corresponding PDF for the droplet for the droplet diameter, i.e., number fraction distribution (NFD), is defined as follows:

$$f_N\left(d_d\right) = \frac{\left(\dfrac{f_v\left(d_d\right)}{d_d^3}\right)}{\displaystyle\int_0^{\infty}\left(\dfrac{f_v\left(\delta\right)}{\delta^3}\right)d\delta}.\tag{7.75}$$

The cumulative number fraction (CNF) is expressed as follows:

$$F_N\left(d_d\right) = \int_0^{d_d} f_N\left(\delta\right)d\delta.\tag{7.76}$$

The sequence of Eqs. (7.73) to (7.76) shows how the CVF, VFD, NFD and CNF are connected to each other.

Figure 7.6 shows an example of the outcome of Eqs. (7.73) and (7.75) for $d_{v,50} = 60\,\mu m$ and two values for the distribution width parameter, $\gamma = 2$ and $\gamma = 4$, using the Rosin–Rammler distribution. More particularly, Figure 7.6b shows that for $\gamma = 2$ (Rosin–Rammler), the curve approaches the ordinate axis asymptotically as the droplet size decreases. This is explained mathematically by combining Eqs. (7.74) and (7.75), which yields $f_N(d_d) \sim d_d^{\gamma-4}$ and means that unimodal number density curves only exist when $\gamma > 4$ [17].

Therefore, when a Rosin–Rammler distribution is prescribed with a width parameter $\gamma < 4$, upper and lower limits for the droplet size should be prescribed ($d_{d,\,max}$ and $d_{d,\,min}$), the $d_{d,\,min}$ being more critical in the NFD curve. The divergence of the integrals at $d_d \to 0$ limit is a known drawback of the Rosin–Rammler distribution. A way to by-pass the problem induced by the lower limit of integration in Eq. (7.75) is to prescribe a skew distribution such as the lognormal function for the small diameters [17]. Hence, a combination of the lognormal and Rosin–Rammler (LN-RR) distributions is used as follows:

$$
F_v(d_d) = \begin{cases} \dfrac{1}{\sigma\sqrt{2\pi}} \displaystyle\int_0^{d_d} \dfrac{1}{\delta} \exp\left(-\dfrac{\left[\ln(d_d) - \ln(d_{v,50})\right]^2}{2\sigma^2} \right) d\delta; & d_d \le d_{v,50} \\[4mm] 1 - \exp\left[-\ln(2)\left(\dfrac{d_d}{d_{v,50}} \right)^\gamma \right] & ; \quad d_d > d_{v,50} \end{cases}
$$
(7.77)

where the width parameters are related by:

$$
\sigma = \dfrac{2}{\sqrt{2\pi}\,[\ln(2)]\,\gamma}
$$
(7.78)

to ensure continuity at $d_{v,50}$.

Figure 7.6b illustrates the unimodal number density curve obtained with the lognormal Rosin–Rammler distribution, even with a spread parameter $\gamma < 4$ ($\gamma = 2$ in this case).

An alternative to Eqs. (7.73) and (7.77) is to solely use the lognormal distribution to characterize the droplet size distribution.

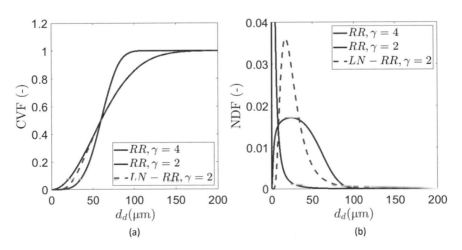

FIGURE 7.6 Droplet size distributions for $d_{v,50} = 60\,\mu m$ in terms of Cumulative Volume Fraction (CVF) (a) and Number Fraction Distribution (b).

Regardless of the type of distribution used, it is important to bear in mind that $d_{v,50}$ and γ are spray-dependent parameters that should be determined experimentally.

Similarly to $d_{v,50}$, other 'average' diameters can be defined based on the volume. For example, $d_{v,90}$ (resp. $d_{v,99}$) is defined such that 90% (resp. 99%) of the volume of the water is contained in droplets with diameter below $d_{v,90}$ (resp. $d_{v,99}$). These definitions are important in classifying water spray systems. For example, in NFPA [18] and Grant et al. [19], a water spray can be considered as mist if the measured $d_{v,99}$ at the coarsest part of the spray in a plane 1 m from the nozzle is <1 mm. In a more detailed classification reported in Grant et al. [19], the $d_{v,90}$ is: (1) lower than 200 μm for 'Class 1' sprays, (2) between 200 and 400 μm for 'Class 2' sprays, and (3) between 400 and 1,000 μm for 'Class 3' sprays.

Another approach in defining a 'mean' droplet diameter consists of considering the following expression, which reads:

$$
d_{mn} = \frac{\displaystyle\int_0^\infty f_N(d_d)\, d_d^m\, d(d_d)}{\displaystyle\int_0^\infty f_N(d_d)\, d_d^n\, d(d_d)}, \tag{7.79}
$$

where n and m are integer number such that $n = m - 1$ and m ranges from 1 to 4.

In a (simplified) discretized form, Eq. (7.79) reads:

$$
d_{mn} = \frac{\displaystyle\sum_i n_i\, d_{d,i}^m}{\displaystyle\sum_i n_i\, d_{d,i}^n}, \tag{7.80}
$$

where n_i is the number of droplets of size $d_{d,i}$.

In heat transfer applications, the d_{32} (i.e., $m = 3$ and $n = 2$), called also the Sauter Mean Diameter (SMD), is used because it takes into account the distribution of volume-to-surface ratio (which is larger for larger droplets). Sometimes use is made of d_{43} in the context of lognormal distributions. This is called the 'Herdan' diameter.

To conclude this section on droplet size distribution, it is important to mention that the droplet size is a key factor in determining the dominant mechanism(s) occurring during the interaction of the water system with smoke and flames. For example, large droplets (as typically encountered in conventional sprinkler systems) have a higher momentum (than small droplets) and can thus penetrate more easily through flames and also wet the fuel, whereas small droplets (as in mist) have the ability to absorb more heat. More details will be given in Section 7.3, 'Applications'.

7.2.6 DILUTE SPRAYS AND DENSE SPRAYS

In dilute sprays, droplets have no influence on their respective trajectories because they are sufficiently far from each other. However, when the inter-particle spacing approaches about ten particle diameters, a reduction in drag may occur and Eqs. (7.6) for isolated particles may no longer be valid. Several drag reduction correlations exist in the literature, such as the correlation of Ramírez-Muñoz et al. [20] used in McGrattan et al. [2]:

$$
\frac{C_D}{C_{D0}} = W\left[1 + \frac{\mathrm{Re}_d}{16}\frac{1}{(L/d_d - 1/2)^2}\exp\left(-\frac{\mathrm{Re}_d}{16}\frac{1}{(L/d_d - 1/2)}\right)\right], \tag{7.81}
$$

$$W = 1 - \frac{C_{D0}}{2} \left[1 - \exp\left(-\frac{\mathrm{Re}_d}{16} \frac{1}{(L/d_d - 1/2)} \right) \right],$$ (7.82)

where C_{D0} and C_D are the drag coefficients of an isolated particle and a trailing particle (i.e., a particle in the wake of another), respectively, and L is the distance between two droplets.

The above correlation has been developed for two identical spherical particles in tandem and positioned in line, compared to the flow direction. More correlations are available in the literature for several configurations and all types of applications.

In practical calculations, dense sprays are defined based on the local liquid volume fraction, α, which is linked to L as follows:

$$L/d_d = \left(\frac{\pi}{6} \alpha \right)^{\frac{1}{3}},$$ (7.83)

A 'critical' inter-particle spacing of $L/d_d = 10$ corresponds thus to $\alpha \simeq 5 \times 10^{-4}$.

More research is required on this topic, especially in water mist applications.

If the volume fraction occupied by the liquid exceeds 10^{-3} [1], droplet collision and coalescence near the nozzle become important, as mentioned, for example, in Jenny et al. [1].

7.2.7 SPRAY-INDUCED MOMENTUM

As mentioned in Section 7.1.1, droplets experience a drag force, Eq. (7.3), as they flow through the gas. Vice versa, through Newton's third law (action–reaction), the droplets exert a drag force onto the gas phase. Thus, any spray will induce momentum onto the gas phase. Very much alike what was described for fire plumes (Section 3.5), smoke plumes (Section 4.2) and momentum-driven jets (Section 2.11), the displacement of gas by the spray will induce a relative under-pressure, by which surrounding gas will be attracted, or 'entrained'. Indeed, surrounding gas will mix into the spray. In case of smoke, this can lead to dangerous situations, as explained in Section 7.3.1.2.

An interesting recent study, presenting CFD results and experimental data on this phenomenon of air entrainment for water mist type nozzles, is Vaari et al. [21]. Whereas model improvement is still possible, the trends are well captured. The spray is divided into two regimes: a 'momentum' regime, close to the nozzle (the first 40 cm in Vaari et al. [21]), and a 'gravitation' regime, where the droplets evolve towards their equilibrium drop velocity (see Section 7.1.1). In the momentum regime, the spray widens and the droplets, injected with high velocity, break up into smaller droplets, due to the high Weber number (see Section 7.2.3). The air currents, induced by the water flow from the nozzles, drag the smaller droplets into the spray core, thereby increasing the drop density inside that region. Indeed, smaller droplets have a lower Stokes number, Eq. (7.13), and thus follow the air currents more easily than the larger droplets. As an order of magnitude, induced velocities of around 10 m/s are reported in Vaari et al. [21], with droplet inlet velocities in the order of 110 m/s.

A similar reasoning can be held for water sprays from sprinklers. The drag forces cause the local surrounding gas to move, and the resulting relative under-pressure causes entrainment of the surrounding gas into the spray.

Obviously, quantitatively the phenomenon will strongly depend on the spray momentum, i.e., on the water flow rate, the droplet size distribution, the velocity distribution and the spray cone angle, as described in Sections 7.2.1–7.2.5. The reader is referred to specialized literature on this.

7.2.8 Experimental Characterization of Water Sprays

It is very important to experimentally characterize water sprays in order to develop a good understanding of their interaction with a fire-driven flow. From a numerical modelling standpoint, an accurate simulation of water sprays in cold conditions (i.e., in the absence of a fire-driven flow) is necessary before delving into more complex aspects of the physics related to the interaction with smoke, flames and fuels, as will be addressed in Section 7.3.

The droplet size distribution is particularly important because it strongly determines the dominant mechanisms that take place when a water system is activated in the event of a fire. For example, small droplets (in the order of $100\,\mu m$), as typically encountered in water mist systems, are very efficient in smoke cooling (because of the higher surface area) and thermal radiation attenuation, whereas larger droplets (in the order of 1 mm), as typically encountered in conventional sprinkler systems, are more efficient in reaching the fire source and, eventually, wetting the fuel because of their higher momentum.

The 'simplest' experimental technique that is used to characterize the droplet size distribution consists of (a) collecting droplets on a petri dish with thinly spread castor oil, (b) taking pictures with a microscope camera and (c) using the three-point method to estimate the diameter of each droplet, as done, for example, in Noda et al. [22]. The data is then displayed in terms of the number of droplets per interval of droplet diameters, called also 'bin'. For example, Figure 7.7 shows the number of droplets per bin of $20\,\mu m$ diameter. By estimating the total volume of the droplets in each bin and dividing that number by the total volume of all the collected droplets, one can obtain the Volume Fraction Distribution (VFD) and the subsequent other related quantities, as explained in Section 7.2.5. A specific functional form (i.e., Rosin–Rammler, lognormal or a combination) is then chosen to fit the experimental data.

A more advanced experimental technique to measure the droplet size distribution is the Phase Doppler Particle Analysis (PDPA), which allows to track the size and number of droplets in a sampling volume using laser. Furthermore, the PDPA allows the measurements of droplet velocity and water volume flux measurements. Further developments of advanced techniques for spray

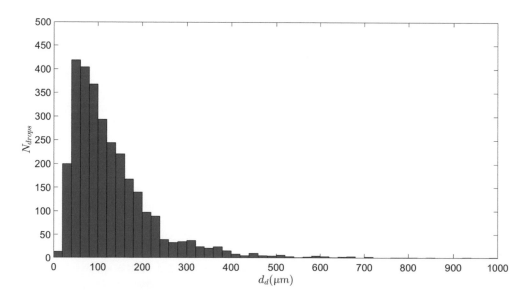

FIGURE 7.7   The number of droplets per bins of $20\,\mu m$ diameter. (Experimental data redrawn from Noda et al. [22].)

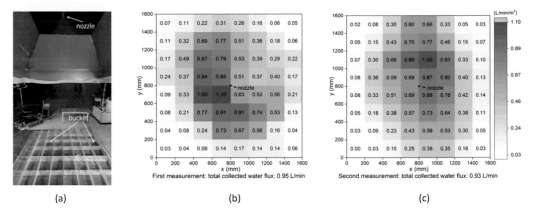

FIGURE 7.8 An example of a bucket test carried out in Liu et al. [24]. (a) Photo showing the positioning of the nozzle and the buckets. (b and c) Two measurements showing the mapping of the water flux along the *x*-axis and *y*-axis.

characterization are carried out in Jordan et al. [23] where a Spatially-resolved Spray Scanning System (4S) has been developed to capture the complete spatio-stochastic nature of the spray at its point of origin.

In addition to the characterization of the spray near the injection and further downstream in terms of droplet size distribution, droplet velocity and water volume flux, it is of high interest to measure the water flux distribution at floor level, which determines the 'footprint' of the spray. This can be done with a 'simple' test called the 'bucket test', an example of which is given in Figure 7.8.

Besides the water flux distribution, the bucket test is a verification test that gives the total water flow rate delivered by the nozzle. The integration of the data in Figure 7.8 gives a total water flow rate of 0.95 lpm (resp. 0.93 lpm) in the first (resp. second) test, which is close to the theoretical flow rate of 1 lpm in the test considered.

7.3 APPLICATIONS OF WATER SPRAYS AND INTERACTION WITH A FIRE-DRIVEN FLOW

Shortly after the activation of a sprinkler or a water mist nozzle, water sprays interact directly with smoke by, primarily, cooling it. Furthermore, they may be designed to form a 'barrier' that blocks the propagation of smoke further downstream from the fire source. However, it is important to realize that the above phenomena are generally accompanied by an adverse effect, namely the loss of stratification which causes, for example, a loss in visibility and difficulties in evacuation.

Besides the interaction with smoke, water is an effective means of controlling and, eventually, suppressing a fire by directly interacting with the flames through different mechanisms such as cooling, dilution and oxygen displacement. Attenuation of thermal radiation from the flames is also an important application of water sprays.

Finally, water may be in direct contact with the fuel in order to prevent ignition. However, special care must be taken regarding liquid fuels.

In this section, several water spray applications (and mechanisms) will be reviewed, making the link with the fundamental aspects of the physics addressed earlier in Sections 7.1 and 7.2.

7.3.1 Interaction with Smoke

7.3.1.1 Smoke Cooling

The popularity of water in firefighting stems essentially from its ability to absorb substantial amounts of heat from the gas phase, mainly by (a) heat-up and (b) evaporation.

In liquid form, the specific heat of water, c_p, varies between 4,186 J/kg K at 20°C and 4,219 J/kg K at 100°C. Therefore, during heating up from 20°C to 100°C (which is the maximum temperature water can attain at atmospheric pressure, in agreement with the law of Clausius and Clapeyron), water absorbs an amount of energy per unit mass equal to approximately:

$$\Delta Q_{\text{heat-up}} = 4,200(100-20) = 336 \text{ kJ/kg}. \tag{7.84}$$

In other words, per kg water, inserted at ambient temperature and heated up to 100°C, an amount of energy equal to 336 kJ is absorbed by the water. Obviously, as long as the water is inside the fire compartment, the energy has not left the compartment. However, as it is taken from the gas phase (smoke and/or flames), the gas temperatures decrease, so that chemistry becomes slower and the fire development is slowed down (see Chapter 5).

If one assumes that the water droplets are mainly heated up by convection, it is of interest to recall Eq. (7.34):

$$\dot{q}_{\text{conv}} = h A_d \left(T_g - T_d \right). \tag{7.85}$$

More specifically, the total area of the droplets, A_d, is particularly interesting to examine. If one considers a specific volume flow rate of water, $\dot{V}_w$, and a constant droplet diameter (i.e., monodisperse spray), d_d, the number of water droplets injected per second is then:

$$n = \frac{\dot{V}_w}{\pi(d_d^3 / 6)}, \tag{7.86}$$

which corresponds to a total surface area per second of:

$$A_d = \left(\frac{\dot{V}_w}{\pi(d_d^3 / 6)} \right)(\pi d_d^2) = \frac{6 \dot{V}_w}{d}. \tag{7.87}$$

This means that the total surface area available for heat-up by convection is inversely proportional to the droplet diameter, which makes water mist systems (small droplet diameters) much more efficient in this regard in comparison to conventional sprinkler systems. Equation (7.87) could also be interpreted as the following: with smaller droplets, less water is needed (i.e., lower $\dot{V}_w$) to provide a similar convective cooling power.

Heat-up (and subsequent convective cooling of the gas) is not the main contribution from the water as heat sink. Indeed, the main contribution is due to evaporation. The latent heat of vaporization for water at atmospheric pressure is approximately $L_v = 2.26$ MJ/kg. In other words, while evaporating at a temperature of 100°C, water absorbs an amount of energy per unit mass equal to approximately:

$$\Delta Q_{\text{evap}} = 2,260 \text{ kJ/kg}. \tag{7.88}$$

This is approximately seven times as much as the value in Eq. (7.84). Again, as long as the steam is inside the fire compartment, the energy has not left the compartment. However, as it is taken from the gas phase (smoke and/or flames), the gas temperatures decrease, so that chemistry becomes slower and the fire development is slowed down (see Chapter 5).

As the steam, which is inert and therefore mainly dilutes the gases, mixes with the environment, its temperature increases further. The specific heat of steam is lower than that of water, but it is still around c_p =1,890 J/(kg K) at 100°C (and increases to around 2,000 J/(kg K) at 300°C and around 2,200 J/(kg K) at 600°C), so that the heat absorption by the steam can still be significant.

As (liquid) water evaporates, steam (water vapour) is formed. The latter can be considered to behave as an ideal gas. Therefore, the volume occupied by steam can be calculated as follows:

$$V_1 = \frac{nRT_1}{P},$$ (7.89)

where R is the universal gas constant ($R = 8.314$ J/K mol), T_1 is the temperature of steam, assumed to be $T_1 = 100°C = 373.15$ K, and P is the pressure taken as $P = 101,325$ Pa. The variable n denotes the number of moles of water which can be calculated as follows:

$$n = \frac{\rho_l}{MW_l} V_0,$$ (7.90)

where ρ_l is the density of liquid water ($\rho_l = 1,000$ kg/m³), MW_l is the molecular weight of water ($MW_l = 18$ g/mol), and V_0 is the volume of liquid water.

Combining the previous two equations gives:

$$V_1 = \frac{\rho_l RT_1}{MW_l P} V_0 \simeq 1,706 V_0.$$ (7.91)

This means that as liquid water evaporates to form steam at $T_1 = 100°C$, it occupies a volume which is about 1,706 larger than that in its liquid state.

Assuming that there are no substantial changes in P and that n is constant, the ideal gas law can be rewritten as:

$$\frac{V}{T} = \text{constant.}$$ (7.92)

which means that if steam heats to a temperature, T, higher than T_1, it will occupy a volume of:

$$V = \frac{V_1}{T_1} T = \frac{1,706 V_0}{376.15} T = 4.572 V_0 T$$ (7.93)

corresponding to a volume expansion of:

$$\frac{V}{V_0} \simeq 4.572 T.$$ (7.94)

It must be appreciated that this volume is added into the compartment of the fire. If it is added to the smoke, it implies an increase in volume of the smoke, due to the formation of steam. This is a disadvantage. On the other hand, if the water takes the heat, required to heat up and evaporate, from the smoke, the smoke volume will shrink. Indeed, the smoke temperature will decrease due to the heat loss, so that the ideal gas law reveals that for a constant pressure, the smoke volume decreases linearly with a decrease in absolute temperature:

$$V_{s,2} = \frac{T_2}{T_1} V_{s,1}.$$ (7.95)

In Eq. (7.95), '1' refers to the situation before water is added into the steam and '2' refers to the situation after addition. The 'well-mixed' assumption is made. Whether or not the addition of water will lead to an increase, decrease or status quo of smoke volume, depends on the combination of Eqs. (7.91) and (7.94). Clearly, if the heat to heat up and evaporate the water is not taken from the smoke (but, e.g., from hot walls or the ceiling), Eq. (7.91) remains in place, while Eq. (7.94) does only indirectly, so that in general an increase in smoke volume is to be expected.

One firefighting technique concerns '3D gas cooling' [25]. With this technique, short pulses (0.5 to 1 s only) of small water droplets are provided, to cool down the hot smoke. As mentioned above, care must be taken that the increase in volume due to steam formation, Eq. (7.94), is at least compensated by the reduction in smoke volume due to smoke temperature reduction, Eq. (7.95). If executed properly, it is indeed possible to reduce the smoke volume with this technique [25]. As such, also no over-pressure is created, so that the smoke is not pushed towards other parts of the building. Cooling down the smoke, the risk for flash-over is reduced (Chapter 5) and the concentration of combustible gases in the smoke layer is reduced. Thus, the steam has an inerting effect and the smoke becomes less flammable (Section 3.1.2). Thus, ignition and rapid premixed combustion in the smoke layer is avoided. In other words, a cooler and safer situation is created for the firefighters. It is recalled that contact of water with hot surfaces (e.g., objects, walls, ceiling) must be avoided with this technique, since then the primary effect would be the formation of steam, Eq. (7.94), increasing the amount of smoke (and/or creating an over-pressure, pushing the smoke to other parts of the building). This explains why short pulses are required in this technique, rather than a continuous water jet. Indeed, much of the water would hit the ceiling, walls and floor, so that the heat for the steam formation is not taken from the smoke. As such, the smoke volume would effectively increase by the steam formation, as explained above.

7.3.1.2 Smoke Logging

The canonical situation concerns the build-up of a hot smoke layer underneath a ceiling, into which, upon activation, a spray from a sprinkler or water mist system is released.

As mentioned above, sprays from sprinklers or water mist systems cool down the smoke layer, through heat transfer from the smoke towards the water droplets (Section 7.1.2). Moreover, the water droplets exert a downward drag force onto the smoke (Section 7.1.1). Both effects lead to "smoke logging", a downward displacement of the smoke layer caused by the water droplets [3]. In addition hereto, smoke will be entrained into the spray due to the spray-induced momentum (Section 7.2.7), so this smoke will also move down. Finally, as explained in Section 7.3.1.1, depending on the heat transfer conditions, the smoke volume can increase due to the additional volume, occupied by steam (Eq. 7.94), if this is not compensated by shrinkage of the smoke layer due to heat losses (Eq. 7.95). Figure 7.9 presents a sketch of the configuration where three zones are distinguished in the model, based on temperature differences:

- *Zone I*: the smoke layer outside the water spray envelope. This zone has the highest average temperature ($T_{s,o}$).
- *Zone II*: there is no smoke in Zone II, and the average temperature remains equal to ambient air temperature ($T_{a,o}$).
- *Zone III*: This zone is occupied by smoke inside the spray envelope ($T_{s,i}$).

It is important to note that the smoke layer considered in Figure 7.9 and in Tang et al. [3] is static/quiescent beneath the ceiling. If the smoke layer is in motion (e.g., in a tunnel configuration or in a ceiling-jet region), the relative droplet velocity $|\vec{u}_d - \vec{u}|$ changes in the drag force and the droplet trajectories are modified.

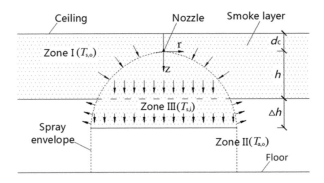

FIGURE 7.9 Schematic sketch of the configuration. (Adapted from Tang et al. [3].)

Whereas the spray has a cooling effect and is primarily intended to interact with the flames and the fire source (Sections 7.3.2 and 7.3.3), this downward displacement of smoke is a disadvantage. Indeed, smoke is hazardous with respect to life safety (and smoke damage), so a downward displacement is a disadvantage. It is thus important to try and quantify this downward displacement.

A classical theory has been presented by Bullen [26], leading to an instability criterion of smoke layer under water spray conditions. The criterion is based on the ratio of the total drag force of water droplets (D) to the 'buoyancy of smoke layer' (B). According to this theory, smoke logging would happen when $D > B$, while otherwise the layer remains stable. However, starting from first principles, it is explained in Tang et al. [3] that this theory needed to be revisited. Indeed, several sets of independent experiments revealed downward smoke displacement in conditions where Bullen's theory predicted a stable situation. As explained in Tang et al. [3], the key reason is that the smoke layer buoyancy is calculated as follows:

$$B = (\rho_a - \rho_s)gV_{\text{smoke}}, \tag{7.96}$$

where ρ_a is the air density, ρ_s is the smoke density and V_{smoke} is the volume of smoke inside the spray envelope.

Obviously, according to the Archimedes Law, this expression is suitable only if V_{smoke} is entirely surrounded by air at ambient temperature. In reality, however, this is never the case, as illustrated in Figure 7.9. On the contrary, the buoyancy force works downwards on smoke that is cooled by the water from the sprinklers/water mist, as it is surrounded by hot smoke. Only the lower part of V_{smoke} is surrounded by ambient air, and therefore, upward buoyancy can only apply to that region. As such, using Eq. (7.96) to calculate the buoyancy force for Bullen's criterion [26] is in general not correct. Only as long as the initial smoke layer is not too thick (and thus most of the spray envelope is in fact surrounded by ambient air), the expression is a reasonable approximation of reality.

Therefore, a model has been developed in Tang et al. [3] and extended (adding heat transfer and entrainment of smoke and air into the spray cone) in Tang et al. [27]. The extended model (Tang et al., 2014) is briefly mentioned here, because it contains the essential ingredients to describe the basic physical phenomena. The reader is referred to Tang et al. [3,27] for more detail.

A first important aspect in Tang et al. [3,27] is the prediction of the spray envelope shape based on the dynamics of a single spherical droplet, as described in Section 7.1.1. It is thus assumed that the spray is monodisperse (i.e., all droplets have the same diameter) and that there is no drag reduction and no droplet deformation, collision and coalescence. Figure 7.10 shows the influence of (a) the spray angle, (b) the water mass flow rate and (c) the droplet diameter on the shape of the spray envelope by varying only one parameter at a time. Obviously, a higher spray angle leads to a wider

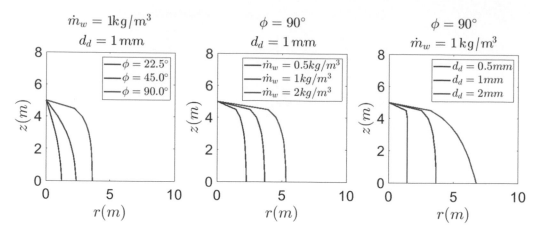

FIGURE 7.10 Spray envelope shape (i.e., vertical distance, z, as a function of the radius, r) under several combinations of the water mass flow rate, $\dot{m}_w$, droplet diameter, d_d, and the elevation angle, ϕ, as defined in Figure 7.4, with $\phi_{max} = \phi$ and $\phi_{min} = 0°$. (Redrawn from Tang et al. [3].)

spray; see Figure 7.10a. Furthermore, a higher mass flow rate leads to a wider spray (see Figure 7.10b), because of the higher initial horizontal momentum. Finally, larger droplets create a wider spray than smaller droplets; see Figure 7.10c. This is due to the higher area-to-volume ratio for the smaller droplets, so that the drag force becomes relatively more important, compared to gravity. The width of the spray (which depends on the spray angle, the mass flow rate and the droplet size, as discussed above) is particularly important in the interaction with the smoke layer because the narrower the spray, the higher the vertically downward momentum of the water droplets and, thus, the more pronounced the downward drag and the subsequent smoke layer displacement.

A second important aspect addressed in Tang et al. [27], besides the momentum exchange between the water droplets and the surrounding smoke, is heat transfer. On the one hand, heat is absorbed by the droplets by convective heat-up and evaporation. On the other hand, the smoke temperature inside the spray, $T_{s,i}$, changes not only by direct cooling from the droplets but also due to either smoke entrainment into the spray envelope or smoke flowing out of it. By virtue of the ideal gas law, the density of smoke, $\rho_{s,i}$, is calculated from $T_{s,i}$ in order to estimate the buoyancy force within Zone III. The properties of air are used, given that smoke mainly consists of entrained air. Furthermore, the temperature increase of water droplets, while travelling through the smoke layer, is ignored. Because the heat exchange by evaporation significantly exceeds the heat exchange related to heat-up (see Section 7.3.1.1), and given the relatively short residence times of water droplets inside the smoke layer, the assumption of constant droplet temperature is acceptable, except for very high smoke layer temperatures [27]. Besides, the temperature difference between water droplets and the surrounding smoke layer is considered as the only driving force for evaporation of water droplets.

The momentum (drag) and heat transfer (buoyancy) aspects briefly discussed above are incorporated in a highly coupled system of equations (for the conservation of mass, momentum and energy) which are solved to give (amongst other quantities) the smoke velocity, $v_{s,i}$, assumed to be vertical and uniform in horizontal planes. The criterion $v_{s,i} = 0\,\text{m/s}$ is used in Tang et al. [3,27] to determine the downward smoke displacement, i.e., the height z is determined where the smoke no longer flows downwards.

An important phenomenon, revealed in Tang et al. [27], is that for high-enough water flow rates and as long as the smoke layer thickness is below a critical value, cool air is entrained into the water spray below the smoke layer, which diminishes the upward buoyancy flow. In such situations, downward smoke layer displacement occurs down to floor level. This is particularly dangerous during the smoke layer build-up phase. This indicates it may be advantageous to opt for moderate water flow rates, as far as possible downward smoke layer displacement is considered. However, obviously the primary purpose is, as mentioned, interaction with flames or fire sources (Sections 7.3.2 and 7.3.3).

As mentioned, the model in Tang et al. [27] has been described for monodisperse sprays, i.e., only 1 droplet diameter (equal to $d_{v,50}$) has been considered. An interesting extension would be the introduction of poly-disperse sprays, e.g., applying droplet size distributions (7.72) or (7.76). Another interesting extension would be the introduction of a range of elevation angles, as illustrated in Section 7.2.2.

While not applied to water mist sprays, the model of Tang et al. [27] contains many of the required ingredients to do so, particularly if the extensions as just mentioned have been added. Yet, droplet collision and coalescence would also need to be introduced for water mist sprays.

7.3.1.3   Smoke Blocking

Smoke, originating from a fire, can be blocked by a water curtain. Indeed, the vertically downward momentum, caused by the water curtain, induces entrainment of air (from the 'protected' side) and smoke (from the 'fire' side). Thus, the horizontal momentum of the smoke is pushed downwards. For sufficiently weak smoke and sufficiently strong water curtains, effective compartmentation can be obtained. Indeed, people could walk through the water curtain from the fire side to the protected side, where they would no longer suffer from high radiation levels.

If the smoke momentum is such that smoke partly flows through the curtain, the possible smoke stratification (see Section 4.5.5) upstream of the water curtain is destroyed, due to the downward drag and the cooling of the smoke. This is disadvantageous in the light of evacuation purposes. Yet, the smoke will be much cooler and diluted by water vapour, so that possibly toxic species concentrations are reduced (although the latter effect may not be substantial).

More research is required to allow for a systematic discussion on the use of water curtains in terms of compartmentation.

7.3.2   Interaction with the Flame

7.3.2.1   Cooling the Flame Zone Directly

In a flame, the reaction zone thicknesses are in the order of 0.1 mm. Droplets from water mist systems have diameters in that order of magnitude (Section 7.2.5) and can thus interact directly with the reaction zone, taking heat from the flames (by heat-up and evaporation) and reducing the temperature. As explained in Section 3.1.1, a reduction in temperature causes a reduction in chemical kinetics. In other words, the Damköhler number (3.13) decreases because the chemical time scale (3.14) increases. The reduction in chemical kinetics can be such that the flame effectively extinguishes, particularly if the temperature drops below a threshold value (Section 3.1.1). For obvious reasons, the water mass flow rate applied is important. This is also called the 'application rate' [19]. There is an optimum application rate, extinguishing the fire with the minimum total amount of water supply required. However, in practice the 'preferred' application rate is higher (up to three or four times higher), because with higher application rate (if applied efficiently), the time to fire extinction is shorter. There is also a critical application rate below which the fire cannot be extinguished. In general, this critical rate increases with increasing 'pre-burn' time (i.e., the time the fire has already had to develop before the water is applied) and with increasing ventilation factor of the compartment, whereas it is lower when the water is applied from the base of the fire upward, rather than vice versa [19].

The flame cooling phenomenon can only happen if the droplets can reach the flames [28]. Indeed, the upward motion caused by the fire plume (Section 3.5.1) and smoke plume (Section 4.2) can push the water mist droplets away, because their Stokes number, Eq. (7.13), is typically very low. Whether or not this will happen, will depend to a large extent on the mass rate of air/smoke entrained into the water spray [28,29].

7.3.2.2 Dilution Effect and Inerting

In addition to the removal of heat from the flames, water (vapour) is inert, so that the gases are also diluted. This hinders the mixing of combustible gases with oxygen, again reducing the combustion intensity. Furthermore, atmospheric inerting is promoted through the production of steam [19]. The above-cited mechanisms, i.e., dilution, inerting, and flame cooling, are the primary fire suppression mechanisms of water mist systems, which are extremely effective (i.e., very rapid extinction) if a large fire is allowed to develop in a confined space, whereby the oxygen is already depleted. With reduced ventilation factor, times to extinction typically reduce significantly, because the steam formation results in strong inerting ('smothering' the fire [19]). It is mentioned in Grant et al. [19], though, that water mist is particularly efficient if the compartment boundaries have a high thermal conductivity, as is typical in industry plants or off-shore applications, because then the compartment boundaries remain hot and impinging water droplets keep on evaporating on the surface. The latter is not guaranteed in residential applications.

7.3.2.3 Blowing Effect and Oxygen Displacement

As steam occupies a much larger volume than liquid water (Eq. 7.94), there is a 'blowing' effect due to the expansion involved. If the steam is formed near the flame, the expansion causes a flow away from the flames, which causes an additional resistance against the flow of combustible gases and oxygen to the flames.

7.3.2.4 Attenuation of Thermal Radiation Using Water Mist

Water curtains can be interpreted as special cases of water sprays. Being typically vertical, under the influence of the gravity field, one can interpret a water curtain as a sprinkler or water mist spray where the azimuthal angle only covers a very narrow range, rather than the entire range $0 \leq \theta \leq 2\pi$ (see Section 7.2.2). The working principle resembles that of an air curtain, as sketched in Figure 6.13, although the momentum is typically higher (due to the much higher mass density of water, compared to air) and water is capable of blocking heat better than air.

A key property of water curtains (besides smoke blocking, as explained in Section 7.3.1.3) is the very strong attenuation of radiation. In that sense, a water curtain can be considered as a means of 'compartmentation', in that it can prevent fire spread by blocking the radiation. The strong attenuation of radiation is due to the combined effect of two phenomena, namely absorption of radiation by water and scattering of radiation. The absorption of radiation is not only important with respect to radiation blocking, but also because droplet heating and evaporation are stimulated. This acts as a heat sink for the smoke (Section 7.3.1.1) or fire (Section 7.3.2.1) and thus relates to the action of water as fire suppressant.

In the present section, we briefly discuss the scattering effect. The Mie scattering theory [30] allows to calculate the monochromatic extinction coefficient β_λ for radiation with wavelength λ as the sum of the monochromatic absorption coefficient κ_λ and the monochromatic scattering coefficient σ_λ:

$$\beta_\lambda = \kappa_\lambda + \sigma_\lambda. \tag{7.97}$$

An important quantity is the size parameter, which is the ratio of the droplet radius to the wavelength of radiation considered (pre-multiplied by 2π):

$$\chi = \frac{2\pi r_d}{\lambda}. \tag{7.98}$$

Indeed, the coefficients can then be computed as follows [2]:

$$\kappa_\lambda = \int_0^\infty Q_a(\chi,\lambda)\pi r_d^2 f(r_d)dr_d, \tag{7.99}$$

$$\sigma_\lambda = \int_0^\infty Q_s(\chi,\lambda)\pi r_d^2 f(r_d)dr_d. \tag{7.100}$$

In Eqs. (7.99) and (7.100), $f(r_d)$ is the distribution function of droplet radius (cf. Eq. (7.75) or Eq. (7.79)) and $Q_a(\chi,\lambda)$ and $Q_s(\chi,\lambda)$ denote 'efficiencies' for absorption and scattering, respectively. Note that the absorption and scattering coefficients (7.99) and (7.100) are local quantities, i.e., their values vary in physical space. Also note that the integrations can be very tedious, as water droplets range in size from below $10\,\mu m$ (evaporating droplets) to well beyond $1,000\,\mu m$ in the case of sprinkler sprays (Section 7.2.5). Therefore, this integration is sometimes simplified in models in CFD simulations. For example, in McGrattan et al. [2], use is made of the Sauter Mean Diameter (SMD) values, Eq. (7.75), to replace the entire integration. The use of SMD is logical in terms of heat transfer, as it is characteristic for the volume–area ratio, as mentioned in Section 7.2.5.

The relevant wavelength spectrum in case of fire relates essentially to the visible and infrared spectra, say $0.4\,\mu m \le \lambda \le 100\,\mu m$. Shorter wavelengths correspond to higher temperatures, so particularly the range of $0.4\,\mu m \le \lambda \le 20\,\mu m$ is typically of relevance as far as radiation is concerned [31]. The higher end of this wavelength range corresponds to emitting hot surfaces (estimated at $T = 700$ K in Modest [30]), or could also correspond to the radiation from a hot smoke layer. Still higher wavelengths correspond to relatively low temperatures, so radiation is not the dominant heat transfer mechanism, while flames (temperatures of 1,100 K and higher) mainly emit at lower wavelengths.

The range of $0.4\,\mu m \le \lambda \le 20\,\mu m$ implies that for typical droplet size distributions, the size parameter (7.98) is much larger than unity, except perhaps for the smallest droplets. This implies that the main type of scattering is 'geometric scattering': radiation is reflected on the surface of the spherical droplets, causing a change in direction. In general, reflection of incident radiation dominates over diffraction (change in direction without contact) and refraction (part of the radiation going through spheres) [32].

In Viskanta and Tseng [33], it is illustrated that in the visible and near-infrared spectra ($\lambda < 2\,\mu m$), water droplets are predominantly scattering. For wavelengths larger than $2\,\mu m$, water absorbs strongly and large droplets are nearly opaque to incident radiation. Indeed, it is illustrated in Viskanta and Tseng [33] that water is essentially transparent in the visible spectrum ($0.4\,\mu m \le \lambda \le 0.75\,\mu m$), semi-transparent in the near-infrared spectrum ($0.75\,\mu m \le \lambda \le 2\,\mu m$) and effectively opaque in the infrared spectrum ($\lambda > 2\,\mu m$). In brief, it is illustrated in Viskanta and Tseng [33] that scattering is dominant for small droplets and short wavelengths, while absorption predominates for larger droplets and shorter wavelengths. This is confirmed in Hostikka and McGrattan [34]: for droplets with radius up to $200\,\mu m$ (which is the highest value tested in Hostikka and McGrattan [34]), the total extinction coefficient is practically equal to the scattering coefficient (Eq. 7.97) for wavelengths $0.1\,\mu m \leq \lambda \leq 1.5\,\mu m$, and even for $\lambda > 3\,\mu m$, the scattering remains as important as the absorption.

To close this section, it is mentioned that the radiation shielding effect of a water curtain is sometimes exploited by firefighters, trying to approach a fire source. The water spray is then not meant as fire suppressant, but creates a water shield, behind which the firefighters can approach the fire source to a distance from where the attack can be organized.

7.3.3   INTERACTION WITH THE FUEL

7.3.3.1   Fuel Wetting and Blanketing

An important suppression mechanism is the cooling of the fuel surface, so that the pyrolysis rate or evaporation rate reduces and the mass flow rate of combustible gases is reduced (see Section 5.2). Indeed, the endothermic heating up and vaporization processes require energy, which is taken by the water as heat coming from the flames, but also as heat coming from the (solid or liquid) fuel, which has a surface temperature higher than the evaporation temperature of water. As such, heat is taken from the endothermic pyrolysis or vaporization of the fuel, required to release combustible gases. If the fuel surface temperature is cooled down sufficiently (e.g., to below the fire point for a liquid fuel), the fire is suppressed. It is also mentioned that pre-wetting of adjacent objects can slow down fire spread, for similar reasons, namely that, heating up and evaporating, the water consumes a portion of the heat, transferred from the flames and the hot surfaces, so that this heat cannot be used for pyrolysis or evaporation of fuel.

Fuel wetting occurs mainly when sprinklers are used because the large water droplets have the ability to essentially flow through the flames with relatively little interaction, reaching thus the surface and wetting it. This results in a reduction in heat feedback to the evaporating liquid fuel or the pyrolysing solid fuel (Section 5.2). The latter effect does not come into play in the case of gaseous fuel (e.g., supplied from a leakage in a pipe).

The total water requirement for the cooling of the fuel surface often depends more on the heat content of the fuel than on the fire heat release rate [19]. The rate of heat absorption from the fuel, required to extinguish the fire, is in general far less than the fire heat release rate [19].

Fuel wetting may result in the formation of a barrier on the fuel, i.e., fuel 'blanketing'. In this case, water acts as additional thermal resistance against convection and as resistance against the flow of combustible gases from the fuel. The water film can also absorb radiation, so that the heat feedback from the flames towards the fuel is attenuated.

7.3.3.2   Special Care with Liquid Fuels

It is noted that water is not always beneficial. Special care is required with liquid fuels, floating on water, particularly with fuel mixtures in which a hot zone can propagate downwards [35]. Indeed, if water is provided on the surface of such burning fuels and the fire is not extinguished, the water can sink to the bottom of the pan with the fuel. If steam is formed at the bottom, below the burning liquid fuel, the very dangerous situation of 'boil-over' can occur: steam eruptions, causing fierce spilling of fuel over the pan edges and very hazardous liquid fuel burning.

Similarly, the supply of water on burning oil can be very dangerous. Indeed, if heavy oil is burning as a liquid pool, the contact area of the liquid fuel to the air is limited to the surface area of the pool. If now the boiling temperature is higher than the evaporation temperature of water (100°C), the supply of water will cause very rapid formation of steam, thereby pushing the liquid fuel out of the pan in the form of many droplets. This creates a much larger contact area for the liquid fuel and causes far more hazardous burning than before. Such fires are typically extinguished by removing the oxygen supply, rather than by water.

7.3.4   Interaction with the Sprinklers: Sprinkler Skipping

To close this section, another type of interaction between smoke and water from a sprinkler spray is mentioned, namely 'sprinkler skipping'. This term refers to irregular sprinkler operating sequence, compared to what would be expected from the plume ceiling-jet behaviour underneath the ceiling. This phenomenon is caused by droplets from the originally activated spray(s), convected by the moving smoke layer in the ceiling-jet region and effectively cooling down neighbouring sprinkler heads, so that the latter are not activated and other sprinkler heads, further away from the fire source, are activated first. This leads to an inefficient use of the water supply to the sprinkler system. From the above, it is clear that primarily small droplets, with the Stokes number (Eq. 7.13) sufficiently small so that they follow the smoke flow, can impinge onto neighbouring sprinkler heads, particularly if they are deflected into a horizontal motion by the sprinkler head. Also in water mist systems, even when the droplets are injected downwards, the skipping of nozzles can be an issue, because the droplets are so small that they are easily pushed upwards and sidewards by the fire and smoke plume.

7.4   INTERACTION OF WATER SPRAYS WITH SMOKE AND HEAT CONTROL SYSTEMS

It has been shown, throughout this chapter, that water sprays (conventional sprinkler systems or mist) can be quite effective in fire suppression and control, thanks to several mechanisms such as cooling and thermal attenuation of radiation. Nevertheless, it is of interest to consider combining sprinklers with Smoke and Heat Extraction Vents (SHEVs). Indeed, SHEVs present the advantage of improving visibility and tenability, which helps the egress of occupants and facilitates the location of the seat of the fire by firefighters in order to proceed with a faster extinguishment, whereas, as mentioned in Section 7.3.1.2, sprinklers can result in smoke logging and a subsequent loss of visibility.

An important question to be considered then is whether the combination of sprinklers and SHEVs can be detrimental to either or both technologies. For example, it is possible that heat extraction by vents delays the activation of sprinklers, causing (potentially) a larger fire. On the other hand, the cooling effect of sprinklers may significantly reduce the buoyancy force and the subsequent amount of smoke extracted through the vents.

The general picture that emerges from the comprehensive literature review carried out in Beyler and Cooper [36], and which is focused on storage and warehouse facilities, is that 'heat venting used in conjunction with sprinklers [...] does not have a negative effect on sprinkler performance'. It is also recommended to rely on venting because its 'early activation [improves visibility] and has no detrimental effect on sprinkler performance'. Although general guidelines and strategies can be followed on the operating conditions of sprinklers in conjunction with SHEVs (e.g., 'ganging the vents and operating them by smoke detector or first sprinkler activation' [36]), it remains of importance to carry out detailed studies on the scenarios and configurations of interest.

In Liu [37], an experimental study has been carried out on the combined effect of water mist and fires. The results show that both longitudinal ventilation and water mist contribute to reduce the gas temperature near the ceiling. Furthermore, water-induced cooling reduces the ventilation velocity required to maintain a certain backlayering distance. These results are interesting, but should be considered with caution because they are strongly scenario-/configuration-dependent. Furthermore, they have been obtained at a reduced scale. Thus, further studies should be carried out to examine the scale effect.

REFERENCES

1. P. Jenny, D. Roekaerts and N. Bijshuizen (2012) "Modeling of turbulent dilute spray combustion", *Progress in Energy and Combustion Science* Vol. 38, pp. 846–887.
2. K. McGrattan, R. McDermott, M. Vanella, S. Hostikka and J. Floyd (2021) *FDS Technical Reference Guide*. http://firemodels.github.io/fds-smv/manuals.html.
3. Z. Tang, J. Vierendeels, Z. Fang, and B. Merci (2013) "Description and application of an analytical model to quantify downward smoke displacement caused by a water spray," *Fire Safety Journal*, Vol. 55, pp. 50–60.
4. V. Novozhilov (2002) "A simple engineering model for sprinkler spray interaction with fire products," *International Journal on Engineering Performance-Based Fire Codes*, Vol. 4, pp. 95–103.
5. D.B. Spalding (1953) "The combustion of liquid fuels", *Proceedings of the Combustion Institute*, Vol. 4, pp. 847–864.
6. M. Chrigui, F. Sacomano, A. Sadiki and A. Masri (2014) 'Evaporation modeling for polydisperse spray in turbulent flow', in B. Merci and E. Gutheil (Eds.), *Experiments and Numerical Simulations of Turbulent Combustion of Diluted Sprays*, ERCOFTAC Series Vol. 19. Springer, New York, 160 p. ISBN 978-3-319–04678-5, doi: 10.1007/978-3-319-04678-5.
7. W.A. Sirignano (1999) *Fluid Dynamics and Transport of Droplets and Sprays*. Cambridge University Press: Cambridge.
8. W.E. Ranz and W.R. Marshall (1952) "Evaporation from drops," *Chemical Engineering Progress*, Vol. 48, (Part 1) pp. 141–146 (Part 2) pp. 173–180.
9. X. Zhou and H.-Z. Yu (2011) "Experimental investigation of spray formation as affected by sprinkler geometry," *Fire Safety Journal*, Vol. 46 (3), pp. 140–150.
10. P.E. Santangelo (2012) "Experiments and modeling of discharge characteristics in water-mist sprays generated by pressure-swirl atomizers," *Journal of Thermal Science*, Vol. 6, pp. 539–548.
11. D.T. Sheppard (2002) *Spray Characteristics of Fire Sprinklers*. NIST GCR 02–838: Gaithersburg, MD.
12. N. Ren, H.R. Baum and A.W. Marshall (2011) "A comprehensive methodology for characterizing sprinkler sprays," *Proceedings of the Combustion Institute*, Vol. 33, pp. 2547–2554.
13. P.H. Dundas (1974) "The scaling of sprinkler discharge: Prediction of drop size", Report No. 10, Factory Mutual Research Corporation.
14. G. Heskestad (1972) "Proposal for studying interaction of water sprays with plume in sprinkler optimization program", Factory Mutual Research Corporation.
15. T.M. Myers and A.W. Marshall (2016) "A description of the initial fire sprinkler spray", *Fire Safety Journal*, Vol. 84, pp. 1–7.
16. T. Beji, S. Ebrahim Zadeh and B. Merci (2017) "Influence of the particle injection rate, droplet size distribution and volume flux angular distribution on the results and computational time of water spray CFD simulations", *Fire Safety Journal*, Vol. 91, pp. 586–595.
17. A.G. Bailey (1983) "The Rosin-Rammler size distribution for liquid droplet ensembles", *Journal of Aerosol Science*, Vol. 14, pp. 39–46. doi: 10.1016/0021-8502(83)90083-6.
18. NFPA (2015) 750 – Standard on Water Mist Fire Protection Systems.
19. G. Grant, J. Brenton and D. Drysdale (2000) "Fire suppression by water sprays," *Progress in Energy and Combustion Science*, Vol. 26, pp. 79–130.
20. J. Ramírez-Muñoz, A. Soria and E. Salinas-Rodríguez (2007) "Hydrodynamic force on interactive spherical particles due to the wake effect," *International Journal of Multiphase Flow*, Vol. 33(7), pp. 802–807.
21. J. Vaari, S. Hostikka, T. Sikanen and A. Pajaanen (2012) "Numerical simulations on the performance of water-based fire suppression systems", *VTT Technology*, Vol. 54, p 150.
22. S. Noda, B. Merci and T. Beji (2021) "Experimental and numerical study on the interaction of a water mist spray with a turbulent buoyant flame", *Fire Safety Journal*, Vol. 120, p. 103033.
23. S.J. Jordan, N.L. Ryder and A.W. Marshall (2017) "Spatially-resolved spray measurements and their implications", *Fire Safety Journal*, Vol. 91, pp. 723–729.
24. Y. Liu, T. Beji, M. Thielens, Z. Tang, Z. Fang and B. Merci (2022) "Numerical analysis of a water mist spray: The importance of various numerical and physical parameters, including the drag force," *Fire Safety Journal*, Vol. 127, p. 103515.
25. K. Lambert and S. Baaij (2011) *Brandverloop – Technisch bekeken, tactisch toegepast*. Sdu Uitgevers: Hague, Netherlands.

26. M.L. Bullen (1974) Fire Research Note 1016, Fire Research Station, Borehamwood, UK, pp. 1–11.

27. Z. Tang, Z. Fang and B. Merci (2014) "Development of an analytical model to quantify downward smoke displacement caused by a water spray for zone model simulations" *Fire Safety Journal*, Vol. 63, pp. 89–100.

28. D.J. Rasbash and Z.W. Rogowski (1957) "Extinction of fires in liquids by cooling with water sprays," *Combustion and Flame*, Vol. 1, pp. 453–466.

29. D.J. Rasbash, Z.W. Rogowski and G.W.V. Stark (1960) "Mechanisms of extinction of liquid fuels with water sprays" *Combustion and Flame*, Vol. 4, pp. 223–234.

30. M.F. Modest (2003) *Radiative Heat Transfer*. Academia Press: Cambridge, MA.

31. M. Försth and K. Möller (2013) "Enhanced absorption of fire induced heat radiation in liquid droplets," *Fire Safety Journal*, Vol. 55, pp. 182–196.

32. F.P. Incropera and D.P. De Witt (1996) *Fundamentals of Heat and Mass Transfer*. John Wiley and Sons Ltd: Hoboken, NJ.

33. R. Viskanta and C.C. Tseng (2007) *Combustion Theory and Modelling*, Vol. 11(1), pp. 113–125. Taylor & Francis Ltd: Oxford.

34. S. Hostikka and K. McGrattan (2006) "Numerical modeling of radiative heat transfer in water sprays," *Fire Safety Journal*, Vol. 41, pp. 76–86.

35. D. Drysdale (2011) *An Introduction to Fire Dynamics*, 3rd Ed. John Wiley & Sons, Ltd: Hoboken, NJ.

36. C.L. Beyler and L.Y. Cooper (2001) "Interaction of sprinklers with smoke and heat vents", *Fire Technology*, Vol. 37, pp. 9–35.

37. Y. Liu (2021) "The combined effect of water mist system and longitudinal ventilation on fire and smoke behavior in tunnel fires", Doctoral dissertation, Ghent University (Belgium) and Wuhan University (China).

8 Introduction to Fire Modelling in Computational Fluid Dynamics

8.1 INTRODUCTION

Computational Fluid Dynamics (CFD) is an advanced technique devoted to the analysis of a wide variety of turbulence, heat transfer and combustion problems. It consists of solving a set of Partial Differential Equations (PDEs) for conservation of mass, momentum and energy in conjunction with several auxiliary models dedicated, for example, to the treatment of turbulence, combustion and boundary conditions (see Chapter 2). The continuous increase in computational power has made CFD a popular technique among practitioners, engineers and researchers in many fields, including fire research and engineering.

Similar to the majority of practical combustion problems, fire is characterized by a complex coupling between several aspects exemplified in turbulence-chemistry interaction (TCI) or turbulence-radiation interaction (TRI). From this perspective, CFD is a very interesting tool that could provide particularly detailed and complex features of the fluid mechanics of fire. The large domain over which a fire might spread, requires nevertheless a subtle compromise between computational capabilities, on the one hand, and reliability and accuracy requirements, on the other.

Fire involves a large range of spatial and temporal scales. Despite the increasing computational power, only a subset of these scales could be simulated, and the rest must be modelled. The increase in CFD capabilities over the years can best be illustrated by the general trend of moving from the RANS (Reynolds-Averaged Navier–Stokes) technique, characterized by time-averaging of all eddies, to the LES (Large Eddy Simulation) technique, where large scales are resolved but subgrid processes are modelled. Furthermore, the current computational resources allow, but only for certain small-scale and academic cases, the exact solution of the flow field. This is called Direct Numerical Simulation (DNS). In terms of resolution, a nice parallelism has been suggested in Ref. [1] between these three techniques: DNS, LES and RANS and three self-portrait paintings of Rembrandt, Van Gogh and Picasso. It is also suggested in Ref. [1] that Rembrandt is not necessarily a better painter than Van Gogh or Picasso. In the same way, a low-resolution technique, such as RANS, should not necessarily be undermined. This brings us to the point of the level of knowledge required to perform reliable CFD simulations.

Performing a CFD simulation and obtaining a set of results for a specific test is *per se* not a difficult task. In most of the available CFD packages, developers strive to render 'the exercise' accessible to potential users that are not necessarily familiar with all the physical and numerical modelling hidden behind colourful and animated contour and isosurface plots. The quality of the available post-processing techniques might confer to the uninitiated user an overly rated sense of confidence. The CFD package then becomes a 'black box'. It is therefore imperative to perform CFD simulations with a good knowledge of the physics involved in the problem to be addressed. The results must be analysed, at best systematically, with a critical attitude towards not only the capabilities of the CFD package itself but also the modelling choices made by the user (cell size, boundary conditions, physical models…).

DOI: 10.1201/9781003204374-8

In general, the quality of a CFD package can only be ensured if it is continuously the subject of a Verification and Validation (V&V) process that goes in line with every new development. As discussed in Ref. [2], several definitions of V&V have been provided by several institutions. However, it is generally agreed that verification activities imply that the *accuracy of a computational solution is primarily measured relative to two types of highly accurate solutions: analytical solutions and highly accurate numerical solutions.* On the other hand, a validation process aims at measuring *how accurately the computational results compare to the experimental data, with quantified error and uncertainty of both.*

In the previous chapters of this book, a detailed description of several fluid mechanics problems of interest to fire research and engineering has been presented. In specific and simplified configurations, analytical and semi-empirical approaches have been exposed. However, for more complex configurations (e.g., induced by the evolving architectural design approaches), the powerful CFD technique becomes a more suitable and attractive approach. This chapter aims at providing the reader first with the basic theoretical knowledge required to understand the fundamentals of CFD simulations. Then, a specific section is devoted to worked-out examples of different complexities based on well-documented experimental tests that have become benchmark tests for CFD validation. Through these examples, a number of 'good practice rules' will be addressed in order to build good-quality CFD models.

8.2   LAMINAR DIFFUSION FLAMES

8.2.1   INSTANTANEOUS TRANSPORT EQUATIONS

The instantaneous transport equations are written as follows:
- **Mass** (see Eq. 2.31)

$$\frac{\partial \rho}{\partial t} + \nabla \cdot (\rho \vec{v}) = 0. \tag{8.1}$$

- **Momentum** (see Eq. 2.38)

$$\left\{ \begin{aligned} \frac{\partial}{\partial t}(\rho v_x) + \rho v_x \frac{\partial v_x}{\partial x} + \rho v_y \frac{\partial v_x}{\partial y} + \rho v_z \frac{\partial v_x}{\partial z} &= -\frac{\partial p}{\partial x} + \frac{\partial \tau_{xx}}{\partial x} + \frac{\partial \tau_{xy}}{\partial y} + \frac{\partial \tau_{xz}}{\partial z} + \rho g_x \\ \frac{\partial}{\partial t}(\rho v_y) + \rho v_x \frac{\partial v_y}{\partial x} + \rho v_y \frac{\partial v_y}{\partial y} + \rho v_z \frac{\partial v_y}{\partial z} &= -\frac{\partial p}{\partial y} + \frac{\partial \tau_{xy}}{\partial x} + \frac{\partial \tau_{yy}}{\partial y} + \frac{\partial \tau_{yz}}{\partial z} + \rho g_y \\ \frac{\partial}{\partial t}(\rho v_z) + \rho v_x \frac{\partial v_z}{\partial x} + \rho v_y \frac{\partial v_z}{\partial y} + \rho v_z \frac{\partial v_z}{\partial z} &= -\frac{\partial p}{\partial z} + \frac{\partial \tau_{xz}}{\partial x} + \frac{\partial \tau_{yz}}{\partial y} + \frac{\partial \tau_{zz}}{\partial z} + \rho g_z \end{aligned} \right. \tag{8.2}$$

- **Species** (see Eq. 2.67)

$$\frac{\partial}{\partial t}(\rho Y_i) + \nabla \cdot (\rho Y_i \vec{v}) = \nabla \cdot (\rho D_i \nabla Y_i) + \rho S_i, \, i = 1, \ldots, n. \tag{8.3}$$

- **Enthalpy** (see Eq. 2.41)

$$\frac{\partial}{\partial t}(\rho h) + \nabla \cdot (\rho h \vec{v}) = \nabla \cdot \left(\frac{\mu}{\mathrm{Pr}} \nabla h + \mu \sum_{i=1}^{N} \left(\frac{1}{\mathrm{Sc}_i} - \frac{1}{\mathrm{Pr}} \right) h_i \nabla Y_i \right) + \rho S_h, \tag{8.4}$$

where the Schmidt number, Sc, and the Prandtl number, Pr, are defined as follows (see Eqs. 2.89 and 2.88):

$$Sc = \frac{v}{D},$$

(8.5)

$$Pr = \frac{v}{\alpha}.$$

(8.6)

For unity Lewis number for all species (Le$_i$ = Sc $\cdot$ Pr^{-1} = 1 for all i, Eq. 2.90), this further simplifies to:

$$\frac{\partial}{\partial t}(\rho h) + \nabla \cdot (\rho h \vec{v}) = \nabla \cdot \left(\frac{\mu}{Pr} \nabla h\right) + \rho S_h.$$

(8.7)

The viscous tensor is given by Eq. (2.3):

$$\tau_{ij} = \mu \left[\left(\frac{\partial v_i}{\partial x_j} + \frac{\partial v_j}{\partial x_i} \right) - \frac{2}{3} \delta_{ij} \frac{\partial v_k}{\partial x_k} \right],$$

(8.8)

where μ is the laminar viscosity and δ_{ij} is the Kronecker symbol.

Species molecular diffusivities are typically described using Fick's law, Eq. (2.7):

$$\vec{J}_i = -\rho D_i \nabla Y_i.$$

(8.9)

8.2.2 COMBUSTION MODELLING

Let us consider the irreversible single-step chemical reaction between fuel (F) and oxidizer (O), which generates products (P). In terms of mass fraction, this chemical reaction may be written as follows:

$$v_F \ Y_F \ + \ v_O Y_O \to v_P Y_P, \text{(R1)}$$

where Y_F, Y_O and Y_P are the mass fractions of the fuel, the oxidizer and the product, respectively, and Y_F, Y_O and Y_P are the corresponding stoichiometric molar coefficients of the reaction. The mass stoichiometric coefficient is then expressed as follows:

$$s = \frac{v_O \ W_O}{v_F \ W_F},$$

(8.10)

where W_O and W_F are the molar weights of the oxidizer and the fuel, respectively.

0.2.2.1 Infinitely Fast Chemistry

The transport equations for fuel and oxidizer mass fractions and temperature can be written as follows [3]:

$$\frac{\partial}{\partial t}(\rho Y_F) + \nabla \cdot (\rho Y_F \vec{v}) = \nabla \cdot (\rho D \nabla Y_F) + \rho S_F,$$

(8.11)

$$\frac{\partial}{\partial t}\left(\rho Y_O\right)+\nabla\cdot\left(\rho Y_O\vec{v}\right)=\nabla\cdot\left(\rho D\nabla Y_F\right)+\rho S_O, \tag{8.12}$$

$$\frac{\partial}{\partial t}\left(\rho T\right)+\nabla\cdot\left(\rho T\vec{v}\right)=\nabla\cdot\left(\frac{\lambda}{c_p}\nabla T\right)+\rho S_T, \tag{8.13}$$

where the fuel and oxidizer are assumed to have equal diffusivities, D, and the heat capacities, c_p, of all chemical species are assumed equal and independent of temperature.

The oxidizer rate and the reaction rate for temperature are related to the fuel reaction rate through:

$$S_O = s\, S_F, \tag{8.14}$$

$$S_T = -\frac{\Delta h_c}{c_p} S_F. \tag{8.15}$$

Therefore, the conservation equations for fuel, oxidizer and temperature become:

$$\frac{\partial}{\partial t}\left(\rho Y_F\right)+\nabla\cdot\left(\rho Y_F\vec{v}\right)=\nabla\cdot\left(\rho D\nabla Y_F\right)+\rho S_F, \tag{8.16}$$

$$\frac{\partial}{\partial t}\left(\rho Y_O\right)+\nabla\cdot\left(\rho Y_O\vec{v}\right)=\nabla\cdot\left(\rho D\nabla Y_O\right)+\rho s S_F, \tag{8.17}$$

$$\frac{\partial}{\partial t}\left(\rho T\right)+\nabla\cdot\left(\rho T\vec{v}\right)=\nabla\cdot\left(\frac{\lambda}{c_p}\nabla T\right)-\rho\frac{\Delta h_c}{c_p} S_F. \tag{8.18}$$

Combining these three equations two by two and assuming a unity Lewis number, i.e., $\mathrm{Le} = \lambda/\rho c_p D = 1$, shows that the three quantities $Z_1 = sY_F - Y_O$, $Z_2 = \left(c_p T/\Delta h_c\right)+Y_F$ and $Z_3 = s\left(c_p T/\Delta h_c\right)+Y_O$ follow the same transport equation (with different boundary conditions) without source terms:

$$\frac{\partial}{\partial t}\left(\rho Z\right)+\nabla\cdot\left(\rho Z\vec{v}\right)=\nabla\cdot\left(\rho D\nabla Z\right). \tag{8.19}$$

Using the appropriate boundary conditions for the three scalars Z_1, Z_2 and Z_3 leads to the same normalized passive scalar with the same boundary condition: 0 on the oxidizer side and 1 on the fuel side.

For instance, the scalar Z_1 is normalized as follows:

$$Z = \frac{1+Z_1}{1+s}. \tag{8.20}$$

- On the fuel side (i.e., $Y_F = 1$ and $Y_O = 0$), $Z_1 = s$ and $Z = 1$
- On the oxidizer side (i.e., $Y_F = 0$ and $Y_O = 1$), $Z_1 = -1$ and $Z = 0$
- At the level of the reaction zone, infinitely fast chemistry implies $Y_F = 0$ and $Y_O = 0$. Thus, $Z_1 = 0$ and

$$Z = Z_{st} = \frac{1}{1+s}, \tag{8.21}$$

where Z_{st} is referred to as the stoichiometric mixture fraction.

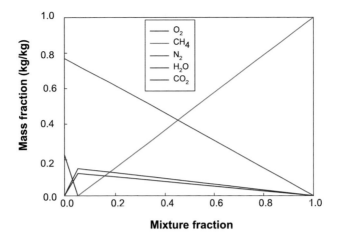

FIGURE 8.1 Species mass fractions as a function of the mixture fraction, in the case of complete combustion of methane at atmospheric conditions.

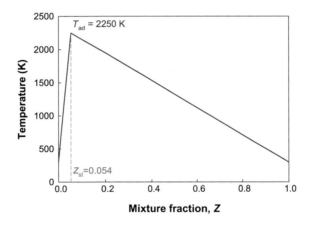

FIGURE 8.2 Temperature profile as a function of the mixture fraction, in the case of complete combustion of methane at atmospheric conditions and without heat losses.

By normalizing for instance Z_2, one can demonstrate that at the reaction zone (in the absence of heat losses), the adiabatic flame temperature is calculated as follows:

$$T_{ad} - T_{amb} = \frac{\Delta h_c}{c_p(1+s)}. \tag{8.22}$$

As discussed in Chapter 2, the mixture fraction concept, under the infinitely fast chemistry assumption, leads to piecewise functions of species concentrations (see Figure 8.1) and temperature (see Figure 8.2) as a function of the mixture fraction.

8.2.2.2 Finite-Rate Chemistry

The rate of the overall irreversible single-step chemical reaction (R1) is usually expressed as follows [4]:

$$k = AT^n \exp(-E_a / RT)[F]^a [O]^b, \tag{8.23}$$

TABLE 8.1

Examples of Single-Step Reaction Parameters [4]

Fuel	A	E_a	a	
CH_4	1.3×10^8	48.4	−0.3	1.3
CH_4	8.3×10^5	30.0	−0.3	1.3
C_2H_6	1.1×10^{12}	30.0	0.1	1.65
C_3H_8	8.6×10^{11}	30.0	0.1	1.65
C_4H_{10}	7.4×10^{11}	30.0	0.15	1.6
C_5H_{12}	6.4×10^{11}	30.0	0.25	1.5

where A is the pre-exponential factor, T is the temperature, E_a is the activation energy and R is the universal gas constant. The variables $[F]$ and $[O]$ are the molecular fractions of the fuel and the oxidizer, respectively. The parameters n, a and b are determined experimentally (see Table 8.1). Table 8.1 provides single-step reaction rate parameters for several hydrocarbons. The oxidizer is molecular oxygen. Units are cm-sec-mole-kcal-Kelvin [4].

The global reaction (R1) is a convenient way to approximate the effects of many elementary reactions. A more detailed description of the chemistry is provided by two-step reaction mechanisms. For instance, the following two-reaction model for methane oxidation is proposed in Ref. [4] to account for the production of carbon monoxide, CO:

$$CH_4 + \frac{3}{2}O_2 \rightarrow CO + 2H_2O, \tag{R2}$$

$$CO + \frac{1}{2}O_2 \rightarrow CO_2. \tag{R3}$$

8.3   TURBULENCE MODELLING

In contrast to laminar flows that are regular and stable, turbulent flows exhibit a chaotic behaviour that involves fluctuations at small scales in time and space. As mentioned in Section 2.8, the transition from laminar to turbulent is characterized by a dimensionless number, the Reynolds number. At low Reynolds numbers, the flow is laminar, whereas it is turbulent at high Reynolds numbers.

In a turbulent flow, structures of different scales, namely eddies, undergo a cascade process where large instable eddies break into smaller ones in a decaying process until the transferred kinetic energy is dissipated at the level of the smallest eddies. Richardson (1922) described the process as follows [5]:

> *Big whirls have little whirls,*
> *Which feed on their velocity;*
> *And little whirls have lesser whirls,*
> *And so on to viscosity.*

The 'big whirls' or largest eddies produce and contain most of the turbulent kinetic energy, E. An eddy is considered as part of the largest eddies (i.e., a 'big whirl') as long as its size ℓ is larger than a minimum size (see Section 2.8.3.2). This is the energy containing range. The transfer of turbulent kinetic energy to eddies that are smaller and smaller ('little whirls which feed on their velocity') occurs over the inertial sub-range. At the level of the smallest eddies, characterized by the Kolmogorov length scale, η (Eq. 2.117), the turbulent kinetic energy is dissipated into heat under the effect of the molecular viscosity, v ('and so on to viscosity'). This range is called the dissipation range.

8.3.1 DNS

As mentioned in Chapter 2 (Section 2.8.3.3), in DNS all the turbulent motions down to the smallest turbulent scales are resolved. The number of grid cells (resp. numerical operations) required for a three-dimensional DNS calculation of a turbulent flow is proportional to $\mathrm{Re}^{9/4}$ (resp. Re^{3}) [6]. For reacting flows (as it is the case in fire-related flow simulations), the chemical time and length (e.g., flame thickness) scales can be even significantly smaller than the smallest turbulent scale. As a consequence, the DNS technique cannot be applied to practical fire-related flows [6].

8.3.2 RANS

As mentioned in Section 2.8.3.3, as opposed to DNS, none of the turbulent motions are resolved in RANS. In Reynolds averaging, a variable Q is split into a mean $\bar{Q}$ and a deviation from the mean Q':

$$Q = \bar{Q} + Q'. \tag{8.24}$$

By definition $\overline{Q'} = 0$.

When the Favre (mass weighted) average $\tilde{Q}$ is used the variable is decomposed into:

$$Q = \tilde{Q} + Q'', \tag{8.25}$$

where

$$\tilde{Q} = \frac{\overline{\rho Q}}{\bar{\rho}}. \tag{8.26}$$

Again, by definition:

$$\tilde{Q}'' = \frac{\overline{\rho(Q - \tilde{Q})}}{\bar{\rho}} = 0. \tag{8.27}$$

The Reynolds-averaged Navier–Stokes equations for low-Mach number flows (weakly compressible) using Favre-averaged quantities are:

- Mass

$$\frac{\partial \bar{\rho}}{\partial t} + \frac{\partial}{\partial x_j} \left(\bar{\rho} \tilde{v}_j \right) = 0. \tag{8.28}$$

- Momentum

$$\frac{\partial (\bar{\rho} \tilde{v}_i)}{\partial t} + \frac{\partial}{\partial x_j} \left(\bar{\rho} \tilde{v}_j \tilde{v}_i \right) = -\frac{\partial}{\partial x_j} \left(\bar{\rho} \widetilde{v_i'' v_j''} \right) - \frac{\partial \bar{p}}{\partial x_i} + \frac{\partial}{\partial x_j} \overline{\tau_{ij}} + \overline{F_i}. \tag{8.29}$$

- Species

$$\frac{\partial}{\partial t}\left(\bar{\rho}\tilde{Y}_k\right) + \frac{\partial}{\partial x_j}\left(\bar{\rho}\tilde{v}_j\tilde{Y}_k\right) = -\frac{\partial}{\partial x_j}\left(\rho\,\widetilde{v_j''Y_k''}\right) - \frac{\partial\overline{J_j^k}}{\partial x_j} + \overline{\dot{\omega}_k}.$$

(8.30)

- Enthalpy

$$\frac{\partial}{\partial t}\left(\bar{\rho}\tilde{h}\right) + \frac{\partial}{\partial x_j}\left(\bar{\rho}\tilde{v}_j\tilde{h}\right) = -\frac{\partial}{\partial x_j}\left(\bar{\rho}\widetilde{v_j''h''}\right) + \frac{\partial\bar{p}}{\partial t} + \frac{\partial}{\partial x_j}\left(\overline{J_j^h} + \overline{v_i\tau_{ij}}\right) + \overline{v_j F_j}.$$

(8.31)

Note that radiation is ignored, for the moment, in the enthalpy equation.

Closures of the following unknown quantities are needed:

- Reynolds stresses $\widetilde{v_i''v_j''}$;
- Species $\widetilde{v_j''Y_k''}$ and enthalpy $\widetilde{v_j''h''}$ turbulent fluxes, usually closed using a gradient hypothesis:

$$\bar{\rho}\widetilde{v_j''Y_k''} = -\frac{\mu_t}{Sc_{k,t}}\frac{\partial\tilde{Y}_k}{\partial x_j},$$

(8.32)

$$\bar{\rho}\,\widetilde{v_j''h''} = -\frac{\mu_t}{Pr_t}\frac{\partial\tilde{h}}{\partial x_j};$$

(8.33)

- Laminar diffusive fluxes $\overline{J_j^k}$ and $\overline{J_j^h}$, which are generally neglected;
- Chemical reaction rates $\overline{\dot{\omega}_k}$: this is the focus of turbulent combustion modelling (see the next section).

The Reynolds stresses are modelled using the Boussinesq hypothesis:

$$-\bar{\rho}\widetilde{v_i''v_j''} = 2\mu_t\tilde{S}_{ij} - \frac{2}{3}\bar{\rho}k\delta_{ij},$$

(8.34)

where μ_t is the turbulent viscosity, $\tilde{S}_{ij}$ is the strain rate tensor and k is the turbulent kinetic energy, i.e., $k = \frac{1}{2}\widetilde{v_k''v_k''}$.

The strain rate tensor is calculated as follows:

$$\widetilde{S}_{ij} = \frac{1}{2}\left[\left(\frac{\partial\tilde{v}_i}{\partial x_j}\right) + \left(\frac{\partial\tilde{v}_j}{\partial x_i}\right)\right] - \frac{1}{3}\left(\frac{\partial\tilde{v}_k}{\partial x_k}\right)\delta_{ij}.$$

(8.35)

The turbulent viscosity is calculated as follows:

$$\mu_t = \rho C_\mu\frac{k^2}{\varepsilon},$$

(8.36)

where C_μ is a constant.

The parameters k and ε are modelled in RANS using a transport equation for each. Depending on the formulation of these transport equations, several models (described hereafter) are available within the RANS approach. It is also interesting to note that the integral time scale computed within the RANS approach is $\tau = k/\varepsilon$, which corresponds to a length scale $\ell = k^{3/2}/\varepsilon$.

8.3.2.1 Standard $k - \varepsilon$ Model

In this model, C_μ is assumed to be constant: $C_\mu = 0.09$.
  The transport equations for k and ε are:

$$\frac{\partial}{\partial t}\left(\bar{\rho}k\right)+\frac{\partial}{\partial x_j}\left(\bar{\rho}k\tilde{v}_j\right)=\frac{\partial}{\partial x_j}\left[\left(\mu+\frac{\mu_t}{\sigma_k}\right)\frac{\partial k}{\partial x_j}\right]+P+G-\bar{\rho}\varepsilon \tag{8.37}$$

and

$$\frac{\partial}{\partial t}\left(\bar{\rho}\varepsilon\right)+\frac{\partial}{\partial x_j}\left(\bar{\rho}\varepsilon\tilde{v}_j\right)=\frac{\partial}{\partial x_j}\left[\left(\mu+\frac{\mu_t}{\sigma_\varepsilon}\right)\frac{\partial \varepsilon}{\partial x_j}\right]+C_{1\varepsilon}\frac{\varepsilon}{k}P-C_{2\varepsilon}\bar{\rho}\frac{\varepsilon^2}{k}+S_{\varepsilon B} \tag{8.38}$$

with $P = -\overline{\rho v_i'' v_j''}\left(\partial \tilde{u}_i / \partial x_j\right)$ the production of turbulence due to shear, G the production of turbulence due to the buoyancy effect and $S_{\varepsilon B}$ a source term that expresses the effect of buoyancy on the turbulent dissipation rate. For the model constants in the ε-equation, the default values are: $C_{1\varepsilon} = 1.44$, $C_{2\varepsilon} = 1.92$, $\sigma_k = 1.0$ and $\sigma_\varepsilon = 1.3$.
  It is important to note that buoyancy modelling is particularly important in RANS [7]. For instance, in the simplest approach called SGDH (simple gradient diffusion hypothesis), the unknown density-velocity fluctuations and the term G are, respectively, modelled as follows:

$$\overline{\rho' v_j'} = -\frac{\mu_t}{\mathrm{Pr}_t}\frac{1}{\bar{\rho}}\frac{\partial \bar{\rho}}{\partial x_j}, \tag{8.39}$$

$$G = -\frac{\mu_t}{\mathrm{Pr}_t}\frac{1}{\bar{\rho}^2}\frac{\partial \bar{\rho}}{\partial x_j}\left(\frac{\partial \bar{p}}{\partial x_j}+\rho_{\mathrm{amb}}\,g_j\right). \tag{8.40}$$

Such approach (i.e., SGDH) is known to substantially underestimate the effect of buoyancy on turbulence [7]. As a consequence, the spreading rate of a vertical buoyant plume is underestimated and the spreading rate of the horizontal ceiling layer is overestimated [7]. The results could be improved when using the GGDH (generalized gradient diffusion hypothesis) approach where the unknown density-velocity fluctuations and the term G are, respectively, modelled as follows:

$$\overline{\rho' v_j'} = -\frac{3}{2}\frac{C_\mu}{\mathrm{Pr}_t}\frac{k}{\varepsilon}\left(\overline{v_j' v_k'}\frac{\partial \bar{\rho}}{\partial x_k}\right), \tag{8.41}$$

$$G = -\frac{3}{2}\frac{\mu_t}{\mathrm{Pr}_t}\frac{1}{\bar{\rho}^2 k}\left(\overline{v_j' v_k'}\frac{\partial \bar{\rho}}{\partial x_k}\right)\left(\frac{\partial \bar{p}}{\partial x_j}+\rho_{\mathrm{amb}}\,g_j\right). \tag{8.42}$$

The main difference is the inclusion of transversal density gradients into the buoyancy production terms.
  More details are provided in Ref. [7].

8.3.2.2 RNG $k - \varepsilon$ Model

This model is based on the Renormalization Group theory. The same transport equations as in the standard model are solved, except that $C_{1\varepsilon}$ parameter is no longer constant but reads:

$$C_{1\varepsilon} = C_{1\varepsilon}^a - C_\eta, \tag{8.43}$$

where $C_{1\varepsilon}^a$ is a model constant and C_η is defined as follows:

$$C_\eta = \frac{\eta(1-\eta/\eta_0)}{1+\beta\eta^3} \quad \text{and} \quad \eta = \sqrt{\frac{P}{\bar{\rho}C_\mu\varepsilon}}, \tag{8.44}$$

where $\eta_0 = 4.38$, $\beta = 0.015$, and $C_\mu = 0.0845$.

8.3.2.3   Realizable $k - \varepsilon$ Model

In this model, a new transport equation for ε is solved:

$$\frac{\partial}{\partial t}(\bar{\rho}\varepsilon) + \frac{\partial}{\partial x_j}(\bar{\rho}\varepsilon\tilde{v}_j) = \frac{\partial}{\partial x_j}\left[\left(\mu+\frac{\mu_t}{\sigma_\varepsilon}\right)\frac{\partial\varepsilon}{\partial x_j}\right] + C_1\bar{\rho}S_\varepsilon - C_2\bar{\rho}\frac{\varepsilon^2}{k+\sqrt{v\varepsilon}} + S_{\varepsilon B} \tag{8.45}$$

with $S = \sqrt{2\tilde{S}_{ij}\tilde{S}_{ij}}$, $C_1 = \max\left(0.43; \eta/(\eta+5)\right)$, $\eta = Sk/\varepsilon$ and v is the kinematic viscosity. The model constants are $C_2 = 1.9$, $\sigma_k = 1.0$ and $\sigma_\varepsilon = 1.2$.

8.3.2.4   Cubic $k - \varepsilon$ Model

In this model developed in Ref. [8], the ε-equation is a combination of Eqs (8.38) and (8.45) that allows for a good prediction of the spreading rate of an axisymmetric turbulent jet and the near-wall behaviour. The turbulent viscosity (Eq. 8.36) is locally flow dependent.

8.3.3   LES

As opposed to RANS, in the LES technique, turbulent motions are resolved to an extent that is defined by a low-pass filter in wave number space. Conceptually, the filter must be in the inertial sub-range. It is typically taken as the cell size (more precisely the cubic root of the cell volume), which is referred to as implicit filtering. In other words, the turbulent motion of the largest eddies is resolved and the effect of the smallest scales is modelled (i.e., subgrid modelling). As a consequence, in LES there is no averaging (i.e., $\bar{\phi}' \neq 0$) but rather filtering.

Filtering the instantaneous transport equations leads to equations similar to the Reynolds-averaged transport equations:

• Mass

$$\frac{\partial\bar{\rho}}{\partial t} + \frac{\partial}{\partial x_j}(\bar{\rho}\tilde{v}_j) = 0. \tag{8.46}$$

• Momentum

$$\frac{\partial(\bar{\rho}\tilde{v}_i)}{\partial t} + \frac{\partial}{\partial x_j}(\bar{\rho}\tilde{v}_j\tilde{v}_i) = -\frac{\partial}{\partial x_j}\left[\bar{\rho}\left(\widetilde{v_iv_j}-\tilde{v}_i\tilde{v}_j\right)\right] - \frac{\partial\bar{p}}{\partial x_i} + \frac{\partial}{\partial x_j}\overline{\tau_{ij}} + \overline{F_i}. \tag{8.47}$$

• Species

$$\frac{\partial(\bar{\rho}\tilde{Y}_k)}{\partial t} + \frac{\partial}{\partial x_j}(\bar{\rho}\tilde{v}_j\tilde{Y}_k) = -\frac{\partial}{\partial x_j}\left[\bar{\rho}\left(\widetilde{v_jY_k}-\tilde{v}_j\tilde{Y}_k\right)\right] + \overline{\dot{\omega}_k}. \tag{8.48}$$

• Enthalpy

$$\frac{\partial(\bar{\rho}\tilde{h})}{\partial t} + \frac{\partial}{\partial x_j}(\bar{\rho}\tilde{v}_j\tilde{h}) = -\frac{\partial}{\partial x_j}\left[\bar{\rho}\left(\widetilde{v_jh}-\tilde{v}_j\tilde{h}\right)\right] + \frac{\partial\bar{p}}{\partial t} + \frac{\partial}{\partial x_j}\left(\overline{J_j^h}+\overline{v_i\tau_{ij}}\right) + \overline{v_jF_j}. \tag{8.49}$$

Note that radiation is ignored, for the moment, in the enthalpy equation.

The unknown LES quantities are as follows:

- Unresolved residual stresses $\widetilde{v_i''v_j''} = \widetilde{v_i''v_j''} - \tilde{v}_i\tilde{v}_j$ that require a subgrid-scale turbulence model,
- Unresolved residual species fluxes $\widetilde{v_j''Y_k''} = \widetilde{v_j''Y_k''} - \tilde{v}_j\tilde{Y}_k$ and enthalpy fluxes $\widetilde{v_j''h''} = \widetilde{v_j''h''} - \tilde{v}_j\tilde{h}$,
- Filtered chemical reaction rate $\bar{\dot{\omega}}_k$.

Several models have been developed to model the 'residual' or 'subgrid' stresses. The most widely used one (because of its simple formulation) is the Smagorinsky model [9]. In this model, the instantaneous velocities are replaced by the resolved (filtered) velocities and the turbulent viscosity is defined as follows:

$$v_t = \left(C_s\Delta\right)^2\left(2\widetilde{S}_{ij}\widetilde{S}_{ij}\right)^{1/2},$$ (8.50)

where C_s is a model constant and Δ is the filter width (that is, as said earlier, often taken as the cell size). For a homogeneous and isotropic turbulence, the model constant is estimated as $C_s \simeq 0.17$. However, it has been reported to depend on the flow configuration in addition to the fact that the Smagorinsky model is too dissipative. A popular method to overcome these limitations is to use a 'dynamic' approach where the model parameter is determined locally. Thanks to its self-adjustment capability, the 'dynamic' model provides locally the 'right' amount of dissipation. The implementation of the dynamic approach (where the coefficient is taken out of the brackets and thus can be in some cases negative) requires, however, special care (e.g., clipping or averaging) to avoid numerical stabilization problems.

Note that Eq. (8.50) has been modified to account for buoyancy as follows [10]:

$$v_t = \left(C_s\Delta\right)^2\left(2\widetilde{S}_{ij}\widetilde{S}_{ij} - \frac{1}{\text{Fr}\,\text{Pr}_t}\frac{\partial T'}{\partial x_i}\delta_{ij}\right)^{1/2}.$$ (8.51)

It has been found, for example, in Ref. [11], that the buoyancy modification has little influence on simulation results.

A second approach to model the turbulent viscosity in the context of LES is the Deardorff or 'one-equation eddy viscosity' model [12–14] in which the turbulent viscosity is calculated as follows:

$$v_t = c_k\Delta\sqrt{k^{\text{SGS}}},$$ (8.52)

where c_k is a model constant.

A transport equation for the subgrid-scale kinetic energy, k^{SGS}, is solved:

$$\frac{\partial\left(\bar{\rho}k^{\text{SGS}}\right)}{\partial t} + \frac{\partial}{\partial x_j}\left(\bar{\rho}\tilde{v}_j k^{\text{SGS}}\right) = \frac{\partial}{\partial x_j}\left[\left(\mu + \frac{\mu_t}{Sc_t}\right)\frac{\partial k^{\text{SGS}}}{\partial x_j}\right] + P - \bar{\rho}\varepsilon,$$ (8.53)

where P is the production term and the dissipation rate is expressed as follows:

$$\varepsilon = c_\varepsilon\left(k^{\text{SGS}}\right)^{3/2}\Delta^{-1},$$ (8.54)

where c_ε is a dimensionless model constant.

Note that Deardorff's model is used, for instance, in Ref. [15]. Also, a modified version is implemented in Refs. [16,17], where k^{SGS} is estimated from the flow rather than calculated using a transport equation.

Additional turbulent viscosity models [18] are available in the literature such as the Vreman model that relies on a Taylor series expansion of the velocity field.

8.4   TURBULENT NON-PREMIXED COMBUSTION

It is explained in many textbooks (e.g., [3]) that simply using expressions like Eq. (8.2.23) with Favre-averaged values does not provide an accurate expression for the mean reaction rate. Indeed, it is not allowed to ignore the large (and correlated) fluctuations in temperature and species mass fractions calculations. Several approaches have been adopted to overcome this issue. Furthermore, in fires, it is very often important to compute not only the combustion reaction rate but also the generation of products of incomplete combustion such as carbon monoxide (CO), oxides of nitrogen (NO_x), hydrogen cyanide (HCN) and soot, which are toxic [19]. The formation of these products is often governed by slow chemical reactions (i.e., kinetically controlled), and therefore the fast chemistry approach is not valid.

8.4.1   Infinitely Fast Chemistry with a Presumed PDF

If infinitely fast chemistry is considered, there is a direct relation between mass fractions or temperature and the local mixture fraction, Eq. (2.72). One example of an infinitely fast chemistry model is the Burke–Schumann model (i.e., piecewise linear expressions). Using Z now as notation for normalized mixture fraction, ξ, the mean quantities can be directly obtained with a PDF ('probability density function'):

$$\tilde{Y}_k = \int_0^1 Y_k\left(Z^*\right)\tilde{P}\left(Z^*\right)dZ^*, \tag{8.55}$$

$$\tilde{T} = \int_0^1 T\left(Z^*\right)\tilde{P}\left(Z^*\right)dZ^*. \tag{8.56}$$

The probability density function $\tilde{P}\left(Z^*\right)$ quantifies the probability to find a mixture fraction Z within the range of $\left[Z^* - \Delta Z/2, Z^* + \Delta Z/2\right]$.

An important question is thus, what the PDF looks like. The PDF is often modelled as a β-function because it can take on many different shapes.

The β-function is defined as follows:

$$\tilde{P}(Z) = \frac{Z^{a-1}\left(1-Z\right)^{b-1}}{\Gamma(a)\Gamma(b)}\Gamma(a+b), \tag{8.57}$$

where Γ is the gamma function and the parameters a and b are defined as follows:

$$a = \gamma\,\tilde{Z}, \tag{8.58}$$

$$b = \gamma\left(1-\tilde{Z}\right), \tag{8.59}$$

$$\gamma = \frac{\tilde{Z}\left(1-\tilde{Z}\right)}{\tilde{Z}''^2} - 1 \geq 0, \tag{8.60}$$

where $\tilde{Z}$ is the mean mixture fraction computed from the transport equation:

$$\frac{\partial}{\partial t}\left(\bar{\rho}\,\tilde{Z}\right)+\frac{\partial}{\partial x_j}\left(\bar{\rho}\,\tilde{v}_j\tilde{Z}\right)=\frac{\partial}{\partial x_j}\left(\overline{\rho D\frac{\partial Z}{\partial x_j}}-\overline{\rho v_j''Z''}\right). \tag{8.61}$$

The laminar diffusive flux, $\overline{\rho D\left(\partial Z / \partial x_j\right)}$, is generally neglected, and the turbulent flux is modelled with a gradient diffusion hypothesis:

$$\overline{\rho v_j''Z''}=\frac{-\mu_t}{Sc_t}\frac{\partial \tilde{Z}}{\partial x_j}. \tag{8.62}$$

• **In RANS models**, the mixture fraction variance $\tilde{Z}''^2$ is typically retrieved from a transport equation:

$$\frac{\partial}{\partial t}\left(\bar{\rho}\tilde{Z}''^2\right)+\frac{\partial}{\partial x_j}\left(\bar{\rho}\,\tilde{v}_j\tilde{Z}''^2\right)=\frac{\partial}{\partial x_j}\left(\frac{\mu_t}{Sc_{t,\tilde{Z}''^2}}\frac{\partial \tilde{Z}''^2}{\partial x_j}\right)+2\frac{\mu_t}{Sc_t}\frac{\partial \tilde{Z}}{\partial x_j}\frac{\partial \tilde{Z}}{\partial x_j}-\bar{\rho}\tilde{\chi}, \tag{8.63}$$

where the scalar dissipation rate, $\tilde{\chi}$, is defined as follows:

$$\tilde{\chi}=2D\frac{\partial \tilde{Z}}{\partial x_j}\frac{\partial \tilde{Z}}{\partial x_j}, \tag{8.64}$$

where D is the molecular diffusivity.

Assuming similarity to the dissipation of the turbulent kinetic energy, a simple model for the scalar dissipation rate is:

$$\tilde{\chi}=c_\phi\frac{\varepsilon}{k}\tilde{Z}''^2=c_\phi\frac{1}{\tau_t^{RANS}}\tilde{Z}''^2, \tag{8.65}$$

where c_ϕ is the so-called time scale ratio, usually taken as $c_\phi = 2$ [20].

• **In LES models**, the scalar dissipation rate is modelled as follows:

$$\tilde{\chi}=2D_t\frac{\partial \tilde{Z}}{\partial x_j}\frac{\partial \tilde{Z}}{\partial x_j}, \tag{8.66}$$

where D_t is the eddy diffusivity (as opposed to the molecular diffusivity used in RANS, Eq. 8.63) and it can be modelled as follows [20]:

$$D_t=\left(c_z\Delta\right)^2\tilde{S}, \tag{8.67}$$

where $\tilde{S}$ is the Favre-filtered strain rate and c_z is a coefficient that can be determined using a dynamic procedure.

Equation (8.65) rewritten in a LES context gives [20]:

$$\tilde{\chi}=c_\phi\frac{1}{\tau_t^{SGS}}\tilde{Z}''^2. \tag{8.68}$$

Combining Eqs. (8.66) to (8.68) and assuming $\tau_t^{SGS} \sim 1/\tilde{S}$ gives [20]:

$$\tilde{Z}''^2 = c_v \Delta^2 \frac{\partial \tilde{Z}}{\partial x_j} \frac{\partial \tilde{Z}}{\partial x_j},\tag{8.69}$$

where the coefficient $c_v = 2c_z^2 / c_\phi$ can be determined using a dynamic procedure.

In alternative models for LES, a transport equation for the mixture fraction variance is solved.

Thus, having accepted that the PDF is a β-function, the local shape can now be determined from Eq. (8.57), having obtained $\tilde{Z}$ from Eq. (8.61) and $\tilde{Z}''^2$ from Eq. (8.63) or (8.69). The missing piece of information is the chemistry model.

8.4.1.1   Flame Sheet Model

In this model, which is also called the Burke–Schumann model or the 'mixed-is-burnt' model, piecewise linear expressions are used to obtain species concentrations as a function of the mixture fraction (see Figure 8.1).

8.4.1.2   Chemical Equilibrium Model

In this model, instantaneous chemical equilibrium is assumed. With this chemistry model, intermediate species and radicals can be taken into consideration. However, it must be noted that often there is not enough time for the complete system to go to local chemical equilibrium.

8.4.1.3   Steady Laminar Flamelet Modelling (SLFM)

In the laminar flamelet concept, the turbulent flame is viewed as an ensemble of thin laminar one-dimensional flames, i.e., flamelets, which are distorted and stretched by turbulent motion.

The flamelet equation for species is calculated as follows:

$$\rho \frac{\partial Y_k}{\partial t} = \dot{\omega}_k + \frac{\rho}{2} \chi \frac{\partial^2 Y_k}{\partial Z^2},\tag{8.70}$$

where the scalar dissipation rate, χ, is defined in Eq. (8.64).

If the response time of the flamelets to changes in the local mixing is assumed to be much smaller than the mixing time scale, i.e., the Stationary Laminar Flamelet Model (SLFM), Eq. (8.70) becomes:

$$\rho \chi \frac{d^2 Y_k}{dZ^2} + 2\dot{\omega}_k = 0.\tag{8.71}$$

The flame structure is provided in the mixture fraction space as a function of the scalar dissipation rate:

$$Y_k = Y_k(Z, \chi).\tag{8.72}$$

Average reaction rates are computed based on a library where χ is paramterized by its value at stoichiometric fraction, χ_s. Such library is generated by computing the counterflow laminar flame for a range of χ_s. The average reaction rates and species concentrations are then expressed as follows:

$$\overline{\dot{\omega}_k} = \int_0^\infty \int_0^1 \dot{\omega}_k(Z, \chi) \overline{P}(Z, \chi) dZ \, d\chi,\tag{8.73}$$

$$\overline{Y_k} = \int\limits_{0}^{\infty}\int\limits_{0}^{1} Y_k(Z,\chi)\overline{P}(Z,\chi)\,dZ\,d\chi, \tag{8.74}$$

where $\dot{\omega}_k(Z,\chi)$ is the laminar flame solution (see Eq. 8.72) and $\overline{P}(Z,\chi)$ is the joint probability density function (PDF) of Z and χ.

The steady flamelet model is often used, especially in LES, because of its simplicity and considerable improvements over fast chemistry. However, the steady-state assumption is inaccurate if slow chemical or physical processes have to be considered such as the formation of pollutants and radiative heat transfer [20]. If the time-dependent term is kept in Eq. (8.70), the unsteady Laminar Flamelet Model (LFM) is retrieved and finite-rate chemistry can be accounted for as discussed in the next section. For more details on LFM, the reader is referred to, e.g., Ref. [21].

8.4.2 FINITE-RATE CHEMISTRY

8.4.2.1 Eddy Break-Up (EBU) Model and Eddy Dissipation Model (EDM)

In the Eddy Break-Up (EBU) model [22], the mean fuel reaction rate is proportional to the fluctuations of fuel concentration:

$$\overline{\dot{\omega}_F} \sim \overline{\rho}\frac{1}{\tau_t}\left(\tilde{Y}_F''^2\right)^{1/2}, \tag{8.75}$$

where $\tilde{Y}_F''^2$ is calculated using a transport equation similar to $\tilde{Y}_F$ and τ_t is the integral time scale for turbulent mixing.

An update of the EBU model has been proposed in Refs. [23,24] and is referred to as the Eddy Dissipation Model (EDM). In EDM, the fuel reaction rate is assumed to be governed by the mean species concentrations (of fuel, oxidizer and products) and is written in the following form:

$$\overline{\dot{\omega}_F} = -C_R\overline{\rho}\frac{1}{\tau_t}\min\left(\tilde{Y}_F;\frac{\tilde{Y}_O}{s};\frac{\tilde{Y}_P}{1+s}\right), \tag{8.76}$$

where s is the mass stoichiometric coefficient (see Eq. 8.10) and C_R is a model parameter that could be taken as constant (typically a value of 4.0 is used [19]) or modelled, for instance, with the viscous mixing model as follows [19]:

$$C_R = 23.6\left(\frac{\nu}{\tau_t k}\right)^{1/4}. \tag{8.77}$$

The EDM is only suitable for one- or two-step global reactions.

- **In RANS models**, the integral time scale for turbulent mixing is calculated as $\tau_t = k/\epsilon$.
- **In LES models**, the integral time scale for turbulent mixing is calculated as follows:

$$\tau_t = \tau_t^{\mathrm{SGS}} = \frac{k^{\mathrm{SGS}}}{\varepsilon^{\mathrm{SGS}}} = \frac{\Delta}{2\sqrt{k^{\mathrm{SGS}}}}, \tag{8.78}$$

where the superscript SGS denotes subgrid-scale modelling.

A more general expression for the turbulent mixing time is given in Refs. [16,17] by:

$$\tau_{\text{mix}} = \max\left(\tau_{\text{chem}}, \min\left(\tau_d, \tau_t^{\text{SGS}}, \tau_g, \tau_{\text{flame}}\right)\right), \tag{8.79}$$

where τ_d is the molecular diffusion time ($\tau_d = \Delta^2/D_F$), τ_g is a gravitational acceleration-based time scale ($\tau_g = \sqrt{2\Delta/g}$) and τ_{flame} is a time scale related to highly turbulent jet flames.

8.4.2.2 Eddy Dissipation Concept (EDC)

In the generalized EDC model [25], the computational cell is divided into two sub-zones: the reacting 'fine structure' and the 'surrounding fluid' [26]. Chemical reactions are assumed to occur in the fine structure, and the reaction rate of any species is expressed as follows:

$$\overline{\dot{\omega}_k} = \overline{\rho}\,\frac{1}{\tau^*}\left(\frac{\gamma^{*2}}{1-\gamma^{*3}}\right)\left(Y_k^* - \tilde{Y}_k\right). \tag{8.80}$$

The size γ^* of the mean structures and the mean residence time (i.e., chemical reaction time), τ^*, are expressed as follows:

$$\gamma^* = 2.1377\left(\frac{v}{\tau_t k}\right)^{1/4}, \tag{8.81}$$

$$\tau^* = 0.4082\left(\frac{v}{\varepsilon}\right)^{1/2}, \tag{8.82}$$

where τ_t is the turbulent mixing time that is calculated depending on whether RANS or LES modelling is used.

The parameter Y_i^* is the fine-scale species mass fraction after reacting over the time scale τ^*, which can be determined through the laminar finite-rate model.

8.4.2.3 Conditional Moment Closure (CMC)

The Conditional Moment Closure (CMC) for non-premixed turbulent combustion modelling has been developed by Klimenko and Bilger [27]. The main concept of this modelling technique relies on the assumption that the fluctuations of the reactive scalars at a given mixture fraction value (i.e., conditional fluctuations) are small; this is referred to as 'first-order' CMC.

The fluctuations of the chemical source term in the mixture fraction space are neglected. The chemical source term is then expressed as follows:

$$\dot{\omega}_k \mid Z = f\left(Y_1 \mid Z, \ldots, Y_N \mid Z, T \mid Z\right). \tag{8.83}$$

The mean fuel mass fraction, $\tilde{Y}_F$, is expressed as follows:

$$\overline{\rho}\tilde{Y}_F = \int_0^1 \overline{\left(\rho Y_F \mid Z\right)}\,\overline{P}(Z)\,dZ, \tag{8.84}$$

where Z represents the mixture fraction space, $\overline{\left(\rho Y_F \mid Z\right)}$ is the conditional moment of the fuel mass fraction, and $\overline{P}$ is a presumed β-PDF.

The CMC model has originally been proposed in a RANS context, but it has been formulated also in a LES context. More details are found in Ref. [21].

8.4.2.4 Transported PDF Models

One disadvantage of the presumed PDF approach described above is that in reality PDFs do not always evolve within one class of functions. Furthermore, for practical reasons, a very limited number of variables is used to parametrize the chemistry. In the transported PDF (TPDF) approach [28], the shape of the joint PDF is not presumed, and the chemical source term appears in closed form (i.e., no modelling is required for the chemistry). However, the modelling problem is shifted to the micro-mixing term [28]. The main disadvantage of the TPDF method is the relatively high computational cost, so that it is not (yet) commonly used in CFD for fire.

When the TPDF approach (initially implemented in RANS) is implemented in LES, it is called Transported Filtered Density Function, TFDF. It is mentioned in Ref. [20] that *because of the inherent unsteadiness in LES, the main challenges in applying Filtered Density Function (FDF) methods in LES are the computational costs and the formulation of robust and consistent algorithms* [to achieve] *statistical convergence.*

8.5 RADIATION MODELLING

Thermal radiation is a very important mode of heat transfer in fires that becomes increasingly dominant with high temperatures. It involves several processes such as absorption and scattering. When radiant energy strikes a surface, part of it could be absorbed. However, there could be particles along the direction of propagation of radiation that continuously abstract energy from the incident wave and re-radiate that energy in all directions. This process is called scattering. Thermal radiation that is neither absorbed nor scattered (or reflected) is transmitted along the path.

In order to quantify the amount of thermal radiation involved in a fire, one must evaluate the radiation intensity or its rate of change along a direction of propagation (of radiation). This is described by the governing equation for the spectral intensity $I_\lambda(r,s)$, known as the Radiative Transfer Equation (RTE) [29]:

$$\frac{dI_\lambda(r,s)}{ds}+(\kappa_\lambda+\sigma_{\lambda s})I_\lambda(r,s)=\kappa_\lambda I_{\lambda b}(r,s)+\frac{\sigma_{\lambda s}}{4\pi}\int_{4\pi}I_\lambda(r,s')\phi(\lambda,s,s')d\Omega', \qquad (8.85)$$

where r is a position in the spatial domain, s is the unit vector along a direction of propagation of radiation, λ is the wavelength, κ_λ is the absorption coefficient (may be positive or negative for absorbing or emitting media), $\sigma_{\lambda s}$ is the scattering coefficient, $I_{\lambda b}$ is the Planck function describing the spectral intensity of blackbody radiation, Ω is the scattering solid angle and ϕ is a phase function for scattering.

There are mainly two issues here:

1. How to solve the RTE (Section 8.5.1)?
2. How to model the coefficients (e.g., absorption coefficients in Section 8.5.2)?

In order to simplify the RTE in the context of fire dynamics, several considerations and assumptions are addressed hereafter. First of all, one can see that the radiation intensity, I_λ, the absorption coefficient, κ_λ, the scattering coefficient, $\sigma_{\lambda s}$, and the blackbody radiation, $I_{\lambda b}(r,s)$, are all dependent on the wavelength, λ. Integrating the intensities I_λ and $I_{\lambda b}$ over the entire spectrum of wavelengths gives:

$$I=\int_0^\infty I_\lambda\, d\lambda, \qquad (8.86)$$

$$I_b = \int_0^\infty I_{\lambda b} \, d\lambda = \frac{\sigma T^4}{\pi}, \tag{8.87}$$

where σ is the Stefan–Boltzmann constant ($\sigma \simeq 5.67 \times 10^{-11} \, \text{kW} / (\text{m}^2 \, \text{K}^4)$).

Furthermore, the incident mean absorption coefficient is defined as follows:

$$\kappa = \frac{\displaystyle\int_0^\infty \kappa_\lambda I_\lambda d\lambda}{\displaystyle\int_0^\infty I_\lambda d\lambda}, \tag{8.88}$$

and the Planck mean absorption coefficient is defined as follows:

$$\kappa_P = \frac{\displaystyle\int_0^\infty \kappa_\lambda I_{\lambda b} d\lambda}{\displaystyle\int_0^\infty I_{\lambda b} d\lambda}. \tag{8.89}$$

Additionally, it is common practice to neglect the scattering effect that is assumed to be negligible (i.e., $\sigma_{\lambda s} \simeq 0$) for the main combustion products (i.e., water vapour and carbon dioxide, and also carbon monoxide).

Under this assumption (i.e., non-scattering medium) and using Eqs. (8.86–8.89), the RTE becomes:

$$\frac{dI}{ds} = \kappa_P I_b - \kappa I, \tag{8.90}$$

where the first term on the right-hand side is the emission term and the second term on the right-hand side is the absorption term.

Under the grey gas assumption, which states that $\kappa_P = \kappa$, Eq. (8.90) reduces to:

$$\frac{dI}{ds} = \kappa_P (I_b - I) = \kappa_P \left(\frac{\sigma T^4}{\pi} - I \right). \tag{8.91}$$

8.5.1   MODELS FOR RADIATIVE TRANSFER

8.5.1.1   The P-1 Radiation Model

The radiative heat flux onto a surface element is expressed as follows [29]:

$$\vec{q}_{\text{rad}} \cdot \vec{n} = \int_{4\pi} I \, \vec{n} \cdot \vec{s} \, d\Omega, \tag{8.92}$$

where $\vec{n}$ is the unit surface normal (pointing away from the surface into the medium), $\vec{s}$ is the unit vector into a given direction and Ω is a solid angle that varies over a total range of 4π (that takes into account the incoming as well as the outgoing contribution).

The incident radiation onto a surface (i.e., direction-integrated intensity) is defined as follows [29]:

$$G = \int_{4\pi} I(r,s)\,d\Omega. \tag{8.93}$$

Integration of Eq. (8.91) over all solid angles using Eq. (8.92) gives the divergence of the radiative heat flux [29]:

$$\nabla \cdot \vec{q}_{\text{rad}} = \kappa_P \left(4\sigma T^4 - \int_{4\pi} I\,d\Omega \right) = \kappa_P \left(4\sigma T^4 - G \right). \tag{8.94}$$

As mentioned in Ref. [29], this equation is a statement of the *conservation of the radiative energy*.

The radiation intensity can be expressed by means of a series of spherical harmonics. The exact solution is obtained if the order of the approximation is taken as infinity. If the order is taken as 1, the model is called the P-1 model and the transport equation for G is written as follows [29]:

$$\nabla \cdot \left(\frac{1}{3\kappa_P} \nabla G \right) - \kappa_P G + 4\kappa_P \sigma T^4 = 0. \tag{8.95}$$

Solving Eqs. (8.95) and then (8.94) allows to calculate the divergence of the radiative heat flux, $\nabla \cdot \vec{q}_{\text{rad}}$, that is used in the energy transport equation to account for radiative heat transfer.

8.5.1.2   The Finite Volume Method (FVM)

In this method, the directions of propagation are discretized using a polar angle, θ, that varies in $[0,\pi]$, and an azimuthal angle, ϕ, that varies in $[0,2\pi]$. The RTE (under the grey gas assumption for a non-scattering medium) is then expressed as follows:

$$\frac{dI(r,\theta,\phi)}{ds} = \kappa_P \left(\frac{\sigma T^4}{\pi} - I(r,\theta,\phi) \right). \tag{8.96}$$

The solution of Eq. (8.96) allows to calculate the incident radiation, G (Eq. 8.93), and then compute the divergence of the radiative heat flux, $\nabla \cdot \vec{q}_{\text{rad}}$.

There are other methods to solve the RTE that are not addressed here such as the Monte Carlo method [29], the discrete ordinates model (DOM) [30] and the discrete transfer radiative model (DTRM) [31].

8.5.2   MODELS FOR THE ABSORPTION COEFFICIENT

In order to save computational time, it is common practice to avoid using spectral calculations for the absorption coefficient, κ, and rely instead on simplified models. It is important first to clearly make the distinction between gas and soot (which consists of solid carbonaceous particles). The absorption of a mixture is then defined as the sum of gas and soot absorption coefficients:

$$\kappa = \kappa_{\text{gas}} + \kappa_{\text{soot}}. \tag{8.97}$$

The gas absorption coefficient could be, for example, expressed as a function of the temperature [32]:

$$\kappa_{\text{gas}} = 0.32 + 0.28\exp(-T/1,135). \tag{8.98}$$

Another, more complex, model is the Weighted-Sum-of-Grey-Gases (WSGG) model that expresses the gas absorption coefficient as follows:

$$\kappa_{\text{gas}} = -\frac{1}{S}(1-\varepsilon), \tag{8.99}$$

where S is the path length (typically taken as the cell size) and ε is the gas emissivity expressed as follows:

$$\varepsilon = \sum_{i=1}^{I} a_i(T)\left[1 - e^{-k_i PS}\right], \tag{8.100}$$

where I is the number of grey gases, a_i is a temperature-dependent coefficient, κ_i is the absorption coefficient of grey gas i and P is the sum of the partial pressures.

The absorption coefficient for soot is generally correlated to the local soot volume fraction, f_v, and temperature, T. Novozhilov [19] suggests the following correlation:

$$\kappa_{\text{soot}} = 1,264\, f_v\, T. \tag{8.101}$$

8.5.3　Turbulence Radiation Interaction (TRI)

In order to account for the turbulence radiation interaction (TRI), the mean radiative heat source must be evaluated by taking the mean of Eq. (8.94):

$$\overline{\nabla \cdot \vec{q}_{\text{rad}}} = 4\sigma\,\overline{\kappa_P T^4} - \overline{\kappa_P G}. \tag{8.102}$$

Two correlations are thus required:
1. The correlations for the emission term between the Planck mean absorption coefficient and the fourth power of temperature:

$$\overline{\kappa_P T^4} = \overline{\kappa_p}\,\overline{T^4} + \overline{\kappa_P'\left(T^4\right)'} \tag{8.103}$$

   where the fluctuations are denoted by the prime and $\overline{T^4} \neq \overline{T}^4$.
2. The correlations for the absorption term between the Planck mean absorption coefficient and the incident radiation:

$$\overline{\kappa_P G} = \overline{\kappa_p}\,\overline{G} + \overline{\kappa_P' G'}. \tag{8.104}$$

Previous studies [29] have suggested that if the mean free path for radiation is much larger than turbulence eddy length scale, ℓ_t (i.e., $\kappa_P \ell_t \ll 1$), then the local radiative intensity is weakly correlated with the local absorption coefficient, i.e., $\overline{\kappa_P G} \simeq \overline{\kappa_p} \overline{G}$. This is called the optically thin fluctuation approximation (OTFA). This approximation is mostly appropriate for non-sooty and small-scale flames.

Similar to turbulence-chemistry interaction, the treatment of TRI varies with respect to the turbulence modelling framework, namely RANS or LES.

For instance, in Ref. [33] in a RANS context, the OTFA has been used and the term $\overline{\kappa_P T^4}$ has been approximated by:

$$\overline{\kappa_P T^4} = \overline{\kappa_P} \overline{T^4} \left(1 + C_{\text{TRI1}} 6 \frac{\overline{T'^2}}{\overline{T}^2} + C_{\text{TRI2}} 4 \frac{\overline{T'^2}}{\overline{\kappa_P T}} \frac{\partial \kappa_P}{\partial T} \bigg|_{\overline{T}} \right), \tag{8.105}$$

where $\overline{T}$ and T' are the average and the fluctuating component of the temperature, respectively (i.e., $T = \overline{T} + T'$, see Eq. 8.24). The latter is solved using a transport equation. The parameters C_{TRI1} and C_{TRI2} are taken as model constants with $C_{\text{TRI1}} = 2.5$ and $C_{\text{TRI2}} = 1.0$.

In LES, turbulent fluctuations in composition and temperature and their contributions to TRI are partially resolved [34]. The contributions of subfilter-scale fluctuations to TRI (in both the absorption and emission terms) are captured, for instance, in Ref. [34] using a coupling between a stochastic method to solve the RTE and a FDF method to solve the species and temperature fluctuations. If only the resolved-scale values are used (i.e., ignoring TRI) in a LES context, Eq. (8.91) becomes:

$$\frac{d\tilde{I}}{ds} = \tilde{\kappa}_P \left(\tilde{I}_b - \tilde{I} \right) \tag{8.106}$$

with

$$\tilde{I}_b = \frac{\sigma \tilde{T}^4}{\pi}. \tag{8.107}$$

In this situation, in the flaming region where combustion occurs, the term $\tilde{I}_b$ can be significantly underestimated due to the typically lower resolved temperature, $\tilde{T}$, in comparison to the maximum temperature in a diffusion flame. In Refs. [16,17], to correct for this problem, Eq. (8.94) is rewritten in terms of a 'corrected' blackbody radiation intensity, $\tilde{I}_{b,\text{corr}}$:

$$\nabla \cdot \vec{q}_{\text{rad}} = \tilde{\kappa}_P \left(4\pi \tilde{I}_{b,\text{corr}} - \tilde{G} \right). \tag{8.108}$$

The radiation source term can also be expressed as follows:

$$\nabla \cdot \vec{q}_{\text{rad}} = \chi_r \dot{q}''', \tag{8.109}$$

where $\dot{q}'''$ (in kW/m³) is the local heat release rate per unit volume in the cells involving combustion (i.e., $\dot{q}''' > 0$) and χ_r a global radiative fraction that is interpreted as the fraction of energy radiated from the flaming region.

Combining Eqs. (8.108) and (8.109) gives:

$$\tilde{I}_{b,\text{corr}} = \frac{1}{4\pi}\left(\frac{\chi_r \dot{q}'''}{\tilde{\kappa}_P} + \tilde{G} \right). \qquad (8.110)$$

The 'corrected' emission term is then computed in Refs. [16,17] based on Eqs. (8.108) and (8.110) as follows:

$$\tilde{I}_{b,\text{corr}} = C\,\tilde{I}_b, \qquad (8.111)$$

where the corrective factor C is calculated as follows:

$$C = \frac{\chi_r \dot{q}''' + \tilde{G}}{4\tilde{\kappa}_P \sigma \tilde{T}^4}. \qquad (8.112)$$

It is important to note that if the TRI are fully resolved (as in DNS), the radiative fraction is expected to vary locally. However, if only the overall structure and radiation of the flame are sought, the 'global radiative fraction' approach described above can be considered as a satisfactory alternative to more advanced TRI modelling as proposed in Ref. [34].

8.6 THE SOOT PROBLEM

The soot problem has received great attention in the combustion community due to its primary importance in many applications, such as combustors and fire dynamics. Such importance lies first in its substantial contribution to thermal radiation. It is, for instance, a key species in combustor wall heat transfer or fire spread, where ignition of combustible materials is triggered by radiant heating. Furthermore, soot is an essential element to take into consideration when addressing pollutant emissions and toxicity concerns because, for example, soot oxidation can lead to the formation of carbon monoxide.

The soot problem remains nowadays a field of very active research due to the high degree of complexity in the associated physical and chemical phenomena. This has led Tieszen in his *Review on the Fluid Mechanics of Fire* [35] to state:

> Assuming that fire catches the leading edge and that we can extrapolate the trend of the last 50 years, future researches can expect to see first-principles solutions of the soot/thermal radiation loop in fires in practical applications in the decades after the start of the twenty-second century.

8.6.1 SOOT NATURE, MORPHOLOGY AND GENERAL DESCRIPTION OF ITS CHEMISTRY

When a fuel molecule breaks in the combustion process, it leads to many species that undergo several chemical reactions. Soot is the result of many intermediate reactions that lead to clusters of carbon particles. The general picture that emerges from the literature is that Polycyclic Aromatic Hydrocarbons (PAHs) are the dominant precursors to soot particles. Their inception may be initiated through channels constituted of non-aromatic species [36]. The chaining of these aromatic rings (i.e., PAHs) leads to higher-order aromatics until a sufficiently high order is reached to make the transition from PAH to soot [37]. This transition could be made by PAH-PAH reactions and/or H-Abstraction-C_2H_2-Addition (i.e., the HACA mechanism).

8.6.2 IMPORTANCE OF SOOT MODELLING

8.6.2.1 Sootiness and Radiation

Radiation is highly influenced by the amount of soot produced. Therefore, the colour of a flame and its luminosity depend on soot concentration. In his early descriptive study of a candle-like flame, Faraday stated [38]: *it is to this presence of solid particles in the candle flame that it owes its brilliancy.* The bluish colour is a witness of soot-free region, while the yellowish luminosity is due to soot. The wide variety of fuels (solid, liquid or gaseous) that might be involved in a fire implies a wide range in flame sootiness. The study conducted in Ref. [39] on flame heat transfer in storage geometries clearly demonstrates that for approximately the same heat release rates, the sootier the flame, the greater the heat fluxes.

8.6.2.2 Interaction of Soot with Carbon Monoxide

Carbon monoxide formation occurs through a complex chemical and burning process involving strong interactions between several species among which soot is one of the most important. In fact, an extensive amount of literature [40–42] demonstrates that the sootier a flame, the higher, the CO levels. This link between CO and soot is even more emphasised in Ref. [41] where the carbon monoxide yield (i.e., mass fraction), Y_{CO}, was directly linked to soot yield, y_s, via the correlation:

$$y_{CO} = \frac{12 n_{c,F}}{W_F s} 0.0014 + 0.37 y_s, \tag{8.113}$$

where W_F and $n_{c,F}$ are, respectively, the molecular weight of the fuel and the number of carbon atoms in a fuel molecule.

Soot participates in CO formation through oxidation. The two main soot oxidizers, responsible thus for CO formation, are the hydroxyl radical, OH, and oxygen, O_2. As mentioned in Ref. [40], soot oxidation by OH radicals may be described by the irreversible global reaction:

$$C_i + OH \rightarrow C_{i-1} + CO + \frac{1}{2} H_2. \tag{R4}$$

Other reaction products may be possible. C_i denotes a soot particle with i carbon atoms.

Soot oxidation by oxygen is described by the following reaction:

$$C + \frac{1}{2} O_2 \rightarrow 2CO. \tag{R5}$$

Carbon monoxide modelling becomes even more important knowing that its formation reaction is much faster than its oxidation reaction by oxygen and the formation of carbon dioxide, as described in the following reaction:

$$CO + \frac{1}{2} O_2 \rightarrow CO_2. \tag{R6}$$

Carbon dioxide may also be reduced at the surface of solid carbon (soot) to form CO:

$$CO_2 + C_i \rightarrow 2CO. \tag{R7}$$

8.6.3 The Sootiness of Fuels

8.6.3.1 The Laminar Smoke Point Height

There is a wide range of fuels' sootiness (i.e. propensity to produce soot). The classification of fuels with respect to their sootiness can be performed using the laminar smoke point (LSP) concept [45,46]. The LSP height, ℓ_{sp}, is defined as the flame height prior to when it starts emitting smoke from its tip. Generally speaking, the sootier the fuel, the lower its ℓ_{sp}. Increasing the fuel flow rate (and therefore the flame height) beyond a critical value leads to an 'overproduction' of soot, which is not fully oxidized and starts escaping from the tip. Table 8.2 provides LSP height values for some fuels.

8.6.3.2 The Threshold Sooting Index (TSI)

Yang et al. [47] pointed out the fact that when fuels have significantly different molecular weights, W_F, there is an offset in the flame height which must be accounted for. This effect is incorporated in the Threshold Sooting Index (TSI) expressed as follows:

$$\mathrm{TSI} = a\left(\frac{W_F}{\ell_{sp}}\right) + b,\tag{8.114}$$

where a and b are apparatus-dependent constants. The TSI ranges from 0 to 100. Table 8.3 provides the ℓ_{sp} [43,44] and TSI [48] values reported in the literature for several fuels.

TABLE 8.2
LSP Height of Some Fuels as Reported in Refs. [43,44]

Fuel	Chemical Formula	ℓ_{sp} (m)
Methane	CH_4	0.290
Propane	C_3H_8	0.162
Ethylene	C_2H_4	0.106
Dodecane	$C_{12}H_{26}$	0.108
Polypropylene	C_3H_6	0.050
Acetylene	C_2H_2	0.019
Polystyrene	C_3H_8	0.015
2-Methylbutane	C_5H_{12}	0.113
Trimethylpentane	C_8H_{18}	0.070
Cyclohexane	C_6H_{12}	0.087

TABLE 8.3
LSP Height [43,44] and TSI [48] of Several Pure Hydrocarbons

Fuel	Chemical Formula	W_F (g/mol) in *g/mol*	ℓ_{sp} (mm)	TSI
Methane	CH_4	16	290	
Propane	C_3H_8	44	162	0.6
Ethylene	C_2H_4	28	106	1.3
Butane	C_4H_{10}	58	160	1.4
Heptane	C_7H_{16}	100	123	2.7
Propylene	C_3H_6	42	29	4.8

The TSI values reveal, for example, that propylene is almost four times sootier than ethylene and eight times sootier than propane. The TSI of methane is not provided, but its LSP height and direct measurements in methane-fuelled flames show that soot concentrations are an order of magnitude lower than in propane flames. The use of the TSI and LSP concepts allows quantifying the sootiness of a fuel and should be systematically referred to instead of (or in addition to) the qualitative classification of *lightly*, *moderately* and *heavily* sooty fuels.

8.6.4   Soot Modelling

A large body of literature has been devoted to several topics related to soot diagnostics and modelling in several applications (e.g., diesel engines or fires). The review of Kennedy [49] provides a general overview of the advances and challenges in soot modelling.

8.6.4.1   Laminar Flames

The conservation equation for soot mass fraction, Y_s, in a laminar flame reads:

$$\frac{\partial}{\partial t}\left(\rho^* Y_s\right) + \nabla\cdot\left(\rho^* Y_s \vec{v}\right) = \nabla\cdot\left(\rho^* Y_s \vec{v}_{th}\right) + \dot{\omega}_s. \tag{8.115}$$

This equation is similar to Eq. (8.3) for species mass fraction, except that the density ρ^* is a two-phase density (that takes into account gas and solid (soot) density) and soot diffusivity is expressed in terms of a thermodiffusion velocity, v_{th}. The variable $\dot{\omega}_s$ is a source term for soot.

The two-phase density ρ^* is expressed as follows:

$$\rho^* = \frac{\rho}{1 - Y_s}, \tag{8.116}$$

where ρ is the gas-phase density.

In Eq. (8.115), it is assumed that the transport of soot particles due to molecular diffusion is negligible. The movement of soot is mainly attributed to thermophoretic forces that are generated from temperature gradients. The thermodiffusion velocity v_{th} is therefore expressed as follows:

$$\left(v_{th}\right)_j = -0.54\frac{\mu}{\rho}\frac{\partial}{\partial x_j}\left(\ln(T)\right) = -0.54\frac{\mu}{\rho}\frac{1}{T}\frac{\partial T}{\partial x_j}. \tag{8.117}$$

Soot mass fraction, Y_s, is related to soot volume fraction, f_v, through:

$$\rho^* Y_s = \rho_s f_v, \tag{8.118}$$

where ρ_s is the soot density, taken as $1{,}800\,\text{kg/m}^3$ [19].

The conservation equation can then be rewritten as follows:

$$\frac{\partial f_v}{\partial t} + \nabla\cdot\left(f_v \vec{v}\right) = \nabla\cdot\left(f_v \vec{v}_{th}\right) + \phi_{f_v}, \tag{8.119}$$

where ϕ_{f_v} is a volumetric source term expressed as follows:

$$\phi_{f_v} = \frac{\dot{\omega}_s}{\rho_s}. \tag{8.120}$$

Soot modelling consists then of providing an explicit expression for ϕ_{f_v} or $\dot{\omega}_s$.

8.6.4.2 Detailed and Reduced Chemistry Models

Several detailed soot chemistry models involve a great number of species and reactions. For example, D'Anna et al. [50] developed a kinetic model based on fuel chemistry and the formation of aromatics and their growth to soot clusters and particle oxidation. The formation of gas-phase compounds was modelled by 60 species and 280 reactions. In Ref. [51], a detailed reaction mechanism of 365 reactions and 62 species was included. In Ref. [36], an even more elaborated model was suggested with a set of 527 reactions involving 99 chemical species.

In order to simplify detailed chemistry models and make them applicable to cases of practical interest with affordable computational requirements, modellers resort to experimental results underlying the importance of specific species. For example, in Ref. [52], it was shown that the most aromatic abundant product in a counterflow laminar ethylene diffusion flame is benzene. Under this consideration, the pathway to soot formation has been reduced to four reactions and nine species. The four reactions describe the major soot mechanisms: nucleation, surface growth, coagulation and oxidation. Using the same assumption of one prevailing species for soot formation, in Ref. [53] another simplified model was proposed based on acetylene concentration. The set of chemical reactions has been reduced to four reactions involving seven species. More details are provided in the following.

8.6.4.3 Models Describing Inception, Coagulation, Surface Growth and Oxidation

In the Moss model [54–57], the presence of carbon black in flames is attributed to four major processes, which control the soot number density, $n\left(\mathrm{m}^{-3}\right)$, and volume fraction, $f_v\left(\mathrm{m}^3/\mathrm{m}^3\right)$. These processes are: (a) nucleation, (b) coagulation, (c) surface growth and (d) oxidation.

- *Nucleation (or inception)* is the first step that initiates soot production through gaseous reactions. Pyrolysis of the initial hydrocarbon produces smaller hydrocarbons. Then, the low-molecular (gaseous) hydrocarbons are converted into solid carbon. The nucleation term acts therefore as a source term for both n and f_v.
- *Coagulation (or collisional coagulation)* between soot particles is described as the physical process that leads to the formation of clusters. The number of soot nuclei decreases, and therefore, the coagulation term serves as a sink term for n. This process occurs in conjunction with a chemical one: surface growth.
- *Surface growth* is generally assumed to be dependent on the available soot area (and therefore the number density n) and the controlling process in soot production. It is modelled as a source term in the f_v equation.
- *Oxidation* of soot is also referred to in the literature as soot combustion, burnout or destruction. It is the competition between soot oxidation and formation (i.e., the result of nucleation coagulation and surface growth) that determines the amount of soot that escapes from the flame because it has not been oxidized.

In the model described above, the soot volume fraction f_v is strongly dependent on the soot number density, n (m^{-3}), that is related to the soot particle concentration, X_n ($\mathrm{mol\,kg}^{-1}$), through:

$$X_n = \frac{n}{\rho^* N_A},\qquad (8.121)$$

where N_A is the Avogadro number ($N_A \simeq 6.02 \times 10^{23}\,\mathrm{mol}^{-1}$).

Equation (8.115) is then solved in conjunction with a conservation equation for X_n, which reads:

$$\frac{\partial}{\partial t}\left(\rho^* X_n\right) + \nabla \cdot \left(\rho^* X_n \vec{v}\right) = \nabla \cdot \left(\rho^* X_n \vec{v}_{\text{th}}\right) + \phi_{X_n},\tag{8.122}$$

where ϕ_{X_n} is the source term.

Using Eq. (8.121), it can be clearly seen that Eq. (8.122) is equivalent to:

$$\frac{\partial}{\partial t}\left(n/N_A\right) + \nabla \cdot \left(\left(n/N_A\right)\vec{v}\right) = \nabla \cdot \left(\left(n/N_A\right)\vec{v}_{\text{th}}\right) + \phi_{X_n}.\tag{8.123}$$

The source terms for Eqs. (8.115) and (8.123) are expressed as follows:

$$\dot{\omega}_s = \dot{\omega}_s^{\text{nuc}} + \dot{\omega}_s^{\text{grow}} - \dot{\omega}_s^{\text{ox}}\tag{8.124}$$

and

$$\phi_{X_n} = \phi_{X_n}^{\text{nuc}} - \phi_{X_n}^{\text{coag}} - \phi_{X_n}^{\text{ox}},\tag{8.125}$$

where the superscripts nuc, grow, coag and ox denote nucleation, surface growth, coagulation and oxidation, respectively.

- *Nucleation*: The nucleation terms are expressed as follows:

$$\dot{\omega}_s^{\text{nuc}} = C_\delta\, C_\alpha\, \rho^2\, T^{1/2} X_c \exp(-T_\alpha / T)\tag{8.126}$$

and

$$\phi_{X_n}^{\text{nuc}} = C_\alpha\, \rho^2\, T^{1/2}\, X_c \exp(-T_\alpha / T),\tag{8.127}$$

where C_δ is a constant taken as 144 and C_α is a fuel-dependent nucleation constant. The activation temperature for nucleation is taken as $T_\alpha = 46,100\,\text{K}$. The variable X_c denotes the molar fraction of the main gaseous species considered for soot formation.

- *Surface growth*: The surface growth term is expressed as follows [54]:

$$\dot{\omega}_s^{\text{grow}} = C_\gamma\, \rho\, T^{1/2}\, X_c \left(\rho_s f_v\right)^{2/3} n^{1/3} \exp\left(-T_\gamma / T\right)\tag{8.128}$$

or [57]

$$\dot{\omega}_s^{\text{grow}} = C_\gamma\, \rho\, T^{1/2}\, X_c\, n\, \exp\left(-T_\gamma / T\right),\tag{8.129}$$

where C_γ is a fuel-dependent surface growth constant and T_γ is the surface growth activation temperature taken as $T_\gamma = 12,100\,\text{K}$.

- *Coagulation*: The coagulation term is expressed as follows:

$$\phi_{X_n}^{\text{coag}} = C_\beta\, T^{1/2} \left(\frac{n}{N_A}\right)^2,\tag{8.130}$$

where C_β is a fuel-dependent coagulation coefficient.

- *Oxidation*: The oxidation terms are expressed as follows:

$$\dot{\omega}_s^{ox} = \left(\frac{36\pi}{\rho_s^2}\right)^{1/3} n^{1/3} \left(\rho_s f_v\right)^{2/3} \tau_{ox}$$

(8.131)

and

$$\phi_{X_n}^{ox} = N_A^{1/3} \rho \left(\frac{\rho_s f_v}{\rho}\right)^{-1/3} \left(\frac{n}{\rho N_A}\right)^{4/3} \tau_{ox},$$

(8.132)

where τ_{ox} is a surface oxidation term. The three commonly employed models for this term are the Nagle–Strickland–Constable (NSC) [58], the Lee–Thring–Beer (LTB) [59] and Fenimore and Jones (FJ) models [60].

Table 8.4 displays a summary of some features that Moss and co-workers have used in soot modelling from 1988 to 2007 [54–57,61].

On the basis of the summary provided in Table 8.4, the following observations can be made:

TABLE 8.4

Summary of Values Used by Moss and Co-Workers, Period 1988–2007 [54–57, 61]

References	Fuel	X_c	$\dot{\omega}_s^{grow}$	$\dot{\omega}_s^{ox}$	$\phi_{X_n}^{ox}$	Coefs.
[54]	Ethylene (C_2H_4)	Parent Fuel	$\propto n$	$= 0$	$= 0$	$C_\alpha = 1.7\times10^8$
						$C_\gamma = 4.2\times10^{-17}$
						$C_\beta = 1\times10^9$
[57]	Methane (CH_4)	Parent Fuel	$\propto \left(\rho_s f_v\right)^{2/3} n^{1/:}$	$\neq 0$	$\neq 0$	$C_\alpha = 6.54\times10^4$
						$C_\gamma = 0.1$
						$C_\beta = 1.3\times10^7$
[55]	16.8% C_2H_4 + 83.2% N_2	Total HC	$\propto n$	$\neq 0$	$= 0$	$C_\alpha = 6\times10^6$
						$C_\gamma = 6.3\times10^{-14}$
						$C_\beta = 2.25\times10^{15}$
[56]	Propylene (C_3H_6)	Total HC	$\propto n$	$\neq 0$	$= 0$	$C_\alpha = 1.5\times10^6$
						$C_\gamma = 8.5\times10^{-13}$
						$C_\beta = 2.0\times10^9$
[56]	MMA ($C_5H_8O_2$)	Total HC	$\propto n$	$\neq 0$	$= 0$	$C_\alpha = 3.68\times10^5$
						$C_\gamma = 8.5\times10^{-13}$
						$C_\beta = 2.0\times10^9$
[61]	Heptane (C_7H_{16})	C_2H_2	$\propto n$	$\neq 0$	$= 0$	$C_\alpha = 6\times10^6$
						$C_\gamma = 6.3\times10^{-14}$
						$C_\beta = 2.25\times10^{15}$

- Except for the modification performed in Ref. [57], the surface growth term has been consistently modelled as proportional to the number density, n. This implies that Moss and co-workers suggest, on the grounds of their studies, that it is the formulation to be retained.
- The oxidation source term for the conservation equation of f_v has been systematically implemented starting from 1990 [57]. The oxidation source term for the conservation equation of n was only implemented in Ref. [57]. The balance between nucleation and coagulation has been considered to be sufficient to model the evolution of the number density n.
- In Ref. [61], soot concentrations in a turbulent heptane flame were predicted using the constants of ethylene calibrated in Ref. [56]. However, according to the TSI measurements displayed in Table 8.3, heptane is twice as sootier as ethylene.
- The main species used in soot formation was taken at first in Refs. [54,57] as the parent fuel, then in Refs. [55,56] as the total hydrocarbons and finally in Ref. [61], acetylene was used. It is not clear to what extent these changes affect the numerical calibration of the constants.

8.6.4.4   The Laminar Smoke Point (LSP) Model

In Ref. [62], an analytical model was developed for a global soot formation rate at the centreline of a laminar diffusion flame, in the form of an Arrhenius-type equation. The results of the model provided a first theoretical framework in accordance with several laminar smoke point (LSP) measurements. The most attractive model feature relates to the fact that the pre-exponential factor for the soot formation rate is inversely proportional to the LSP height, ℓ_{sp}. This allows the extension of the model from one fuel to another to be based on a standardized experimental measure of ℓ_{sp} instead of a numerical basis (which is the case in the Moss model). The second attractive feature lies in the fact that the smoke point (SP) model does not explicitly account for nucleation, coagulation and surface growth and thus does not require the computation of the number density, n, or the soot surface area. In [43,45] a similar approach was followed by implementing a smoke point model for soot in the CFD code Fire Dynamics Simulator (FDS) [16,17]. However, instead of Arrhenius-type equations, a soot formation and oxidation mapping (as suggested in Ref. [63]) was used based on polynomials with coefficients calculated as a function of the fuel stoichiometry and LSP height. Encouraging results were obtained for ethylene, propylene and propane by calibrating third-order polynomials. The work of Ref. [62] has been implemented in a CFD code in Ref. [64], with some adjustments. In addition to the implementation of a simplified soot oxidation model, these adjustments included the bounding of the soot formation region not only near stoichiometry as in Ref. [62], but also at the fuel-rich side. The source term for soot mass fraction is expressed as follows:

$$\dot{\omega}_s = \dot{\omega}_s^f - \dot{\omega}_s^{ox}, \tag{8.133}$$

where the superscripts f and ox denote, respectively, soot formation and soot oxidation terms.

The soot formation source term is expressed as follows:

$$\dot{\omega}_s^f = A\,\rho^2\,Y_F\,T^\gamma\,e^{-T_a/T}, \text{ if } Z_{s,ox} \leq Z \leq Z_{s,f}, \tag{8.134}$$

where A is a fuel-dependent pre-exponential factor, Y_F is the fuel mass fraction, γ is a temperature exponent taken as 2.25 and T_a is an activation temperature taken as 2,000 K. The variables $Z_{s,f}$ and $Z_{s,ox}$ denote, respectively, the upper (incipient) and lower bounds for soot formation in the mixture fraction, Z, space. The pre-exponential factor was numerically calibrated for ethylene in Ref. [64] and taken as $A_{C_2H_4} = 4 \times 10^{-5}$ (m^3/kg K$^\gamma$). It is calculated for other fuels using the LSP height according to:

$$\frac{A_{\text{fuel}}}{A_{C_2H_4}} = \frac{\ell_{\text{sp},C_2H_4}}{\ell_{\text{sp,fuel}}}. \tag{8.135}$$

8.6.4.5 Fuel Conversion Model

This is the simplest approach for soot modelling, and the same approach could be applied for carbon monoxide as well. Let us consider the following single-step mixing-controlled reaction for gas-phase combustion:

$$\nu_F C_n H_m + \nu_{O_2} O_2 \to \nu_{H_2O} H_2O + \nu_{CO_2} CO_2 + \nu_{CO} CO + \nu_s C.$$

where ν_i are the stoichiometric molar fractions of each of the species i. The following molar fractions are calculated:

$$\nu_F = 1,$$

$$\nu_{H_2O} = \frac{m}{2},$$

$$\nu_{O_2} = n + \frac{m}{4} - \frac{\nu_{CO}}{2} - \nu_s,$$

$$\nu_{CO_2} = n - \nu_{CO} - \nu_s.$$

The molar fractions of CO and soot are calculated from their yields y_i according to the following formula:

$$\nu_i = \frac{W_F}{W_i} y_i \tag{8.136}$$

where W_F is the molecular weight of the fuel. Values for soot and CO yields are provided in Ref. [65] for several fuels.

Similar to the fuel and oxidizer, a state relationship can be obtained for the soot mass fraction:

$$Y_s = \frac{\nu_s W_c}{\nu_F W_F} \left(Y_{F,o} Z - Y_F \right) \tag{8.137}$$

This approach has been experimentally verified in Ref. [66] for soot concentrations in the overfire region (i.e., the fuel-lean region where *no flame luminosity was ever seen*). It has been indeed demonstrated in Ref. [66] (for a number of turbulent diffusion flames fuelled with acetylene, propylene, ethylene and propane) that soot behaves in this region like a passive scalar, resulting in an excellent correlation between f_v or Y_s and the mixture fraction, Z (see Eq. 8.7.25). In the flame region, soot formation is governed by slow chemical reactions. Thus, soot concentrations cannot be predicted by infinitely fast chemistry (i.e., using the mixture fraction approach).

8.6.4.6 Turbulent Flames

Obviously, turbulent flames have a more practical interest in fire dynamics. In order to develop soot models suitable for turbulent flows, two approaches are available.

The first approach consists of using turbulent combustion techniques to average source terms that were validated in laminar flows. For instance, in Ref. [67], the Alternative Conditional Source-term Estimation (A-CSE) technique was used to apply and validate the LSP model for turbulent pool fires. The Moss model is generally applied using the same formulation as in the laminar case by using local average values within each cell. In Ref. [68], a transported PDF approach has been applied to two turbulent ethylene flames at high Reynolds numbers ($Re = 11,800$ and $Re = 15,600$). The soot model used in Ref. [68] has been extensively validated in laminar counterflow diffusion flames [69]. A LES/PDF-based modelling has also been used in Ref. [70]. It is important to mention that in this first approach (i.e., averaging the soot source term using turbulent combustion modelling), it is always important to verify the outcome of the soot model parameters in laminar flames (with the fuel of interest) before proposing an implementation in a turbulent flow because, in a turbulent flow, uncertainties in soot modelling (i.e., soot chemistry) may be masked by other modelling uncertainties like the treatment of turbulence and/or radiation.

A second approach consists of developing models that are directly calibrated for turbulent flames such as the Khan model [71] or the Magnussen model [24]. For instance, in Ref. [24], two transport equations are solved for $\tilde{Y}_s$ and $\tilde{X}_n$ (similar to the laminar flames, see Eqs. 8.115 and 8.123). However, the mean reaction rates for soot oxidation are not modelled as in Eqs. (8.131) and (8.132) but rather based on the turbulent time scale k/ε:

$$\overline{\dot{\omega}_s^{ox}} = C_R \bar{\rho} \frac{\varepsilon}{k} \tilde{Y}_s \min\left(1, \frac{\tilde{Y}_O}{\tilde{Y}_s s_s + Y_F s}\right), \tag{8.138}$$

$$\overline{\phi_{X_n}^{ox}} = C_R \bar{\rho} \frac{\varepsilon}{k} \tilde{X}_n \min\left(1, \frac{\tilde{Y}_O}{\tilde{Y}_s s_s + Y_F s}\right), \tag{8.139}$$

where s_s is the mass stoichiometric ratio of soot oxidation calculated as follows:

$$s_s = \frac{W_C}{W_{O_2}} \simeq 2.667. \tag{8.140}$$

8.7 BASICS OF NUMERICAL DISCRETIZATION

8.7.1 Discretization Schemes

8.7.1.1 Description of a 1-D Example

Let us consider the simplified transient convection-diffusion equation of a property ϕ in a given 1-dimensional flow field u:

$$\frac{\partial \phi}{\partial t} + u \frac{\partial \phi}{\partial x} = \alpha \frac{\partial^2 \phi}{\partial x^2} + S_\phi, \tag{8.141}$$

where α is the diffusion coefficient and S_ϕ is a source term responsible for the generation of the property ϕ.

If time and space are discretized as follows:

$$t(n) = n\Delta t, \tag{8.142}$$

$$x(i) = i\Delta x, \tag{8.143}$$

then the first time and space derivatives at $t(n) = n\Delta t$ and $x(i) = i\Delta x$ can be expressed as follows:

$$\left(\frac{\partial \phi}{\partial t}\right)_i^n = \frac{\phi_i^{n+1} - \phi_i^n}{\Delta t} \tag{8.144}$$

and

$$\left(\frac{\partial \phi}{\partial x}\right)_i^n = \frac{\phi_{i+1}^n - \phi_i^n}{\Delta x}. \tag{8.145}$$

The approximation of the derivatives in the previous equations is called forward differencing in, respectively, time and space. Equation (8.145) is very rarely used in CFD.

A more general approach to estimate the derivative $(\partial \phi / \partial x)_i^n$ is the following:

$$\left(\frac{\partial \phi}{\partial x}\right)_i^n = \frac{\left(\phi_{i+\frac{1}{2}}^n\right)^{FL} - \left(\phi_{i-\frac{1}{2}}^n\right)^{FL}}{\Delta x}, \tag{8.146}$$

where the suffixes $\pm\frac{1}{2}$ indicate a face value for the cell i and the superscript FL denotes a flux limiter.

The quantity $\left(\phi_{i+\frac{1}{2}}^n\right)^{FL}$ can be expressed as follows:

$$\left(\phi_{i+\frac{1}{2}}^n\right)^{FL} = \begin{cases} \phi_i^n + \dfrac{1}{2}B\left(\phi_{i+1}^n - \phi_i^n\right) & \text{if } u_i^n > 0 \\[2mm] \phi_{i+1}^n - \dfrac{1}{2}B\left(\phi_{i+1}^n - \phi_i^n\right) & \text{if } u_i^n < 0 \end{cases}, \tag{8.147}$$

where B is a limiter function.

A value of $B = 0$ corresponds to the Godunov scheme. For example, in the case $u_i^n > 0$, one obtains:

$$\left(\phi_{i+\frac{1}{2}}^n\right)^{FL} = \phi_i^n \quad \text{and} \quad \left(\phi_{i-\frac{1}{2}}^n\right)^{FL} = \phi_{i-1}^n. \tag{8.148}$$

Inserting Eqs. (8.7.8) into Eq. (8.7.6) gives the following backward differencing equation:

$$\left(\frac{\partial \phi}{\partial x}\right)_i^n = \frac{\phi_i^n - \phi_{i-1}^n}{\Delta x}. \tag{8.149}$$

A value of $B = 1$ corresponds to the central differencing scheme. For example, in the case $u_i^n > 0$, one obtains:

$$\left(\frac{\partial \phi}{\partial x}\right)_i^n = \frac{\phi_{i+1}^n - \phi_{i-1}^n}{2\Delta x}. \tag{8.150}$$

In more sophisticated schemes, the limiter function B depends on the upstream-to-local data ratio. For example, in [16,17] the Superbee flux limiter [72] is used in the LES mode.

The second space derivative in Eq. (8.141) can be expressed as follows:

$$\left(\frac{\partial^2 \phi}{\partial x^2}\right)_i^n = \frac{\phi_{i+1}^n - 2\phi_i^n + \phi_{i-1}^n}{\Delta x^2}. \tag{8.151}$$

Similar to the first derivative, there are also more options here.

8.7.1.2 Explicit Scheme

As seen above, there are several options to discretize Eq. (8.141). By combining Eqs. (8.144), (8.150) and (8.151) (forward differencing for time and central differencing for space), Eq. (8.141) can be written in a discretized form as follows:

$$\frac{\phi_i^{n+1} - \phi_i^n}{\Delta t} + u \frac{\phi_{i+1}^n - \phi_{i-1}^n}{2\Delta x} = \alpha \frac{\phi_{i+1}^n - 2\phi_i^n + \phi_{i-1}^n}{\Delta x^2} + S_{\phi,i}^n. \tag{8.152}$$

Rearranging Eq. (8.152) gives:

$$\phi_i^{n+1} = \phi_i^n - \frac{u\Delta t}{2\Delta x}\left(\phi_{i+1}^n - \phi_{i-1}^n\right) + \frac{\alpha\Delta t}{\Delta x^2}\left(\phi_{i+1}^n - 2\phi_i^n + \phi_{i-1}^n\right) + \Delta t S_{\phi,i}^n. \tag{8.153}$$

The quantity ϕ at time $t = n+1$ depends then explicitly on values at time $t = n$. Equation (8.153) is then said to be solved using an explicit scheme. The explicit method has the advantage of an easy implementation in a CFD code and very fast computations, since all equations (one for each node here) can be solved independently of each other.

8.7.1.3 Implicit Scheme

Let us consider now backward differencing for the time and first space derivatives and keep the same discretization for the second space derivative. This is referred to as the backward Euler method. The following discretized equations are obtained:

$$\left(\frac{\partial \phi}{\partial t}\right)_i^n = \frac{\phi_i^n - \phi_i^{n-1}}{\Delta t}, \tag{8.154}$$

$$\left(\frac{\partial \phi}{\partial x}\right)_i^n = \frac{\phi_i^n - \phi_{i-1}^n}{\Delta x}, \tag{8.155}$$

$$\left(\frac{\partial^2 \phi}{\partial x^2}\right)_i^n = \frac{\phi_{i+1}^n - 2\phi_i^n + \phi_{i-1}^n}{\Delta x^2}. \tag{8.156}$$

The discretized differential equation could then be written as follows:

$$\frac{\phi_i^n - \phi_i^{n-1}}{\Delta t} + u\frac{\phi_i^n - \phi_{i-1}^n}{\Delta x} = \alpha\frac{\phi_{i+1}^n - 2\phi_i^n + \phi_{i-1}^n}{\Delta x^2} + S_{\phi,i}^n. \tag{8.157}$$

Rearranging Eq. (8.157) gives:

$$a_1\phi_{i+1}^n + a_2\phi_i^n + a_3\phi_{i-1}^n = \phi_i^{n-1} + S_{\phi,i}^n, \tag{8.158}$$

where

$$a_1 = -\frac{\alpha\Delta t}{\Delta x^2}, \tag{8.159}$$

$$a_2 = 1 + \frac{u\Delta t}{\Delta x} + \frac{2\alpha\Delta t}{\Delta x^2}, \tag{8.160}$$

$$a_3 = -\frac{u\Delta t}{\Delta x} - \frac{\alpha\Delta t}{\Delta x^2}. \tag{8.161}$$

Note that as opposed to the explicit method, a system of algebraic equations (millions of equations if millions of nodes are used) must be solved in order to find the value of ϕ^n at each of the spatial nodes from the values at the previous time step, ϕ^{n-1}. Therefore, the considered numerical scheme here is said to be constructed using an implicit method.

The problem can be written in the form of a matrix multiplication. For instance, for $i = 1,2,3,4,5,6$, one obtains:

$$AX = B, \tag{8.162}$$

where

$$A = \begin{pmatrix} a_2 & a_1 & 0 & 0 & 0 & 0 \\ a_3 & a_2 & a_1 & 0 & 0 & 0 \\ 0 & a_3 & a_2 & a_1 & 0 & 0 \\ 0 & 0 & a_3 & a_2 & a_1 & 0 \\ 0 & 0 & 0 & a_3 & a_2 & a_1 \\ 0 & 0 & 0 & 0 & a_3 & a_2 \end{pmatrix}, \tag{8.163}$$

$$X = \begin{pmatrix} \phi_1^n \\ \phi_2^n \\ \phi_3^n \\ \phi_4^n \\ \phi_5^n \\ \phi_6^n \end{pmatrix}, \tag{8.164}$$

and

$$B = \begin{pmatrix} \phi_1^{n-1} + S_{\phi,1}^n + a_3\phi_0^n \\ \phi_2^{n-1} + S_{\phi,2}^n \\ \phi_3^{n-1} + S_{\phi,3}^n \\ \phi_4^{n-1} + S_{\phi,4}^n \\ \phi_5^{n-1} + S_{\phi,5}^n \\ \phi_6^{n-1} + S_{\phi,6}^n - a_1\phi_7^n \end{pmatrix}. \tag{8.165}$$

The values ϕ_0^n and ϕ_7^n denote the boundary conditions that can be estimated from a Dirichlet or a Neumann boundary condition as will be explained in the next section. An alternative implicit discretization scheme to the backward Euler method is the Crank–Nicolson method where each component of the partial differential equation is written as the following:

$$\left(\frac{\partial \phi}{\partial t}\right)_i^n = \frac{\phi_i^{n+1} - \phi_i^n}{\Delta t}, \tag{8.166}$$

$$\left(\frac{\partial \phi}{\partial x}\right)_i^n = \frac{1}{2}\left[\left(\frac{\phi_{i+1}^{n+1} - \phi_{i-1}^{n+1}}{2\Delta x}\right) + \left(\frac{\phi_{i+1}^n - \phi_{i-1}^n}{2\Delta x}\right)\right], \tag{8.167}$$

$$\left(\frac{\partial^2 \phi}{\partial x^2}\right)_i^n = \frac{1}{2(\Delta x)^2}\left[\left(\phi_{i+1}^{n+1} - 2\phi_i^{n+1} + \phi_{i-1}^{n+1}\right) + \left(\phi_{i+1}^n - 2\phi_i^n + \phi_{i-1}^n\right)\right]. \tag{8.168}$$

8.7.2   INITIAL AND BOUNDARY CONDITIONS

Further information is required to find the solution of the partial differential equation:

- the initial conditions, ϕ_i^0
- the boundary conditions: If the physical domain is discretized with I number of nodes (i.e., $i = 1,\ldots,I$), the values at the two boundaries can be denoted by ϕ_0 and ϕ_{I+1}. These values (or one of them) can be prescribed in time. This is referred to as the Dirichlet boundary condition. Another option is to prescribe a zero value for the gradients $\left(\partial \phi / \partial x\right)_{i=1}$ and/or $\left(\partial \phi / \partial x\right)_{i=I}$. This is referred to as the Neumann boundary condition. Other boundary conditions are possible of course.

8.7.3   PROPERTIES OF NUMERICAL METHODS

In this section, five important properties of numerical methods (in terms of performance) are discussed: consistency, stability, convergence, conservativeness and boundedness.

8.7.3.1   Consistency

A numerical scheme is consistent if truncation errors vanish as the mesh size and time step tend to zero. For example, by taking $\alpha = 0$ and S_ϕ in Eq. (8.141), a pure convection equation is obtained:

$$\frac{\partial \phi}{\partial t} + u\frac{\partial \phi}{\partial x} = 0. \tag{8.169}$$

Using central differencing in space and forward Euler in time and considering the Taylor series expansions, Eq. (8.169) becomes:

$$\underbrace{\frac{\phi_i^{n+1}-\phi_i^n}{\Delta t}+u\frac{\phi_{i+1}^n-\phi_{i-1}^n}{2\Delta x}}_{\text{Difference scheme}}-\underbrace{\left(\frac{\partial\phi}{\partial t}+u\frac{\partial\phi}{\partial x}\right)_i^n}_{\text{Differential terms}}+\varepsilon_\tau=0,\tag{8.170}$$

where the residual of the difference scheme, which is expressed as follows:

$$\varepsilon_\tau=-\frac{\Delta t}{2}\left(\frac{\partial^2\phi}{\partial t^2}\right)_i^n-u\frac{(\Delta x)^2}{6}\left(\frac{\partial^2\phi}{\partial x^3}\right)_i^n+O\left[(\Delta t)^2,(\Delta x)^4\right]\tag{8.171}$$

tends to zero as the mesh size and time step tend to zero.

8.7.3.2 Stability

A numerical scheme is stable if the numerical errors that are generated during the solution of discretized equations are not magnified (i.e., small perturbations of initial data lead to small variations in the solution).

8.7.3.2.1 Example of a Stability Analysis

As mentioned above, the explicit method has the advantage of an easy implementation in a CFD code. However, there are restrictions on the time and space steps that need to be met in order to find a stable solution. This is examined here for Eq. (8.153) using the Von Neumann stability analysis. Such analysis consists of considering a perturbation of the solution, ϕ, that could be expressed as follows:

$$\phi_i^n=\phi_{0,i}^n+\varepsilon_i^n,\tag{8.172}$$

where ϕ_0 satisfies the discretized differential equation and ε_i^n is a perturbation at a discrete location i and a time step n that is expressed in the Fourier terms as follows:

$$\varepsilon_i^n=e^{jk(i-1)\Delta x}=\cos\left(k(i-1)\Delta x\right)+j\sin\left(k(i-1)\Delta x\right).\tag{8.173}$$

The linearized difference equation for the perturbation without the source term becomes:

$$\varepsilon_i^{n+1}=\varepsilon_i^n-\frac{u\Delta t}{2\Delta x}\left(e^{jk\Delta x}-e^{-jk\Delta x}\right)\varepsilon_i^n+\frac{\alpha\Delta t}{\Delta x^2}\left(e^{jk\Delta x}-2+e^{-jk\Delta x}\right)\varepsilon_i^n.\tag{8.174}$$

Knowing that:

$$\frac{e^{jk\Delta x}+e^{-jk\Delta x}}{2}=\cos(k\Delta x)\tag{8.175}$$

and

$$\frac{e^{jk\Delta x}-e^{-jk\Delta x}}{2}=j\sin(k\Delta x)\tag{8.176}$$

Equation (8.174) becomes:

$$\varepsilon_i^{n+1}=G\varepsilon_i^n,\tag{8.177}$$

where

$$G = 1 - \frac{2\alpha\Delta t}{\Delta x^2}\left[1 - \cos(k\Delta x)\right] - j\frac{u\Delta t}{\Delta x}\sin(k\Delta x). \tag{8.178}$$

The Von Neumann stability criteria states that the magnitude $|G|$ must be less or equal to 1:

$$|G| = \left[1 - \frac{2\alpha\Delta t}{\Delta x^2}\left[1 - \cos(k\Delta x)\right]\right]^2 + \left[\frac{u\Delta t}{\Delta x}\sin(k\Delta x)\right]^2 \leq 1. \tag{8.179}$$

In general, the extreme values for G occur when the sine and cosine are at their extreme values of 1, 0 or −1. Therefore, the following three particular values for $k\Delta x$ are examined hereafter:

- When $k\Delta x = 0$, $G = 1$: this mode is neutrally stable, neither increasing nor decreasing with increasing values of n.
- When $k\Delta x = \pi$, $|G| = \left(1 - \frac{4\alpha\Delta t}{\Delta x^2}\right)^2$. In order for the numerical scheme to be stable, the following stability condition must be met:

$$\frac{2\alpha\Delta t}{\Delta x^2} \leq 1. \tag{8.180}$$

- When $k\Delta x = \pi/2$, the stability condition reads $|G| = \left(1 - \frac{2\alpha\Delta t}{\Delta x^2}\right)^2 + \left(\frac{u\Delta t}{\Delta x}\right)^2 \leq 1$. The left side of the inequality is the sum of squares. Therefore, neither of the sides can exceed 1, which leads to:

$$\frac{2\alpha\Delta t}{\Delta x^2} \leq 2 \tag{8.181}$$

and

$$\frac{u\Delta t}{\Delta x} \leq 1. \tag{8.182}$$

The first condition, (8.181), is a less restrictive condition than (8.180). The second condition, (8.182), is a new stability condition that is called the CFL (Courant–Friedrichs–Levy) condition.

Rearranging the main inequality, i.e., $|G| = \left(1 - \frac{2\alpha\Delta t}{\Delta x^2}\right)^2 + \left(\frac{u\Delta t}{\Delta x}\right)^2 \leq 1$ by expanding the first term, leads to an additional stability criterion:

$$\frac{u^2\Delta t}{2\alpha} \leq 2 - \frac{2\alpha\Delta t}{\Delta x^2}. \tag{8.183}$$

The minimum value on the right side of the inequality above, i.e., (Eq. 8.183), according to (Eq. 8.180) is 1, which leads to:

$$\frac{u^2\Delta t}{2\alpha} \leq 1. \tag{8.184}$$

In total, three stability conditions must be satisfied in order to ensure that no Fourier modes will grow with time (i.e., as n grows) and therefore no unbounded instability develops:

- Condition 1 (rearranged inequality (8.184))

$$\alpha \geq \frac{u^2 \Delta t}{2}. \tag{8.185}$$

- Condition 2 (CFL condition)

$$\frac{u \Delta t}{\Delta x} \leq 1 \tag{8.186}$$

- Condition 3 (inequality (8.180))

$$\frac{2 \alpha \Delta t}{\Delta x^2} \leq 1 \tag{8.187}$$

The first condition in particular is interesting because it implies that solutions will be unstable for any non-zero time step unless there is a sufficient amount of diffusion. The remaining two conditions are restrictions on the time step size.

As opposed to the explicit method, a low CFL number is not required for stability in the implicit method.

8.7.3.2.2 Numerical Dissipation

By taking $\alpha = 0$ and S_ϕ in Eq. (8.141), a pure convection equation is obtained:

$$\frac{\partial \phi}{\partial t} + u \frac{\partial \phi}{\partial x} = 0, \tag{8.188}$$

Using backward discretization in space and forward in time and considering Taylor series expansions, Eq. (8.188) becomes:

$$\frac{\partial \phi}{\partial t} + u \frac{\partial \phi}{\partial x} = \frac{u \, \Delta x}{2} \left(1 - \frac{u \, \Delta t}{\Delta x}\right) \frac{\partial^2 \phi}{\partial x^2}, \tag{8.189}$$

where the right-hand side is called a numerical dissipation (diffusion) term. The numerical scheme can be stabilized by means of artificial diffusion. Upwind schemes are usually suited for RANS, and central differencing is suited for LES, in order to limit excessive numerical dissipation [73]. For the sake of completeness, it is mentioned that upwinding is particularly necessary for mixture fraction/species transport equations.

8.7.3.3 Convergence

A numerical method is said to be convergent if the numerical solution approaches the exact solution as the step size tends to zero. Very often in complex problems where the Navier–Stokes equations are solved (in conjunction with several sub-models), it is practically impossible to know beforehand the choice of time and space steps that ensure convergence. Therefore, convergence is investigated in practice by performing a grid sensitivity study: one starts on a coarse mesh (i.e., relatively large cell size) and refines gradually until the solution becomes 'mesh insensitive'. It is important to note that if the mesh is used as a filter in LES calculations, the solution is by definition grid-dependent. That is why we prefer the term 'mesh insensitive'.

8.7.3.4 Conservativeness

In order to understand conservativeness, let us consider first two discretization methods: the Finite Difference Method (FDM) and the Finite Volume Method (FVM).

The FDM is based on a discretization of the differential form of the conservation equations such as the following conservation equation for the mixture fraction:

$$\frac{\partial}{\partial t}(\rho Z) + \nabla \cdot (\rho Z \vec{v}) = \nabla \cdot (\rho D \nabla Z). \tag{8.190}$$

The solution is obtained using directly the schemes described above.

The FVM is based on a discretization of the integral forms of the conservation equations. Equation (8.190) is then expressed as follows:

$$\int_{CV} \frac{\partial}{\partial t}(\rho Z) dV + \int_{CV} \nabla \cdot (\rho Z \vec{v}) dV = \int_{CV} \nabla \cdot (\rho D Z) dV, \tag{8.191}$$

where CV is a control volume.

Using the Gauss-divergence theorem, we obtain:

$$\frac{\partial}{\partial t} \int_{CV} (\rho Z dV) + \int_{A} \vec{n} \cdot \vec{v} \rho Z dV = \int_{A} \vec{n} \cdot \rho D \nabla Z \, dV. \tag{8.192}$$

Equation (8.192) represents the flux balance of the quantity (Z) in the control volume CV. As opposed to FDM where a property ϕ is solved directly at each node i, in FVM the problem consists of computing (e.g., convective and/or diffusive) fluxes across the boundaries (i.e., the areas A).

Let us consider now the steady-state diffusion problem that is expressed as follows:

$$\int_{A} \vec{n} \cdot \rho D \nabla Z \, dV = 0 \tag{8.193}$$

and restrict ourselves to the 1-D problem.

Figure 8.3 shows a volume that is divided into three control volumes with nodes positioned in the centre of each and numbered from 1 to 3. Diffusive fluxes occur from the left to the right (see arrows). The inlet area to volume 1 is called A_{in}, and the outlet area from volume 3 is called A_{out}. The area between volumes 1 and 2 (resp. volumes 2 and 3) is called A_{1-2} (resp. A_{2-3}). The fluxes across the boundaries of the total volume are called q_{in} and q_{out}. Diffusive fluxes across volumes 1 to 3 are expressed as follows:

$$q_1 = (\rho D) A_{1-2} \frac{(Z_2 - Z_1)}{\Delta x} - q_{in}, \tag{8.194}$$

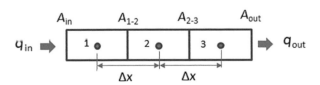

FIGURE 8.3 A sketch of 1-D diffusion problem.

$$q_2 = (\rho D) A_{2-3} \frac{(Z_3 - Z_2)}{\Delta x} - (\rho D) A_{1-2} \frac{(Z_2 - Z_1)}{\Delta x}, \tag{8.195}$$

$$q_3 = q_{out} - (\rho D) A_{2-3} \frac{(Z_3 - Z_2)}{\Delta x}, \tag{8.196}$$

where the flux is interpolated using a central difference formulation.

By summing Eqs. (8.194–8.196), one obtains:

$$q_1 + q_2 + q_3 = q_{out} - q_{in}. \tag{8.197}$$

Equation (8.197) shows the overall conservation of the mixture fraction for the 1-D diffusive problem at hand. In general, the FVM is said to be conservative by construction as long as the flux interpolation remains consistent [74]. On the contrary, the FDM does not necessarily ensure conservativeness because it relies on the Taylor series expansions where high-order terms are neglected.

8.7.3.5 Boundedness

The boundedness of a numerical scheme can be first seen as its ability to prevent quantities such as temperatures and concentrations from being negative. Furthermore, according to Ref. [74], the boundedness criterion states that: *In the absence of sources, the internal nodal values of property ϕ should be bounded by its boundary values.* This is illustrated in Ref. [74] by the example of a steady-state conduction problem without sources and with temperatures of 500°C and 200°C. In this case, all interior values of the temperature should be <500°C and >200°C. Another essential requirement for boundedness [74] is that *an increase in the variable ϕ at a node i should result in an increase in ϕ at neighbouring cells.* If this is not the case, the solution might not converge, or if it does, it contains 'wiggles' or spurious undershoots and overshoots in the vicinity of steep gradients.

In Refs. [16,17], a flux limiter (see Eq. 8.7.7) is used along with a flux correction that adds the minimum amount of numerical diffusion required to maintain boundedness.

8.7.4 Pressure-Velocity Coupling

The numerical methods described above are applied to the conservation equations discussed earlier in order to compute the flow field. However, a solution cannot be readily found (with the equations and numerical techniques presented so far) because there is no (transport or other) equation for pressure. Several procedures have been developed to solve the pressure-velocity linkage, such as the SIMPLE (Semi-Implicit Method for Pressure-Linked Equations) algorithm [75,76] or the PISO (Pressure Implicit with Splitting of Operators) algorithm [77]. The reader is referred to CFD textbooks (e.g., [74]) for more details.

8.7.5 The Importance of the Computational Mesh

The choice of the mesh resolution depends on the relevant types of length scales to the problem to be simulated with CFD: flame scales, vent flow scales and wall flow scales [78].

For fire plumes, it is suggested in Ref. [79] that 'there must be at least 10×10 cells over the burner surface in order to correctly capture its influence'. Another approach consists of defining a mesh resolution index as follows [80]:

$$R^* = \frac{\max(\Delta x, \Delta y, \Delta z)}{D^*}, \tag{8.198}$$

where D^* is the characteristic plume length scale defined as follows:

$$D^* = \left(\frac{\dot{Q}}{\rho_\infty T_\infty c_p \sqrt{g}} \right)^{2/5}. \qquad (8.199)$$

Numerical simulations have shown that plume dynamics can be simulated accurately only if the resolution limit is about $R^* = 0.1$ or smaller [80].

For vent flows, the length scale is typically based on the dimensions of the opening (i.e., vent). It could be, for instance, the hydraulic diameter or the ventilation factor, $AH^{1/2}$. A good starting point for the mesh resolution (i.e., cell size) is to divide the length scale by 10.

It is more difficult to estimate the 'right' resolution for wall flows, especially if there is pyrolysis. The best approach in this is to perform 'numerical tests' on simplified configurations.

8.8 BOUNDARY CONDITIONS

As in all CFD problems, computing a fire dynamics problem requires a correct definition and set-up of its boundary conditions. The most common boundary conditions in CFD are [74]: inlet, outlet, wall, pressure, symmetry and periodicity.

Each condition requires a specific treatment with several available techniques in order to connect the nodes of the domain (internal nodes) to additional nodes requiring the physical boundary (or near-boundary internal nodes). The intent of this chapter is not to describe the numerical implementation for general-purpose CFD but rather to address the BCs in the context of a fire problem from a physical standpoint.

Figure 8.4 illustrates several boundary conditions that need to be addressed when solving a fire problem with CFD. They are categorized in this chapter as follows:

1. *Fire source*: This is a very influential BC in fires. The size of the fire (in terms of geometry and heat release rate), in conjunction with other BCs determined by the geometry, controls the generated thermal environment as well as the amounts of smoke and toxic products produced.

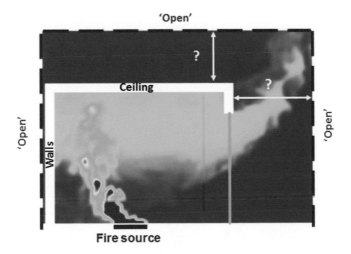

FIGURE 8.4 Boundary conditions for an enclosure fire with an open doorway.

2. *Walls*: Walls constitute the 'bounding envelope' of the fire environment in an enclosure fire. They lead to the containment of smoke and heat in a given volume. The accumulation of 'hot smoke' may result in a substantial heat feedback to the fire source depending on the heat losses to the walls, enhancing thus mass and heat transfer at the fire level and contributing to fire growth (see Chapter 5). In addition to the thermal effects of a wall, computing correct flow velocities at its vicinity (boundary) is particularly important for applications such as the activation of ceiling-mounted heat detectors (or sprinklers or fusible-link roof vents) where the activation time depends (among other factors) on the ceiling-jet velocity.

3. *Open boundary conditions* (*natural ventilation*): Limiting the computational domain only to the enclosure where the fire started may often not be sufficient to compute the flow correctly at the level of the openings. For instance, in case of an open doorway, there will be a vent flow (see Chapter 5) with (at the early stages of the fire) an upper hot smoke layer and a layer of incoming fresh air underneath (see Figure 8.4). The most suitable approach is then to extend the computational domain to a region away from geometrical openings so that the BC does not cause a disturbance of the flow through the opening. In the context of a fire, an 'open' boundary condition is associated with natural ventilation because it has the effect of conveying fresh air to the enclosure fire and/or extracting smoke and heat through vertical (open doorways and windows) or horizontal (roof vents) openings by buoyancy. The open BC is therefore essential in problems dealing with smoke and heat control via natural ventilation.

4. *Mechanical ventilation and pressure effects*: Another approach for smoke and heat control is to rely on mechanical ventilation for different types of buildings such as atria, car parks or tunnels with several possible design options for smoke extraction (see Chapter 6). Mechanical ventilation is also extensively used in Nuclear Power Plants (NPPs) where rooms are generally sealed and connected to each other with a mechanical ventilation network. The latter is set up to ensure a pressure cascade that prevents the release of toxic radioactive materials. If a fire occurs in one of these confined rooms, the induced pressure build-up alters significantly the normal operating conditions of the fans. The pressure effect is in this case an essential parameter in the set-up of the fan boundary condition. Another pressure effect to consider in the set-up of BCs is related to leakages.

8.8.1 Fire Source

8.8.1.1 Gaseous Fuel

A fire source can be viewed, from a mass transfer standpoint, as the release of one or several gaseous species, i, with (see Eq. 2.38):

$$\sum_i Y_{i,f} = 1, \tag{8.200}$$

where $Y_{i,f}$ is the mass fraction of species i at the fire source boundary (denoted with the subscript f).

The gas density at the level of the fire source can be calculated from the ideal gas law as follows:

$$\rho_f = \frac{\bar{p}}{RT_f \sum_i \frac{Y_{i,f}}{W_i}}, \tag{8.201}$$

where $\bar{p}$ is the pressure, R is the ideal gas constant, T_f is the surface temperature at the fire source surface and W_i is the molecular weight of species i.

The mass loss rate of each species i can then be expressed as follows:

$$\dot{m}_i'' = \rho_f w_f Y_{i,f} - \rho_f D_{i,f} \left(\frac{\partial Y_i}{\partial z} \right)_{z=0} , \tag{8.202}$$

where w_f is the gas velocity at the fire source surface and $D_{i,f}$ is the mass diffusion coefficient of species i.

The relation between $\dot{m}_i''$, ρ_f and w_f is:

$$\sum_i \dot{m}_i'' = \rho_f w_f . \tag{8.203}$$

The system of the three previous Eqs. (8.201-8.203) must be solved.

If one homogeneous species is considered at the source (typically one fuel), the system reduces to:

$$\dot{m}'' = \rho_f w_f \tag{8.204}$$

and

$$\rho_f = \frac{W_F \bar{p}}{R T_f} . \tag{8.205}$$

The mass loss rate per unit area can be either prescribed directly or indirectly by setting the heat release rate per unit area, which is expressed as follows:

$$\dot{Q}_f'' = \dot{m}'' \Delta h_c , \tag{8.206}$$

where Δh_c is the heat of combustion of the fuel.

8.8.1.2   Liquid Fuel

If the fire source is a liquid fuel, the liquid must first undergo an evaporation process that leads to the release of gaseous species (see Chapter 2), which is then computed following the previously described equations. An estimate of the Mass Loss Rate Per Unit Area (MLRPUA) (as could be prescribed, for instance, in Refs. [16,17]) can be given by the empirically derived correlation [81]:

$$\dot{m}'' = \dot{m}_\infty'' \left(1 - \exp(-k\beta D) \right), \tag{8.207}$$

where $\dot{m}_\infty''$ is a limiting burning rate, $k\beta$ is an extinction coefficient and D is the diameter of the pool fire. Average values of $\dot{m}_\infty''$ and $k\beta$ for steady-state burning of several liquids are given in Table 8.5.

There is, however, a disadvantage in prescribing a fixed MLRPUA. Whereas in reality when there is a heat build-up in an enclosure, the radiative feedback to the fuel surface may enhance substantially the fuel vaporization process and thus varies the MLRPUA taking into account heat transfer considerations. A model for the evaporation rate of the fuel is required in this case.

TABLE 8.5
Data for Estimating the Burning Rate of Large Pools [81]

Liquid	$\rho\left(\text{kg/m}^3\right)$	$\dot{m}''_\infty\left(\text{kg/m}^2\,\text{s}\right)$	$k\beta\left(\text{m}^{-1}\right)$
Liquid methane	415	0.078	1.1
Liquid propane	585	0.099	1.4
Ethanol	796	0.017	-
Methanol	794	0.015	-
Butane	573	0.078	2.7
Benzene	874	0.085	2.7
Hexane	650	0.074	1.9
Heptane	675	0.101	1.1
Acetone	791	0.041	1.9
Gasoline	740	0.055	2.1
Kerosene	820	0.039	3.5
Crude oil	830-880	0.022-0.045	2.8

TABLE 8.6
Flammability Parameters for Solid Fuels [82]

Combustible	$L_v\,(\text{kJ/g})$	$\dot{q}''_F\left(\text{kW/m}^2\right)$	$\dot{q}''_L\left(\text{kW/m}^2\right)$	$\dot{m}''_{\text{ideal}}\left(\text{g/m}^2\text{s}\right)$
Polyethylene	2.32	32.6	26.3	14
Polypropylene	2.03	28.0	18.8	14
Wood (Douglas fir)	1.82	23.8	23.8	13
Polystyrene	1.76	61.5	50.2	35
Polyurethane foam (rigid)	1.52	68.1	57.7	45
Polyurethane foam (flexible)	1.22	51.2	24.3	32
FR plywood	0.95	9.6	18.4	10

8.8.1.3   Solid Fuel

Similar to liquid fuels, the mass loss rate per unit area (MLRPUA) for a solid fuel can be either prescribed or computed via the following equation:

$$\dot{m}'' = \frac{\dot{q}''_F + \dot{q}''_E - \dot{q}''_L}{L_v}, \tag{8.208}$$

where $\dot{q}''_F$ and $\dot{q}''_E$ refer to heat fluxes to the surface from the flame and from external radiation, respectively. The variables $\dot{q}''_L$ and L_v refer to heat losses at the surface and heat of pyrolysis of the solid (see values from Table 8.6 for small-scale tests).

Equation (8.208) can be also written as follows:

$$\dot{m}'' = \dot{m}''_{\text{ideal}} + \frac{\dot{q}''_E - \dot{q}''_L}{L_v}, \tag{8.209}$$

where $\dot{m}''_{\text{ideal}} = \dot{q}''_E / L_v$.

8.8.1.4   Turbulence Inflow Boundary Conditions

Turbulence inflow boundary conditions are primarily important for momentum-driven flows, where the inlet BCs are blown far into the domain. The aim is to reproduce turbulent fluctuations in a realistic manner. In fire dynamics problems, the inlet could be, for example, a gaseous burner or a localized area of pyrolysed gases from a liquid or a solid fuel bed. It could also be the injection area of a mechanical device such as a fan, a booster or an air curtain. In the latter cases, a *good* modelling of inflow boundary conditions can help reproduce realistic flow patterns that are required to study the influence of these mechanic devices on smoke dynamics.

In RANS simulations, the turbulent kinetic energy, k, and the dissipation rate, ε, are computed by solving a transport equation. Therefore, inlet boundary conditions are required for k and ε. The following expressions can be used [13]:

$$k_{\text{inlet}} = \frac{3}{2}\left(U_{\text{inlet}}I\right)^2,\tag{8.210}$$

$$\varepsilon_{\text{inlet}} = C_\mu^{3/4}\,\frac{k_{\text{inlet}}^{3/2}}{\ell},\tag{8.211}$$

where U_{inlet} is the mean velocity at the inlet, I is the turbulence intensity at the inlet, C_μ is a constant and ℓ is a characteristic length scale (that is taken sometimes as $\ell = 0.07\,L$, where L is the characteristic diameter of the inlet [74]).

In Ref. [83], an overview is provided of several methods used to generate realistic inflow turbulence boundary conditions for LES and DNS. These methods are classified into three categories: recycling methods, synthetic turbulence methods and forcing techniques.

- *Recycling methods*: The general approach consists of generating a database of turbulent velocity fluctuations using a *precursor simulation*. The inflow data for the *main simulation* is taken from a location in the *precursor simulation* where a fully turbulent flow is developed.
- *Synthetic turbulence*: This approach consists of applying a stochastic procedure that generates at the inlet a velocity signal that resembles turbulence. Several methods can be used:

  1. *Algebraic methods*: The instantaneous velocity field is decomposed into a mean value and a fluctuation component, i.e., $u_i = U_i + u_i'$. The fluctuation component is expressed as follows:

$$u_i' = r_i\sqrt{\frac{2}{3}k},\tag{8.212}$$

  where k is the turbulence kinetic energy and r_i are independent random variables for each velocity component at each point and time step. This method reproduces the target mean velocity and kinetic energy profiles but does not reproduce correlations in time and space, which is not realistic. An improvement of this method consists of expressing the fluctuation as follows:

$$u_i' = r_j a_{ij},\tag{8.213}$$

  where a_{ij} is referred to as the Cholesky decomposition of the Reynolds stress tensor R_{ij} when the random data satisfies $\langle r_i r_j\rangle = \delta_{ij}$ and $\langle r_i\rangle = 0$.
  Another algebraic method is the digital filtering procedure developed in Ref. [84].

2. *Spectral methods*: These methods rely on the decomposition of the signal into Fourier modes to initialize the flow domain with a three-dimensional homogeneous and isotropic synthetic velocity field.
3. *Mixed methods*: These methods combine algebraic and spectral approaches.
4. *The Synthetic Eddy Method (SEM)*: In this model [83], the velocity fluctuation, u_i', is viewed as the average velocity of moving eddies generated at the inlet. It is then expressed as follows:

$$u_i' = \frac{1}{\sqrt{N}} \sum_{k=1}^{N} a_{ij} \varepsilon_j^k f_{\sigma_{ij}(x)} \left(x - x^k \right),$$ (8.214)

where N is the number of eddies, a_{ij} is the Cholesky decomposition of the Reynolds stress tensor R_{ij}, ε_j^k are independent random variables taken from any distribution with zero mean and unit variance, σ_{ij} is the eddy length scale and f is a shape function which provides the velocity distribution of the eddies located at x^k. The subscripts i and j denote, respectively, the velocity component and the spatial direction. Expression (8.214) allows the generation of non-isotropic velocity fluctuations at the inlet. If isotropic structures are produced, the three-dimensional length scale σ_{ij} which defines the structure of turbulent eddies for each velocity component i in each direction j is replaced by one value, i.e., $\sigma_{ij} = \sigma$. In order to generate synthetic velocity fluctuations at the inlet with the SEM, four parameters must be prescribed: (a) the velocity components, (b) the Reynolds stresses, (c) the eddy length scale(s) and (d) the Reynolds stress tensor components.

- *Forcing techniques*: An example of such techniques is addressed in Ref. [85] where a sinusoidal forcing is imposed at the inlet boundary to generate the turbulence for an early transition to a turbulent state. The fluctuation axial velocity at the inlet is expressed as follows:

$$u' = AU(r) \sum_{n=1}^{N} \sin\left(\frac{2\pi f t}{n} + \theta \right),$$ (8.215)

where A is the amplitude of the forcing, N is the number of modes, f is the passage frequency of the vertical structures at the end of the potential core (calculated from St = fD/U = 0.3 where St is the Strouhal number) and θ is the azimuthal angle.

8.8.2 WALLS

8.8.2.1 Velocity

In order to solve the flow accurately near a wall (e.g., ceiling jet), it is important to have a very small cell size. Such cell size is often very small in comparison with the cell size away from walls. The main approach is then to prescribe wall functions that rely on the concept of a dimensionless local wall distance defined as follows:

$$y^+ = \frac{y u_\tau}{v},$$ (8.216)

where y is the normal distance from the wall, u_τ is the wall friction velocity and v is the kinematic viscosity.

The wall friction velocity is expressed as follows:

$$u_\tau = \sqrt{\frac{\tau_w}{\rho}}, \qquad (8.217)$$

where τ_w is the wall shear stress and ρ is the density.

A local dimensionless velocity near the wall can be defined as follows:

$$u^+ = \frac{\tilde{u}}{u_\tau}, \qquad (8.218)$$

where $\tilde{u}$ is the local velocity.

Very close to the wall ($y^+ < 5$), the fluid is considered to be mainly dominated by viscous shear stress. Turbulent shear stress effects are neglected. It is assumed also that the shear stress is almost equal to the wall shear stress, τ_w, throughout the layer. This is expressed as follows:

$$\mu \frac{\partial u}{\partial y} = \tau_w, \qquad (8.219)$$

where μ is the dynamic viscosity ($\mu = \rho v$).

By discretizing Eq. (8.219), one obtains:

$$\mu \frac{\tilde{u} - u_w}{y - y_w} = \tau_w, \qquad (8.220)$$

where y_w is the wall coordinate taken as $y_w = 0$.

Applying the no-slip boundary condition at the wall level, i.e., $u_w = u_{(y=y_w=0)} = 0$, along with Eq. (8.220) gives:

$$\tilde{u} = \frac{y\,\tau_w}{\mu} = \frac{y}{\mu} \rho u_\tau^2. \qquad (8.221)$$

By using the definition of the dimensionless numbers used above, one obtains [13]:

$$u^+ = y^+. \qquad (8.222)$$

Given the linear expression provided in Eq. (8.222), the fluid layer adjacent to the wall is called a linear sub-layer.

At some distance from the wall ($30 < y^+ < 500$), a logarithmic expression prevails, leading to what is called a log-law layer:

$$u^+ = \frac{1}{\kappa} \ln\left(y^+\right) + B, \qquad (8.223)$$

where κ is the von Karman constant taken as $\kappa = 0.41$ and $B = 5.3$ [61].

In the region where ($5 < y^+ < 30$), called the buffer layer, both viscous and inertial stresses are important.

The most common approach to approximate the velocity u^+ in the buffer layer (e.g., in Refs. [86,16]) is to match the viscous and log regions at a cross-over point y_0^+ (i.e., intersection between the viscous sub-layer and the logarithmic region). The wall functions are then expressed as follows:

- Linear sub-layer

$$u^+ = y^+ \quad \text{if } y^+ < y_0^+.$$ (8.224)

- Log-law layer

$$u^+ = \frac{1}{\kappa} \ln\left(y^+\right) + B \quad \text{if } y^+ \geq y_0^+.$$ (8.225)

where y_0^+ is computed based on the upper root of:

$$y_0^+ = \frac{1}{\kappa} \ln\left(y_0^+\right) + B.$$ (8.226)

Depending on the exact values of κ and B that are considered, y_0^+ is around 11.5 (in Ref. [86] $y_0^+ = 11.225$ and in Ref. [16] $y_0^+ = 11.81$).

It is worth noting that these functions are based on small pressure gradients, local equilibrium of turbulence and constant near-wall stress layer. Non-equilibrium wall functions that are recommended for complex flows are not addressed here.

8.8.2.2 Temperature

As mentioned in the introduction, heat conduction through walls (ceiling and floor) is an important component in the heat balance of an enclosure fire. The one-dimensional heat conduction equation through the solid boundary of a wall (without any additional source term, e.g., no combustible wall) reads:

$$\rho c \frac{\partial T}{\partial t} = \frac{\partial}{\partial x}\left(k \frac{\partial T}{\partial x}\right).$$ (8.227)

Two boundary equations are required to solve this equation: the first one on the front surface (surface subjected to heat from the flames and the smoke layer) and the second one at the back side.

On the front surface of the wall, the boundary condition reads:

$$-k\left(\frac{\partial T}{\partial x}\right)_{x=x_0} = \dot{q}_c'' + \dot{q}_r'',$$ (8.228)

where $\dot{q}_c''$ and $\dot{q}_r''$ are, respectively, the convective and radiative fluxes per unit area.

The convective heat flux at the surface of the wall is expressed as follows:

$$\dot{q}_c'' = h\left(T_g - T_{x=x_0}\right),$$ (8.229)

where T_g is the temperature in the centre of the first gas cell.

The radiative heat flux of an opaque wall is expressed as follows:

$$\dot{q}_r'' = \dot{q}_{r,\text{in}}'' - \dot{q}_{r,\text{out}}'',$$ (8.230)

where $\dot{q}_{r,\text{in}}''$ is the incoming radiation and $\dot{q}_{r,\text{out}}''$ is the outgoing radiation ($\dot{q}_{r,\text{out}}'' = \varepsilon \sigma T_w^4$).

If the wall at the front or at the back is assumed to be perfectly insulated, the following boundary condition applies:

$$-k\left(\frac{\partial T}{\partial x}\right)_{x=x_0} = 0. \tag{8.231}$$

8.8.3 OPEN BOUNDARY CONDITIONS (NATURAL VENTILATION)

Consider the configuration of an enclosure fire illustrated in Figure 8.4. There are mainly three types of boundary conditions (BC):

- A BC at the fire source,
- A BC at the walls and ceiling,
- A BC called 'open boundary' condition.

The third one, addressed here, allows fresh air and smoke to flow freely. In reality, the full physical domain corresponding to Figure 8.4 extends indefinitely in all the three directions of space. However, due to computational constraints (finite available resources), the computational domain is truncated. Two questions open up at this level: (a) How to choose the size of the computational domain? and (b) What conditions to apply at its boundaries? The objective is to *use smaller, less computationally expensive domains while maintaining accuracy* [87].

8.8.3.1 Velocity and Scalars

The most common approach is to apply a Von Neumann boundary condition:

$$\frac{\partial \phi}{\partial n} = 0, \tag{8.232}$$

where n is the direction normal to the boundary. For example, FDS [16,17] uses this boundary condition for the tangential velocity, temperatures and mass fractions.

8.8.3.2 Pressure

- *Bernoulli's principle*: This boundary condition has been developed in FDS [16,17]. It relies on the Dirichlet conditions for pressure and assumes that the quantity $H = p/\rho + (u_i u_i)/2$ remains constant along a streamline.
  If we consider, for instance, the boundary $x = x_{\min}$, the boundary value for H reads:
- For outgoing flows:

$$H_{1/2,jk} = \frac{\tilde{p}_{\text{ext}}}{\rho_{1,jk}} + \frac{1}{2}\left(\bar{u}_{1,jk}^2 + \bar{v}_{1,jk}^2 + \bar{w}_{1,jk}^2\right). \tag{8.233}$$

- For incoming flows it is expressed as follows:

$$H_{1/2,jk} = \frac{\tilde{p}_{\text{ext}}}{\rho_{\text{amb}}} + \frac{1}{2}\left(\bar{u}_{\text{amb}}^2 + \bar{v}_{\text{amb}}^2 + \bar{w}_{\text{amb}}^2\right). \tag{8.234}$$

The subscript amb denotes user-specified far-field velocity and density values. The far-field velocity is typically set to zero except for situations involving an external wind. The subscript 1/2 replacing the coordinate i (in the x-direction) denotes that the total pressure is interpolated at the boundary with streamlines normal to the x-direction. In Ref. [87], it is noted that if streamlines are not normal to the boundary, this BC could work poorly.

- *The OpenFOAM buoyant pressure condition*: This BC for pressure in buoyancy-driven flows is expressed as follows:

$$\frac{\partial p}{\partial n} = \frac{\partial \rho}{\partial n} g \Delta z, \tag{8.235}$$

which represents the spatial derivative of the pressure field written as $p = p_0 + \rho g \Delta z$.

In Ref. [87], a more detailed discussion is provided on open boundary conditions.

The modelling approaches described in this section provide an answer to the question asked earlier: 'What conditions to apply at the boundaries'? but, do not explicitly answer the former question on how to choose the size of the truncated domain. Obviously, the larger the domain, the better the sense that less impact is to be expected at the openings.

For example, one can see from Figure 8.4 the outgoing flow of hot smoke and the incoming flow of fresh air. Therefore, although the primary interest might be what happens inside the enclosure, the open boundary condition must not be set at the level of the open doorway but should be extended further downstream. This will allow smoke and fresh air to flow freely. Obviously, if external flaming is expected to occur, it is even more important to extend the domain in order to avoid the computation of combustion at the level of the boundary (which may result in a substantial inaccuracy).

A second simpler case illustrating the importance of the size of the domain is the case of a turbulent buoyant plume in open atmosphere conditions. More specifically, the size of the domain in the horizontal directions is particularly important with respect to the entrainment of fresh air within the fire plume. It is difficult to know beforehand the size that provides the best compromise between accuracy and computational time. That is why very often a sensitivity analysis is performed to select the optimal size. For instance, it is recommended in Ref. [88] for simulations of a medium-scale methanol pool fire to use a domain with a width/depth of more than four times the burner diameter. In Ref. [89], a domain is used with a depth/width 10 times the burner diameter in their simulations of McCaffrey's experiments [90].

In a nutshell, there are no specific quantitative guidelines for the positioning of open boundary conditions, but the CFD user must consider this aspect in the set-up of his model and analysis of its results.

8.8.4 MECHANICAL VENTILATION AND PRESSURE EFFECTS

Often in CFD simulations of fire safety engineering problems, mechanical ventilation must be modelled. The latter could be part of either the building design in normal (non-fire) operating conditions or a heat and smoke control system (see Chapter 6) designed for fire conditions. One example for mechanical ventilation in normal operating conditions is an HVAC (Heat Ventilation and Air Conditioning) system for indoor air quality and comfort. A second example is the set-up of a ventilation network in a nuclear facility in order to create a pressure cascade that prevents the release of hazardous materials in case of an accident [91,92]. In all cases, the outcome of a CFD study could strongly depend on the set-up of the mechanical ventilation.

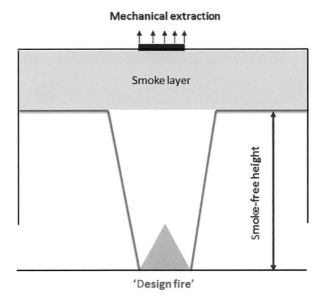

FIGURE 8.5   Mechanical smoke extraction in an atrium.

8.8.4.1   Fixed Velocity

Consider, for example, the case of an atrium (see Figure 8.5) with a mechanical ventilation system (i.e., one fan) placed at its roof in order to extract smoke and ensure a given smoke-free height based on a specified fire size (called a design fire). A simple model would be to assign a fixed velocity, v_{ex}, at the level of the fan, which corresponds to an extraction volume flow rate of:

$$\dot{V}_{\text{ex}} = v_{\text{ex}} A_{\text{ex}}, \tag{8.236}$$

where A_{ex} is the extraction area.

8.8.4.2   Fan Curves and Pressure Effects

Consider now a similar configuration to Figure 8.5 except that the enclosure is:

- much smaller,
- well confined and practically without leaks,
- equipped with an inlet and an exhaust fan as illustrated in Figure 8.6.

In the event of a fire for such a configuration (typically encountered in nuclear facilities [92]), the release of hot combustion products in a closed vessel will substantially increase the overall room pressure. A force is then created and directed towards the outside, decreasing thus the supply of fresh air (by acting in an opposite direction to the supply fan) and increasing the extraction of smoke [91]. The volume flow rates delivered at the inlet and extraction are not constant anymore (as for the case in Figure 8.5), but they rather depend on the discontinuous pressure rise (or decrease) at the level of the fans, as expressed in the following expression:

$$\dot{V}_{\text{fan}} = f\left(\Delta p\right). \tag{8.237}$$

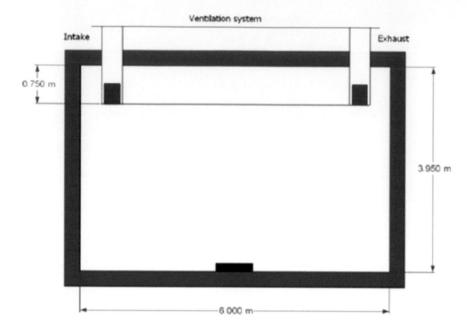

FIGURE 8.6   A sketch of mechanical ventilation in a well-confined room.

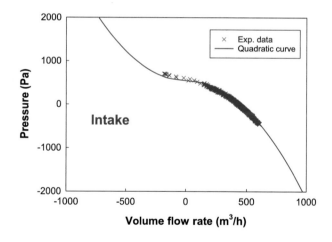

FIGURE 8.7   Variation of intake volume flow rates with pressure (test case examined in Ref. [93]).

The function f could be a polynomial, a piecewise linear function, or piecewise-polynomial function, or a user-defined function and is in fact a fan characteristic.

Figures 8.7 and 8.8 show examples of fan volume flow rate dependence on pressure in the event of a fire in a confined multi-compartment configuration [93]. The volume flow rates in Figures 8.7 and 8.8 are modelled using quadratic fan curves that correspond to the following expression [16]:

$$\dot{V} = \dot{V}_0 \operatorname{sign}\left(\Delta p_{\max} - \Delta p\right) \sqrt{\frac{\left|\left(\Delta p_{\max} - \Delta p\right)\right|}{\Delta p}}, \tag{8.238}$$

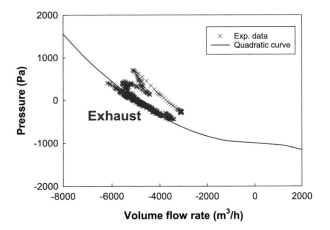

FIGURE 8.8   Variation of extraction volume flow rates with pressure (test case examined in Ref. [93]).

where $\dot{V}_0$ is the volume flow rate at $\Delta p = 0$ and Δp_{max} is called the stall pressure of the fan. For example, in Figure 8.7, $\dot{V}_0 = 450$ m^3/h and $\Delta p_{max} = 550$ Pa.

One can clearly see that when the pressure increases, the volume flow rate at the inlet decreases until it reaches $\dot{V}_0 = 0$ m^3/h at $\Delta p = \Delta p_{max}$. If the pressure increases further, the volume flow rate becomes negative, which means (by convention) that the fan does not behave as an inlet fan anymore but rather as an exhaust. This is called a reverse flow. On the other hand, for the exhaust fan, the higher the pressure, the higher the extraction volume flow rate (see Figure 8.8).

A more complex approach to model flow rates in ventilation branches is to use a Heat Ventilation and Air Conditioning (HVAC) network as implemented in Refs. [16,17] where conservation equations for mass, momentum (including the additional momentum induced by a fan) and energy are solved at the level of each duct. This set of equations is coupled with a consistency condition for pressure. Such a condition ensures [17] that *blowing air or starting a fire within a sealed compartment leads to an appropriate decrease in the divergence within the volume.* In a similar approach used in the CFD code ISIS ('Incendie SImulé pour la Sûreté') [94], the general Bernoulli equation is solved for each branch:

$$L_j \frac{\partial(\rho_j u_j)}{\partial t} = p(t) - p_{ext}(t) - f_j, \tag{8.239}$$

where p is the thermodynamic pressure, p_{ext} is the external pressure (at the end of the branch), ρ_j is the density, u_j is the velocity, L_j is the length of the branch, and f_j is the friction coefficient defined as follows:

$$f_j = \rho_j R_j A_j^2 |u_j| u_j, \tag{8.240}$$

where R_j (in m^{-4}) is the aeraulic resistance of the duct and A_j is the area of the ventilation.

If L_j is not specified, the stationary Bernoulli equation is considered:

$$p(t) - p_{ext}(t) = f_j. \tag{8.241}$$

8.9 EXAMPLES OF CFD SIMULATIONS

In this section, several examples of CFD applications in fire dynamics are discussed. The intent here is to:

- provide the reader with an overview of the physical and numerical aspects explained throughout the book with clear examples,
- illustrate the state-of-the-art of CFD in fire dynamics,
- highlight the main modelling challenges.

Furthermore, the well-documented test cases examined in this section are sorted with an ascending level of complexity (non-reacting plumes → free-burning flames → fuel-controlled enclosure fire → underventilated enclosure fire), which is suitable for CFD beginners.

8.9.1 Non-Reacting Buoyant Plume

The main physical aspects related to non-reacting buoyant plumes were discussed in Chapter 4. An illustration of these aspects is provided here through the CFD simulation of a round turbulent buoyant plume examined in Ref. [95].

8.9.1.1 Test Case Description

A comprehensive set of hot-wire measurements of a round buoyant turbulent plume is reported in Ref. [95]. The plume was generated by forcing a jet of hot air vertically upwards into a quiescent environment at a temperature $T_0 = 292 \pm 1°C$ through a nozzle with a diameter $D = 0.0635$ m. The exit velocity measured by a two-wire probe was $U_0 = 0.98$ m/s. The velocity exit profile was uniform outside the wall boundary layer (of the nozzle) to within 2%. The calculated buoyancy and kinematic momentum at the source are $B_0 = 0.0127$ m^4/s^3 and $M_0 = 0.0030$ m^4/s^2, respectively. Measurements of mean flow (temperature and velocity) and turbulence properties (second- and third-moment profiles) are provided at heights between 10 and 20 diameters.

8.9.1.2 Simulation Set-Up

The simulations shown here have been performed using FDS 6 [16,17]. The round plume source has been modelled with an equivalent square source yielding the same buoyancy flux at the inlet. The computational domain is set with the following dimensions: 0.616 m × 0.616 m × 1.200 m (width × depth × height) and with 'open' boundary conditions at the sides and the top. The width/depth of the computational domain is 11 times the inlet length, which is believed to minimize the boundary condition effects on the simulations and allow a complete description of air entrainment. A mesh with two refinement regions is generated. The smallest cells are 4 × 4 × 4 mm^3 and uniformly placed in a 0.168 × 0.168 × 1.200 m^3 domain. The mesh resolution in this region corresponds to $R^* = 0.08$ (see Eq. 8.198). Indeed, for this case, R^* is the relevant mesh criteria. The cells in the remaining region of the computational region of the domain are 8 × 8 × 8 mm^3. The turbulent viscosity has been modelled using either the constant or the dynamic Smagorinsky model. A first set of simulations were performed without prescribing turbulence inflow boundary conditions. In the second set of simulations, isotropic inflow turbulence was generated at the inlet using the SEM model (described in Section 8.8.1) with a turbulence intensity $I = 30\%$, a number of eddies $N = 1,000$ and an eddy length scale of $\ell_{\text{eddy}} = 0.01$ m.

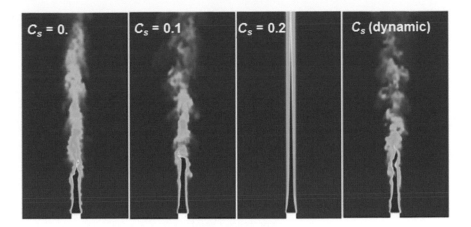

FIGURE 8.9 Effect of the Smagorinsky constant on the instantaneous temperature field (the SEM is not used here).

8.9.1.3 Results

The results are not examined here in detail through quantitative comparison between the numerical and experimental data. We rather focus on a qualitative description on the influence of two modelling aspects: (a) the turbulence viscosity and (b) the turbulence inflow boundary condition.

Figure 8.9 shows instantaneous temperature fields for the values of the Smagorinsky constant ($C_s = 0$, 0.1 and 0.2) in the constant Smagorinsky approach. The most striking result is obtained using a constant Smagorinsky coefficient $C_s = 0.2$, causing a complete suppression of turbulence. This can be explained by the too high turbulent viscosity, which causes excessive dissipation, leading to a damping of the turbulence disturbances. The difference between the three other flow fields is not visible at first sight. However, a closer look at, for example, the main centreline values for vertical velocity (not shown here) between $C_s = 0$ and $C_s = 0.1$ reveals that an increased C_s causes (in addition to turbulence damping) a delay in the jet break-up and results in an overestimation of the velocity. Figure 8.9 illustrates the self-adjustment capability of the dynamic approach. The value of C_s increases from zero near the inlet to a value of 0.14 at around $z = 3.6D$ and then slowly decreases in an asymptotic manner to the value of $C_s = 0.12$ at the furthest positions downstream (Figure 8.10).

Figure 8.11 illustrates the influence of the turbulence inflow boundary conditions on the temperature field. One can note the earlier jet break-up. The hot-air jet breaks up sooner (closer to the inlet) due to the added disturbances at the inlet by creating synthetic eddies in motion to emulate the turbulence boundary condition where hot air is released into the ambient atmosphere.

8.9.2 HOT-AIR PLUME IMPINGING ON A HORIZONTAL PLATE

8.9.2.1 Test Case Description

The test case examined here has been performed in the framework of a study that addresses the interaction between a water spray and a fire plume [96]. Prior to the plume-spray interaction tests, a series of three hot-air jet plumes impinging onto a horizontal ceiling have been conducted and documented in Ref. [97].

The fire plume was simulated by forcing a jet of hot air vertically upwards into a quiescent environment. The hot-air source is a 72 mm diameter circular nozzle issuing from a 254 mm long

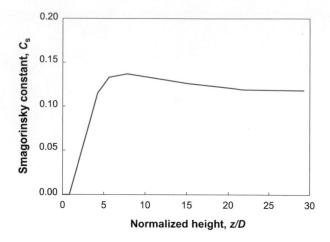

FIGURE 8.10 Evolution of the calculated Smagorinsky constant on the centreline for the dynamic Smagorinsky simulation.

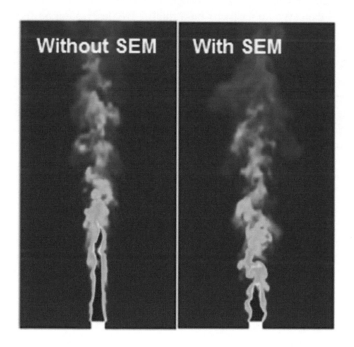

FIGURE 8.11 Effect of the SEM model on the instantaneous temperature field using the dynamic Smagorinsky model.

steel tube. The hot-air exit temperature has been maintained at 205°C at 30 mm above the nozzle exit. Three exit velocities have been tested: 3.3, 4.2 and 5.3 m/s. Only the last test is considered here. A 3 mm thick horizontal aluminium ceiling plate of 1.22 × 1.22 m² has been placed at a distance of 590 mm above the hot-air nozzle. Mean velocities (radial and vertical directions) and velocity fluctuations (vertical and horizontal fluctuations, as well as turbulent shear stresses) of

FIGURE 8.12 Schematic of the experimental set-up for the hot-air plume impinging on a horizontal surface [97].

the vertical plume and the ceiling jet have been measured using the laser-based Particle Image Velocimetry (PIV) technique. Temperature measurements have not been performed, except at 30 mm above the nozzle exit. More details on the experimental set-up can be found in Ref. [97].

8.9.2.2 Simulation Set-Up

The CFD package FireFOAM 2.2.x [15] is used here. Turbulence is modelled by the one-equation eddy viscosity approach. The turbulent viscosity is expressed as follows:

$$\mu_t = \rho c_k \Delta k^{1/2}, \tag{8.242}$$

where c_k is a model constant taken as $c_k = 0.03$. The subgrid-scale kinetic energy, k, is solved using a transport equation where the dissipation rate is expressed as follows:

$$\varepsilon = c_\varepsilon k^{3/2} \Delta^{-1}, \tag{8.243}$$

where c_ε is a dimensionless model coefficient taken as $c_\varepsilon = 1.048$.

The computational domain is 2 m×2 m×0.734 m. It is extended by 0.4 m at each side of the ceiling. An open boundary condition is employed at the sides, bottom and the extended parts on top of the computational domain. At the ceiling, zero-gradient boundary condition is applied for the subgrid-scale viscosity in addition to no-slip boundary condition for the velocity. An unstructured Cartesian computational mesh is used in the simulations. It has been generated using OpenFOAM's mesh generation utility *snappyHexMesh*. The total number of the cells is 7 million, with 14 cells across the jet inlet (5 mm). Mesh refinement is applied in the region of the thermal plume and the ceiling layer flow. The grid is also refined in the ceiling with a minimum wall-normal spacing of $\Delta z = 2$ mm. The numerical simulations are run for 20 seconds of real time. The averaging is done over the period of the last 17 seconds.

A velocity profile is imposed at the inlet using experimental data. Turbulence is generated at the inflow using the *decaying-TurbulenceInflowGenerator* that creates random spots in motion. The velocity fluctuation at the nth time instant is calculated as the sum of fluctuations produced by each spot:

$$u^n(x) = \sum_{i=1}^{M} r_i^n f\left(x - x_{ri}^{(n)}\right), \tag{8.244}$$

where r_i are random numbers and x_{ri} is the position of a random spot i. The variable M denotes the number of spots.

More details can be found in Ref. [98].

8.9.2.3 Results

Figure 8.13 shows the mean vertical velocity along the plume axis. The centreline velocity increases due to buoyancy acceleration before starting to decrease due to turbulent mixing with the surrounding air.

Figure 8.14 displays the ceiling-jet predictions for the mean horizontal velocities. These results show an overall good agreement. However, there are significant deviations (with the experimental data) in the predictions of the viscous sub-layer thickness. In the experiments, the locus of the peak horizontal velocity has been measured for all the radial positions from plume axis at $z = 587$ mm, which corresponds to a momentum thickness of $\delta_{v_{max}} = 3$ mm, whereas the predictions show an increase in the viscous sub-layer thickness from $\delta_{v_{max}} = 4$ mm at a radial position $r = 57.6$ mm from the plume axis to $\delta_{v_{max}} = 12$ mm at $r = 552.6$ mm. A complementary simulation using FDS [16] and prescribing a wall function (see Section 8.8.2) with a cell size of 2 mm near the ceiling showed a similar behaviour to the OpenFOAM results (i.e., an increase in the viscous sub-layer thickness with increased radial distance from the plume axis). These discrepancies (between the simulations results and the experimental data) need to be further investigated. This example highlights the importance of the mesh size near the ceiling in addition to the mesh resolution at the fire source diameter.

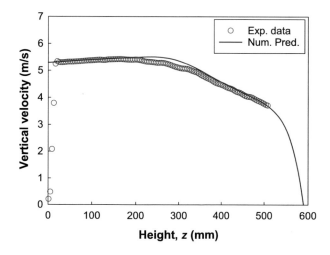

FIGURE 8.13 Comparison of the experimental and numerical vertical velocity profiles in the axis of the jet.

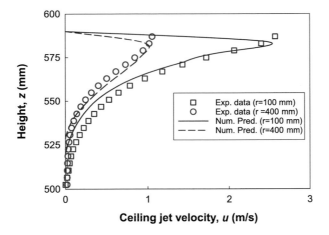

FIGURE 8.14 Comparison of the experimental and numerical profiles of the ceiling-jet horizontal velocity two distances from the plume axis: $r = 100$ mm and $r = 400$ mm.

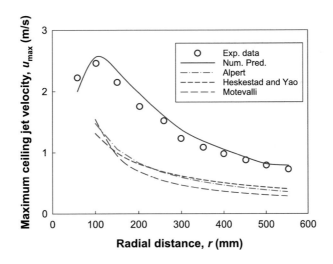

FIGURE 8.15 Comparison of the experimental and numerical profiles of the ceiling-jet horizontal velocity two distances from the plume axis: $r = 100$ mm and $r = 400$ mm.

Finally, it is important to note that the experimental measurements and numerical predictions of the maximum ceiling-jet velocities are significantly higher than the values predicted by the Alpert or Heskestad correlations (see Figure 8.15). This is mainly explained by the fact that the latter correlations apply low Froude number flows (as typically encountered in fires), whereas in the experiment described here, the flow is rather momentum-driven at the inlet (high velocities).

8.9.3 FREE-BURNING TURBULENT BUOYANT FLAME

The added complexity here, with respect to the two previous test cases, is the occurrence of combustion at the source. The flame (i.e., near-field) region is characterized by an entrainment rate, and more generally, a structure that is different from the plume (i.e., far-field region of the flow).

The main properties of turbulent buoyant diffusion flames have been investigated experimentally by many researchers, e.g., [90] and [99]. The flame that is simulated here is part of a test series reported in Ref. [100] on medium-scale propane flames with a heat release rate ranging from 16 to 38 kW.

8.9.3.1   Test Case Description

A 37.9 kW propane flame is generated from a 30 cm diameter burner made of a cylindrical envelope containing about 2,000 vertical tubes of alumina placed near each other and supported by a honeycomb structure. Sampled gases were analysed either by chromatography or using specific infra-red analysers. Temperature was measured by means of 50 μm butt-welded chromel-alumel thermocouples. The error due to soot deposition is estimated to reach a maximum value of 60 K in the region where the mean temperature is maximum. The velocity field was measured using a Laser Doppler Analysis (LDA) system.

8.9.3.2   Simulation Set-Up

The simulation shown here has been performed using the FDS (version 6) [16,17]. The computational domain is set with the following dimensions: $1.50 \, \text{m} \times 1.50 \, \text{m} \times 1.95 \, \text{m}$ (width × depth × height) and with 'open' boundary conditions at the sides and the top. The circular shape of the burner is modelled using the stair-stepping method. A uniform and structured was used with a cell size of 1.5 cm, which gives a mesh resolution of $R^* = 0.06$, which is again here a relevant criterion.

The gas-phase combustion reaction is specified as a single-step mixing-controlled reaction expressed as follows:

$$v_F C_n H_m + v_{O_2} O_2 \rightarrow v_{H_2O} H_2O + v_{CO_2} CO_2 + v_{CO} CO + v_s C,$$

where v_i are the stoichiometric molar fractions of each of the species i. For propane ($n = 3$ and $m = 8$), the following molar fractions are calculated:

$$v_F = 1,$$

$$v_{H_2O} = \frac{m}{2} = 4,$$

$$v_{O_2} = n + \frac{m}{4} - \frac{v_{CO}}{2} - v_s,$$

$$v_{CO_2} = n - v_{CO} - v_s.$$

The molar fractions of CO and soot are calculated from their yields y_i according to the following formula:

$$v_i = \frac{W_F}{W_i} y_i, \tag{8.245}$$

where W_F is the molecular weight of the fuel.

The CO and soot yields are taken as $y_{CO} = 0.005$ and $y_{soot} = 0.024$ [65].

A heat release rate per unit area (HRRPUA) of $\dot{Q}'' = 577 \, \text{kW/m}^2$ is prescribed at the level of the inlet. This value has been calculated by taking into account the slight difference in the area of

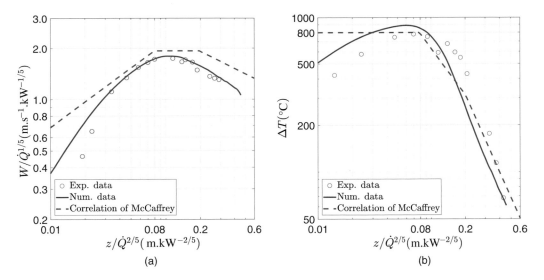

FIGURE 8.16 Comparison of the experimental and numerical centreline profiles of vertical velocity (a) and temperature (b).

the burner due to the stair-stepping method. Alternatively to the HRRPUA, a mass loss rate per unit area (MLRPUA) could have been specified with $\dot{m}'' = \dot{Q}''/\Delta h_c$ where the heat of combustion of propane is taken as $\Delta h_c = 46,000\,\mathrm{kJ/kg}$ [65]. The turbulence model used here is the dynamic Smagorinsky model. The turbulent combustion model used is the Eddy Dissipation Concept (EDC) model (see Section 8.4.2). The RTE equation is solved using the FVM method in conjunction with a prescribed radiative fraction of $\chi_r = 0.286$ as suggested by the experimental database of Tewarson [65]. As explained in Section 8.5.3, TRI is partially accounted for by performing a correction in the emission term.

8.9.3.3 Results

The simulation results displayed in Figure 8.16 show that the centreline mean vertical velocity is well predicted. A lower level of agreement has been obtained for the temperature profile. This could be attributed to several factors such as the simple soot model employed here and other additional simplifications in the computation of thermal radiation (e.g., TRI is only partially accounted for).

Figure 8.17 shows also that the simple conversion model for combustion products provides a good prediction for carbon dioxide. However, carbon monoxide centreline molar concentrations are significantly underestimated (the peak experimental value is 4.13%, and the peak numerical value is 0.02%). This is attributed to the fact that the slow chemistry of CO formation is not accounted for here.

8.9.4 OVER-VENTILATED ENCLOSURE FIRE WITH NATURAL VENTILATION

8.9.4.1 Test Case Description

In Ref. [101], a set of 55 full-scale experiments are described, studying enclosure fire characteristics at the developing stage (i.e., with low heat release rates between 30 and 160 kW). Particularly the mass flow rate through the door or window opening and the effects of opening geometry, fire strength and fire location on the fire plume entrainment rate were examined. The well-documented

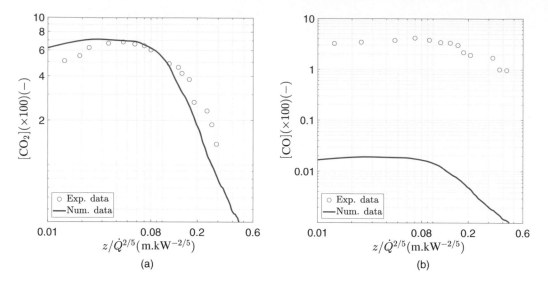

FIGURE 8.17 Comparison of the experimental and numerical centreline profiles of molar concentrations of carbon dioxide (a) and carbon monoxide (b).

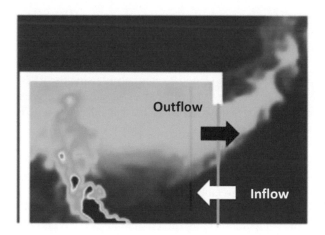

FIGURE 8.18 The Steckler case configuration.

measurements made this series of experiments (known as 'the Steckler case') very popular for code validation.

The description of the experiment provided in Ref. [101] is as follows.

The steady-state flow experiments were conducted in the room shown in Figure 8.18 [2.8 m × 2.8 m × 2.18 m]. The lightweight walls and ceiling were covered with a ceramic fibre insulation board to establish near-steady conditions within 30 minutes following ignition of the 30 cm diameter porous plate diffusion burner. The burner was supplied with commercial-grade methane at a fixed rate.

Several fire strengths, door and window openings and fire locations were examined. However, only one test is considered here for CFD simulations. For this test, the burner was placed in the centre of the room flush to floor level, the fire strength was 62.9 kW, and the door was 0.74 m wide and 1.83 m high (see Figure 8.18).

8.9.4.2 Simulation Set-Up and Results

Similar to the previous example (i.e., free-burning propane flame), gas-phase combustion reaction has been specified as a single-step mixing-controlled reaction (without soot and CO production due to the very low levels produced by methane) along with the same turbulence (i.e., dynamic Smagorinsky), turbulent combustion (i.e., EDC) and radiation (with $\chi_r = 0.14$) models. As shown in Figure 8.18, the computational domain has been extended in front of the doorway in order to minimize the effect of the 'open' boundary condition at the level of the opening on smoke outflow and air inflow. A uniform-structured mesh is used with a cell size of 32 mm, which corresponds to a mesh resolution index of $R^* = 0.1$. Heat loss to the boundaries was computed by specifying the thermal properties of lightweight concrete and ceramic fibre and solving the 1-D Fourier equation for conduction.

The results displayed in Figures 8.19 and 8.20 show an overall very good agreement of the temperature and horizontal velocity profiles at the doorway as well as the temperature profile within

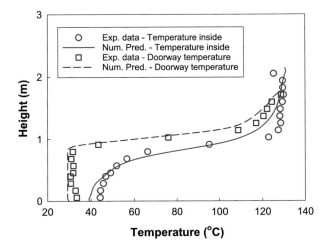

FIGURE 8.19 Temperature results for the Steckler case.

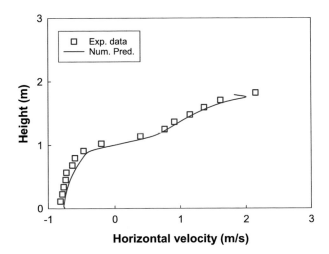

FIGURE 8.20 Horizontal velocity (at the doorway) results for the Steckler case.

the room. One can clearly see the two-layer structure (hot upper layer and cold lower layer) typical for an enclosure fire in its early stages. The interface between the two layers is characterized by a high temperature gradient as observed in Figure 8.19. A two-zone model can thus be viewed for similar type of situations as a good alternative to CFD.

8.9.5   OVER-VENTILATED ENCLOSURE FIRE WITH MECHANICAL VENTILATION

8.9.5.1   Test Case Description

The test case considered here has been examined experimentally in Ref. [102]. The configuration consists of a reduced-scale closed room of 1.50 m×1.26 m×0.99 m height and made of a 2 mm thick steel structure with a 25 mm thick inner layer of calcium silicate. The ventilation system consists of one intake branch and one exhaust branch. Within the latter, an extraction fan is positioned, delivering a volume flow rate of about 40 m³/h prior to the start of the fire. The intake branch is left open. The fire source is a propane square burner with a side length of 18 cm. The mass flow rate of propane has been controlled in order to prescribe (a) a pre-ignition, (b) a fire growth, (c) a steady-state, (d) a decay, and (e) an extinction period, as highlighted by the vertical lines in Figure 8.21. More details are provided in Ref. [102].

8.9.5.2   Simulation Set-Up and Results

The test case described here has been simulated with the FDS (version 6), using mainly the default settings for turbulence, combustion and thermal radiation. A uniform-structured mesh with a cell size of 3 cm has been used. Although this mesh is coarse (and do not allow to solve the fire plume accurately), it is deemed to be sufficient to capture the overall features of the fire dynamics in terms of transient profiles of pressure and ventilation flow rates, as displayed in Figure 8.21.

Figure 8.21 shows that during the fire growth period, the room pressure rises, leading to an increase in the extraction rate and a decrease in the inflow rate of fresh air. As soon as a

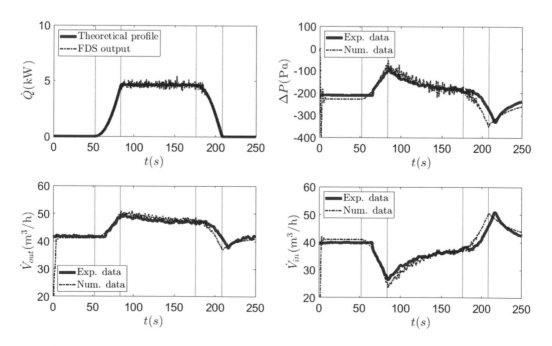

FIGURE 8.21   Experimental and numerical profiles of the fire scenario examined in Li et al. [102].

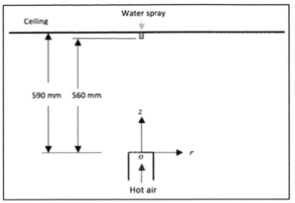

FIGURE 8.22 Schematics of experimental set-up: upward air flow, downward water spray, horizontal ceiling and four measurement windows (A, B, C and D) used for PIV measurements [103].

steady-state heat release rate (HRR) is established, the pressure (which has reached its peak) starts to decrease asymptotically towards a quasi-steady-state value. This value is the result of an equilibrium between the fire, the ventilation and the heat losses to the boundaries. The decrease in the HRR during the decay period causes a decrease in pressure and a subsequent decrease (resp. increase) in the extraction (resp. inlet) ventilation flow rate. A typical fire 'signature' at extinction is the under-pressure peak. The pressure increases then towards its value in the pre-ignition period.

It is important here to note that a detailed characterization of the ventilation network (e.g., in terms of fan curves and resistances) is a key in the prediction of the fire scenario at hand because of the subsequent amounts of fresh air delivered to the fire and smoke extracted from the room. The second important aspect is the modelling of heat losses to the boundaries, which has a direct impact on the pressure levels.

8.9.6 Interaction of a Hot-Air Plume with a Water Spray

In this case, the complexity arises from the need to compute a two-phase flow. Often in the CFD calculations of such a scenario, the conservation equations of mass, momentum and energy are solved using an Eulerian formulation for the continuous phase, i.e., gas phase. The dispersed phase, i.e., liquid phase, is solved using a Lagrangian formulation, where computational droplets resemble the statistics of real droplets.

Since the simulation of the set-up that is described hereafter is an ongoing work in our group, the intent is not to discuss in a comprehensive way the modelling aspects but rather to simply report the experimental configuration and results accompanied by preliminary FireFOAM [15] simulation results reported in Ref. [103].

8.9.6.1 Test Case Description

The test case considered here has been performed in the framework of a study that addresses the interaction between a water spray and a fire plume [96]. Details about the hot-air plume have been described earlier in Section 8.8.2. The water spray is discharged from a Delavan CT-1.5-30°C B full cone nozzle installed at 560 mm above the hot-air nozzle (i.e., 30 mm below the centre of the aluminium ceiling plate). Water spray droplet size, number density and velocity were measured at

TABLE 8.7

Experimental Data [96] and Numerical Predictions [103] of the Stagnation Plane Height

Jet Inlet Velocity (m/s)	Experimental Data (mm)	Numerical Predictions (mm)
5.3	60	78
4.2	320	193
3.3	445	282

mainly two heights above the fire source (530 and 260 mm) using a Shadow Imaging System (SIS) [96]. The water spray nozzle was operated on one pressure of 750 kPa, and the measured water flow rate was 0.084 Lpm (Litre per minute). The objective here is to study the water flux penetration through the fire plume for three inlet velocities 3.3, 4.2 and 5.3 m/s.

8.9.6.2   Simulation Set-Up

The simulation set-up using FireFOAM [15] is similar to Section 8.8.2, since the experimental set-up is essentially the same. A Lagrangian particle injection model is added here in order to track in space and time the discrete spray particles originating from the sprinkler nozzle. The injection mass flow rate was 1.4 mg/s with a spray angle of 30°. The velocity was set to 24.4 m/s at an injection pressure of 750 kPa. A Rosin–Rammler distribution was assumed with a volume mean diameter of 0.06 mm and a spread factor of 2 in order to match the locally measured droplet distribution.

8.9.6.3   Results

Prior to the simulation of the spray-plume interaction, it is important to evaluate the capability of the CFD calculations to reproduce the spray pattern in normal ambient conditions, in the absence of a hot plume. An example of hot-plume calculations without a spray, which are also important to validate, has been described in Section 8.9.2. The isolated spray injection results reported in Ref. [103] show: (a) an underprediction of the particle volume mean diameter in the far field, (b) an overprediction (resp. underprediction) of the centreline liquid volumetric flow rate at the near field (resp. far field) and (c) a similar behaviour for the particle velocity. It has been concluded thus that further adjustments of the spray injection model are needed. Consequently, as displayed in Table 8.7, only qualitatively good predictions were obtained for the water flux penetration (measured by the height of the stagnation plane where the vertical velocity is null) through the fire plume. The latter depends on the ratio of the initial vertical component of the spray thrust to the maximum upward thrust of the fire plume [104].

8.9.7   Underventilated Enclosure Fire with Mechanical Ventilation

8.9.7.1   Test Case Description

The case considered here has been examined in Ref. [94] in the context of large collaborative research project on fire dynamics in Nuclear Power Plants (NPPs) called PRISME [92]. The configuration (see Figure 8.23) consists of a closed room of 5 m × 6 m × 4 m height. The walls are 30 cm thick made of concrete. The ceiling is insulated with 5 cm thick rockwool panels. The ventilation system includes intake and exhaust branches of 0.3 m × 0.6 m that provide a renewal rate of 4.7 h^{-1} (that is equivalent to a volume flow rate of ~ 564 m^3/h). More details are provided in Ref. [94].

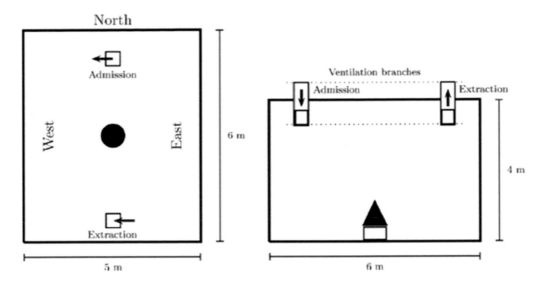

FIGURE 8.23 Experimental set-up in Ref. [94].

8.9.7.2 Simulation Set-Up

The case described above has been simulated [94] using the ISIS open-source software developed by the Institut de Radioprotection et de Sûreté Nucléaire (IRSN). The fire-field model and associated documentation are available at https://gforge.irsn.fr/gf/project/isis. The standard $k - \varepsilon$ model has been used along with the EBU combustion model. The boundary conditions at the intake and exhaust openings are solved using Eqs. (8.240) and (8.241) with the following pressure differences and flow resistances: $\Delta P_{adm} = 126$ Pa, $\Delta P_{ext} = -866 Pa$, $R_{adm} = 3,367 \text{ m}^{-4}$ and $R_{exh} = 32,370 \text{ m}^{-4}$. The prescribed mass loss rate of the fuel (hydrogenated tetrapropylene, HTP, that is equivalent to dodecane) is calculated as follows:

$$\dot{m}'' = \dot{m}''_{open} \left(10 X_{O_2} - 1.1 \right) + \frac{\sigma \left(1 - \varepsilon_f \right) \left(T_g^4 - T_s^4 \right)}{L_v}, \tag{8.246}$$

where $\dot{m}''_{open}$ is the burning rate in open atmosphere conditions, X_{O_2} is the mean volume fraction on a region near the flame, σ is the Stefan–Boltzmann constant, ε_f is the flame emissivity, L_v is the fuel vaporization heat, T_s is the fuel surface temperature and T_g is the mean gas temperature over a part of the domain. The first term on the right-hand side of Eq. (8.246) represents the vitiation effect (i.e., reduced oxygen levels) on the fuel burning rate. The second term represents the radiative heat feedback from smoke and hot walls on the fuel surface. It must be noted that this approach is a simpler alternative way to the detailed pyrolysis modelling and heat transfer at the fuel surface. However, the influence of the size of the domains over which to find the mean oxygen concentration and the mean gas temperature has not been investigated.

For the case at hand, the radiative feedback effect has not been included because of the relatively low gas temperatures obtained (around 200°C at the steady-state stage).

A uniform cell size of 10 cm has been used with a refined meshing of 5 cm around the fire source.

More details about the numerical modelling are provided in [94].

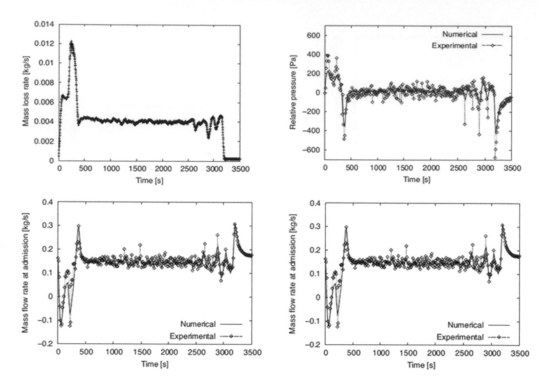

FIGURE 8.24   Experimental and numerical results for the fuel MLR, pressure and volume flow rates in Ref. [94].

8.9.7.3   Results

The experimental data for the fuel MLR displayed in Figure 8.24 shows an initial transient phase that is similar to open atmosphere conditions. At around $t = 400$ s, the MLR decreases abruptly before reaching a steady-state value of 4 g/s that is mainly controlled by the ventilation conditions in the room. At $t = 3,000$ s, the MLR drops to zero due to lack of fuel. The pressure profile (see Figure 8.24) shows an initial increase due to the release of hot combustion products in a well-confined environment. As a consequence, the inlet ventilation flow rate decreases and the extraction rate increases (see Figure 8.24). A reverse flow is observed at the inlet branch between $t = 100$ s and $t = 300$ s. The extinction stage is typically characterized by a high (resp. low) volume flow rate at the inlet (resp. exhaust). The overall behaviour described above is well captured by the numerical simulations (see Figure 8.24). Wall and gas temperatures are also well predicted as shown in Figure 8.25.

The most important issues to carefully examine in cases similar to the case described here are as follows:

1. The fuel response model to the environmental conditions in the room, i.e., oxygen concentrations and gas temperatures. This is addressed, for instance, using Eq. (8.246). However, more complex approaches (including, for instance, the computation of the fuel vaporization rate with the Clausius-Clapeyron equation) can be considered. The formulation of Eq. (8.246) has the advantage of being rather simple to implement in a field model. However, it relies upon mean quantities (i.e., X_{O_2} and T_g) that are averaged over a specified value of the domain, an input variable that has not been carefully investigated

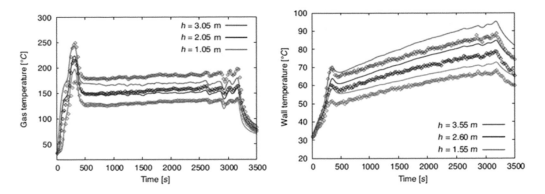

FIGURE 8.25 Results of gas and wall temperatures in Ref. [94].

yet. Furthermore, this approach does not allow solving some important flame features such as partial of full displacement from the pool surface area. Extinction is also a key issue here.

2. The ventilation network behaviour. An accurate modelling of the network response to the pressure rise induced by the release of hot combustion products in a well-confined room is also essential because it determines the amount of oxygen delivered to the room (and thus, the fire).

The strong coupling between the fire and the ventilation network can lead in some cases to combustion and pressure instabilities, reported in Ref. [105], that could be explained as follows. As the fire develops initially, the amount of oxygen present inside the room prior to ignition gets consumed. Furthermore, due to the induced pressure build-up in the closed vessel (i.e., the room) due to gas expansion, the volume flow rate of fresh air delivered by the fan at the inlet duct decreases. Consequently, the oxygen concentration feeding the flame decreases. The flame becomes weak to the level of near-extinction. Partial quenching, in conjunction with heat losses to the walls, induces a pressure drop and subsequently an increased supply of fresh air delivered by the inlet fan. This fresh air revitalizes the flame. The enhanced burning causes then the consumption of most of the oxygen. Hence, the process repeats itself.

8.9.8 Fire Spread Modelling

As opposed to the previous examples of the CFD simulations described above, the case examined here has not been examined experimentally. It is a *fictitious* worked-out case that is mainly meant to illustrate several fire spread modelling strategies and their implications. Fire spread is mainly associated with an increase in the fuel bed surface area (i.e., fire area). As a consequence, the HRR increases with time (as opposed to the previous cases discussed above).

In order to model flame spread, there are essentially three strategies:

1. *Prescribed spread rate model*: In this model, the solid and gas phases are completely decoupled. This means that no heat transfer calculation is performed at the surface of the solid to predict when it starts burning locally. Flame spread is controlled by a prescribed spread rate, v_f. At the level of each cell where the flame is present, the burning rate is calculated according to a prescribed transient profile of HRRPUA, $\dot{Q}''(t)$ or MLRPUA, $\dot{m}''(t)$.

2. *Surface-pyrolysis model*: In this model, the profiles of $\dot{Q}''(t)$ or $\dot{m}''(t)$ are also prescribed. However, the ignition time of a virgin fuel surface is not prescribed, but it is rather computed from a heat transfer calculation through the solid, which requires the knowledge of its thermal properties. When the surface temperature reaches a predefined surface ignition temperature, $T_{s,\text{ign}}$, the burning takes place according to the profiles of $\dot{Q}''(t)$ or $\dot{m}''(t)$.

3. *In-depth pyrolysis model*: In this model, the solid degradation (through the thickness) is computed using an Arrhenius equation that indicates the amount of mass that is being consumed in the burning process, and thus provides as an output the profile $\dot{m}''(t)$. Such equation for a material, i, is expressed as follows:

$$\frac{dY_i}{dt} = -AY_i^{n_i} e^{-\frac{E}{RT_i}} X_{O_2}^{n_{O_2}}, \tag{8.247}$$

where Y_i is the mass fraction, t is the time, A is the pre-exponential factor, n_i and n_{O_2} are the reaction orders for i and oxygen, E is the activation energy, R is the universal gas constant, T_i is the temperature and X_{O_2} is the oxygen volume fraction. Equation (8.247) describes solid degradation for one material using one chemical reaction. In practice, solid fuels in fires are composed of several species that may undergo several chemical reactions.

Table 8.8 summarizes the main parameters required for each fire spread modelling strategy.

In order to illustrate the differences between the models described above, let us consider the simple example of a 10 cm thick horizontal slab made of one homogeneous material and insulated at its bottom. The ignition is located at its top surface in its centre (see Figure 8.26). A first series of three simulations, without external heat flux (see Figure 8.26a), has been performed with the Fire Dynamics Simulator [16] corresponding to the three methods described above. The selected parameters in each case have been chosen such that a similar HRR profile is obtained (see Figure 8.27).

It is interesting to note at this stage the differences in the computational time between the three methods. The first method required a computational time of 40 minutes; for the second, it

TABLE 8.8

Required Parameters in Each Fire Spread Modelling Strategy

Model	Required Parameters
1- Prescribed Spread Rate	Spread rate, v_f
	profiles of $\dot{Q}''(t)$ or $\dot{m}''(t)$
	Material properties not required
2- Surface Pyrolysis	profiles of $\dot{Q}''(t)$ or $\dot{m}''(t)$
	Thermal properties of the solid
	Surface ignition temperature, $T_{s,\text{ign}}$
	Heat of vaporization
3- In-Depth Pyrolysis	Thermal properties of the solid
	Pre-exponential factor, A
	Activation energy, E
	Heat of vaporization

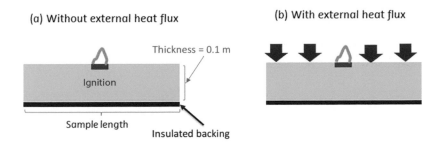

(a) Without external heat flux

Thickness = 0.1 m

Ignition

Sample length

Insulated backing

(b) With external heat flux

FIGURE 8.26 Geometrical configuration for flame spread modelling.

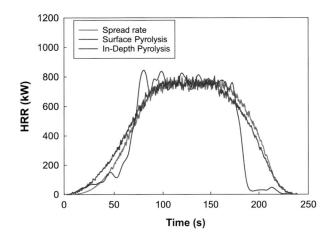

FIGURE 8.27 HRR profiles for the three fire spread modelling strategies.

was 51 minutes; and for the third, 2 hours and 40 minutes were needed. This result renders the first approach more appealing in terms of computational cost. However, this comes at the price of neglecting the coupling between the gas and solid phases. Such *deficiency* is illustrated in the following, applying an external heat flux at the surface of the slab (see Figure 8.26b), which may result in external radiation from smoke and walls (in the case of an enclosure fire).

The following results are obtained:

- As expected, there is no influence on the HRR profile for the first method because the solid phase is decoupled from the gas phase (i.e., there is no excess of pyrolysed fuel due to additional heat).
- The second method shows a faster growth (see Figure 8.28). Indeed, the ignition temperature is reached faster due to the external heat flux. However, the peak and steady-state value remains the same because of the imposed HRRPUA.
- The third method results in faster fire spread and a higher peak HRR, and produces as such a more complete response to the additional external flux (see Figure 8.29). However, it must be noted that such an approach is very difficult to put in place due to the associated complex chemical reactions and heterogeneity of real fuels in fires, which render the estimation of parameters such as the pre-exponential factor and activation energy quite tedious. This is typically done using optimization strategies via, for instance, Genetic Algorithms (e.g., [106]).

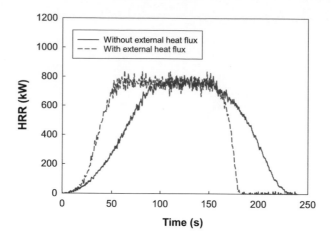

FIGURE 8.28   Influence of the external heat flux on flame spread using the surface-pyrolysis model.

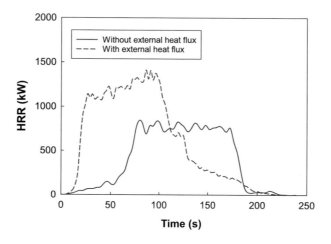

FIGURE 8.29   Influence of the external heat flux on flame spread using the in-depth pyrolysis model.

The test case discussed above shows clearly some of the advantages and limitations of several approaches dealing with fire spread modelling. The CFD user must therefore bear them in mind during both the set-up and analysis phases of a given test case.

8.9.9   CONCLUSIONS

Significant advances have been achieved in CFD simulations of fire dynamics. There is, for example, a relatively good level of confidence when attempting to predict the general features of a flame (see Section 8.9.3) and the smoke dynamics from a well-ventilated fire (see the Steckler case in Section 8.9.4).

However, a lot of challenges still lie ahead. For example, the prediction of carbon monoxide (see Section 8.9.3) and soot concentrations in the flaming region remains difficult (partially) because of the need to account for finite-rate chemistry and its interaction with turbulence and thermal radiation. It is even more so the case when the fire undergoes partial extinction and becomes underventilated. If the fuel is in the condensed phase (i.e., liquid or solid), additional difficulties arise with respect to the heat-up, thermal degradation and pyrolysis for solids (see Section 8.9.8) or evaporation for liquids (see Section 8.9.7). Modelling the interaction between the gas and condensed phases, which is required for predictive simulations as opposed to prescribed simulations (where the HRR or the fuel MLR is an input), remains thus highly challenging. Furthermore, the simulations of water sprays and their interaction with fire-driven flows have gained in maturity, but more effort is needed to improve the current level of predictive capabilities (see, for example, Section 8.9.6). Nevertheless, prior to improving the current level of CFD predictive capabilities, the users of the current CFD tools (whether it is for research or design purposes) need to make a well-motivated choice of several numerical parameters such as the gas-phase cell size, the number of radiation angles or the number of computational droplets per second, which implies comprehensive sensitivity studies that can be quite tedious. These numerical parameters can be indeed as important as the physical parameters.

Finally, to conclude this section, it is important to underscore the key role of experimentalists in the development of reliable and more advanced CFD models by designing well-instrumented and, sometimes, tailor-made experiments. A comprehensive and well-documented experimental dataset allows to (a) assess with confidence the current level of modelling capabilities, (b) identify the current limitations and (c) foster model development.

REFERENCES

1. K. McGrattan (2005) Fire modeling: Where are we? Where are we going? *8th International Symposium on Fire Safety Science*, September 18–23, Beijing, China.
2. W.L. Oberkampf and T.G. Trucano (2002) "Verification and validation in computational fluid dynamics", *Progress in Aerospace Sciences*, Vol. 38, pp. 209–272.
3. T. Poinsot and D. Veynante (2005) *Theoretical and Numerical Combustion*. R.T. Edwards Inc: Australia.
4. C.K. Westbrook and F.L. Dryer (1981) "Simplified reaction mechanisms for the oxidation of hydrocarbon fuels in flames", *Combustion Science and Technology*, Vol. 27, pp. 31–43.
5. L.F. Richardson (1922) *Weather Prediction by Numerical Process*, 1st Ed. Cambridge Mathematical Library Series. Cambridge University Press: Cambridge
6. B. Merci, E. Mastorakos and A. Mura (2010) "Modeling of turbulent combustion", in: M. Lackner et al. (Ed.) *Handbook on Combustion*. Wiley-VCH: Weinheim, Germany, pp. 175–203.
7. K. Van Maele and B. Merci (2006) "Application of two buoyancy-modified k-ε turbulence models to different types of buoyant plumes", *Fire Safety Journal,* Vol. 41, pp. 122–138.
8. B. Merci and E. Dick (2003) "Heat transfer predictions with a cubic k-ε model for axisymmetric turbulent jets impinging onto a flat plate", *International Journal of Heat and Mass Transfer*, Vol. 46, pp. 469–480.
9. J. Smagorinsky (1963) "General circulation experiments with the primitive equations", *Monthly Weather Review*, Vol. 91, pp. 99–164.
10. D.K. Lilly (1962) "On the numerical simulation of buoyant convection", *Tellus*, Vol. XIV(2), p. 144.
11. Z.H. Yan (2007) "Large eddy simulations of a turbulent thermal plume" *Heat Mass Transfer,* Vol. 43, pp. 503–514.
12. S.B. Pope (2000) *Turbulent Flows*. Cambridge University Press: Cambridge.
13. J.W. Deardorff (1972) "Numerical investigation of neutral and unstable planetary boundary layers", *Journal of Atmospheric Sciences*, Vol. 29, pp. 91–115.
14. U. Schumann (1975) "Subgrid scale model for finite difference simulations of turbulent flows in plane channels and annuli", *Journal of Computational Physics*, Vol. 18, pp. 376–404.
15. FireFOAM, https://github.com/fireFoam-dev/.
16. K. McGrattan, et al. (2013) Fire dynamics simulator, User's Guide, NIST Special Publication 1019, Sixth Edition, National Institute of Standards and Technology (NIST), Gaithersburg, MD.
17. K. McGrattan, et al. (2013) "Fire dynamics simulator", Technical Guide, NIST Special Publication 1019, Sixth Edition, National Institute of Standards and Technology (NIST), Gaithersburg, MD.

18. Vreman (2004) "An eddy-viscosity subgrid-scale model for turbulent shear flow: Algebraic theory and applications", *Physics of Fluids*, Vol. 16(10), pp. 3670–3681.
19. V. Novozhilov (2001) "Computational fluid dynamics modeling of compartment fires", *Progress in Energy and Combustion Science*, Vol. 27, pp. 611–666.
20. H. Pitsch (2006) "Large-eddy simulation of turbulent combustion", *Annual Review of Fluid Mechanics*, Vol. 38, pp. 453–482.
21. T. Echekki and E. Mastorakos (2011) *Turbulent Combustion Modeling: Advances, New Trends and Perspectives.* Springer: Berlin/Heidelberg, Germany.
22. D.B. Spalding (1971) "Mixing and chemical reaction in steady confined turbulent flames", *Thirteenth Symposium (International) on Combustion*, The Combustion Institute, Vol. 13(1), pp. 649–657.
23. B.F. Magnussen and B.H. Hjertager (1976) "On mathematical modelling of turbulent combustion with special emphasis on soot formation and combustion", *Sixteenth Symposium (International) on Combustion.* The Combustion Institute, Vol. 16(1), pp. 719–729.
24. I.S. Ertesvag and B.F. Magnussen (2000) "The eddy dissipation turbulence energy cascade model", *Combustion Science and Technology*, Vol. 159, pp. 213–235.
25. B.F. Magnussen (1981) "On the structure of turbulence and a generalized eddy dissipation concept for chemical reaction in turbulent flow", *Nineteenth AIAA Aerospace Science Meeting.*
26. E. De Oldenhof, P. Sathiah and D. Roekaerts (2011) "Numerical simulation of Delft-Jet-in-Hot-Coflow (DJHC) flames using the eddy dissipation concept model for turbulence chemistry interaction", *Flow Turbulence and Combustion*, Vol. 87, pp. 537–567.
27. A.Y. Klimenko and R.W. Bilger (1999) "Conditional moment closure for turbulent combustion," *Progress in Energy and Combustion Science*, Vol. 25(6), pp. 595–687.
28. S.B. Pope (1985) "PDF methods for turbulent reactive flows", *Progress in Energy and Combustion Science*, Vol. 11, p. 119.
29. M.F. Modest (2003) *Radiative Heat Transfer*, 2nd Ed. Academic Press: Cambridge, MA.
30. S. Chandrasekhar (1960) *Radiative Transfer.* Dover: New York.
31. F.C. Lockwood and N.C. Shah (1981) "A new radiation solution method for incorporation in general combustion prediction procedures", *Eighteenth Symposium (International) on Combustion*, The Combustion Institute, Vol. 18(1), pp. 1405–1414.
32. D.F. Fletcher, J.H. Kent, V.B. Apte and A.R. Green (1994) "Numerical simulation of smoke movement from a pool fire in a ventilated tunnel", *Fire Safety Journal*, Vol. 23, pp. 305–325.
33. A.Y. Snegirev (2004) "Statistical modelling of thermal radiation transfer in buoyant turbulent diffusion flames", *Combustion and Flame*, Vol. 136, pp. 51–71.
34. A. Gupta, D.C. Haworth and M.F. Modest, "Turbulence-radiation interactions in large-eddy simulations of luminous and nonluminous nonpremixed flames", *Proceedings of the Combustion Institute*, Vol. 34, pp. 1281–1288.
35. S.R. Tieszen (2001) "On the fluid mechanics of fires". *Annual Review of Fluid Mechanics,* Vol. 33, pp. 67–92.
36. H. Wang and M. Frenklach (1997) "A detailed kinetic modeling study of aromatics formation in laminar premixed acetylene and ethylene flames", *Combustion and Flame*, Vol. 110(1–2), pp. 173–221.
37. F. Tao, V.I. Golovitchev and J. Chomiak (2004) "A phenomenological model for the prediction of soot formation in diesel spray combustion", *Combustion and Flame*, Vol. 136(3), pp. 270–282.
38. M. Faraday (1988) *The Chemical History of a Candle* 1st Ed. Chicago Review Press: Chicago, IL.
39. H. Ingason and J. De Ris (1998), "Flame heat transfer in storage geometries", *Fire Safety Journal,* Vol. 31(1), pp. 39–60.
40. I.M. Kennedy, C. Yam, D.C. Rapp and R.J. Santoro (1996) "Modeling and measurements of soot and species in a laminar diffusion flame", *Combustion and Flame,* Vol. 107(4), pp. 368–382.
41. U.O. Koylu and G.M. Faeth (1991) "Carbon monoxide and soot emissions from liquid-fueled buoyant turbulent diffusion flames". *Combustion and Flame*, Vol. 87(1), pp. 61–76.
42. R.R. Skaggs and J.H. Miller (1995) "A study of carbon monoxide in a series of laminar ethylene/air diffusion flames using tunable diode laser absorption spectroscopy", *Combustion and Flame,* Vol. 100(3), pp. 430–439.
43. C.W. Lautenberger (2002) CFD simulation of soot formation and flame radiation, MSc Thesis, Worcester Polytechnic Institute.

44. Tewarson (1986) "Prediction of fire properties of materials-part I: Aliphatic and aromatic hydrocarbons and related polymers", Factory Mutual Research Corporation J.I. 0K3R3.RC, (NBS Grant 60NANB4D-0043).

45. G.H. Markstein (1985) "Relation between smoke point and radiant emission from buoyant turbulent and laminar diffusion flames", *Proceedings of the Combustion Institute*, Pittsburgh, Vol. 20(1), pp. 1055–1061.

46. C.W. Lautenberger, J.L. De Ris, N.A. Dembsey, J.R. Barnett and H.R. Baum (2005) "A simplified model for soot formation and oxidation in CFD simulation of non-premixed hydrocarbon flames", *Fire Safety Journal*, Vol. 2, pp. 141–176.

47. Y. Yang, A.L. Boehman and R.J. Santoro (2007) "A study of jet fuel sooting tendency using the threshold sooting index (TSI) model", *Combustion and Flame*, Vol. 149, pp. 191–205.

48. H.M. Calcote and D.M. Manos (1983) "Effect on molecular structure on incipient soot formation", *Combustion and Flame*, Vol. 49(1–3), pp. 289–304.

49. I.M. Kennedy (1997) "Models of soot formation and oxidation", *Progress in Energy and Combustion*, Vol. 23, pp. 95–132.

50. A. D'Anna, and J.H. Kent (2006) "Modeling of particulate carbon and species formation in coflowing diffusion flames of ethylene", *Combustion and Flame*, Vol. 144(1–2), pp. 249–260.

51. A. Beltrame, P. Porshnev, W. Merchan-Merchan, A. Saveliev, A. Fridman, L.A. Kennedy, O. Petrova, S. Zhandok, F. Amouri and O. Charon (2001) "Soot and NO formation in methane–oxygen enriched diffusion flames," *Combustion and Flame*, Vol. 124(1–2), pp. 295–310.

52. N. Olten and S. Senkan (1999) "Formation of polycyclic aromatic hydrocarbons in an atmospheric pressure ethylene diffusion flame", *Combustion and Flame*, Vol. 118(3), pp. 500–507.

53. A. Kronenburg, R.W. Bilger and J.H. Kent (2000) "Modeling soot formation in turbulent methane–air jet diffusion flames", *Combustion and Flame*, Vol. 121(1–2), pp. 24–40.

54. J.B. Moss and C.D. Stewart (1998) "Flamelet-based smoke properties for the field modelling of fires", *Fire Safety Journal*, Vol. 30, pp. 229–250.

55. K.J. Syed, C.D. Stewart, and J.B. Moss (1990) "Modelling soot formation and thermal radiation in buoyant turbulent diffusion flames", *Proceedings of the Combustion Institute*, Vol. 23(1), pp. 1533–1541.

56. J.B. Moss, C.D. Stewart and K.J. Young (1995) "Modelling soot formation and burnout in a high temperature laminar diffusion flame burning under oxygen-enriched conditions", *Combustion and Flame*, Vol. 101, pp. 491–500.

57. J.B. Moss and C.D. Stewart (1998) "Flamelet-based smoke properties for the field modelling of fires", *Fire Safety Journal*, Vol. 30, pp. 229–250.

58. J. Nagle, R.F. Strickland-Constable (1962) "Oxidation of carbon between 1000–2000 C", *Proceedings of the Fifth Carbon Conference*, London, Vol. 1, p. 154.

59. K.B. Lee, M.W. Thring, and J.M. Beer (1962) "On the rate of combustion of soot in a laminar soot flame", *Combsution and Flame*, Vol. 6, p. 137.

60. C.P. Fenimore and G.W. Jones (1967) "Oxidation of soot by hydroxyl radicals", *Journal of Physical Chemistry*, Vol. 71, p. 593.

61. J.B.M. Pierce and J.B. Moss (2007) "Smoke production, radiation heat transfer and fire growth in a liquid-fuelled compartment fire", *Fire Safety Journal*, Vol. 42, pp. 310–320.

62. M.A. Delichatsios (1994) "A phenomenological model for smoke-point and soot formation in laminar flames", *Combustion Science and Technology*, Vol. 100, pp. 283–298.

63. J.H. Kent and D.R. Honery (1994) "Soot mass growth in laminar diffusion flames-parametric modelling", in: Bockhorn, H. (Ed.) *Soot Formation in Combustion Mechanics and Models*. Springer: Berlin, Heidelberg, p. 199.

64. T. Beji, J.P. Zhang, W. Yao and M. Delichatsios (2011) "A novel soot model for fires: Validation in a laminar non-premixed flame", *Combustion and Flame*, Vol. 158, pp. 281–290.

65. A. Tewarson (2002) "Generation of heat and chemical compounds in fires", in: P.J. Di Nenno (Ed.), *SFPE Handbook of Fire Protection Engineering*, 3rd Ed., Section 3, Chapter 3.4. National Fire Protection Association: Quincy, MA.

66. Y.R. Sivathanu and G.M. Faeth (1990) "Soot volume fractions in the overfire region", *Combustion and Flame*, Vol. 81, pp. 133–149.

67. W. Yao, J. Zhang, A. Nadjai, T. Beji and M.A. Delichatsios, (2011) "A global soot model developed for fires: Validation in laminar flames and application in turbulent pool fires", *Fire Safety Journal*, Vol. 46(7), pp. 371–387.

68. R.P. Lindstedt and S.A. Louloudi (2005) "Joint-scalar transported PDF modeling of soot formation and oxidation", *Proceedings of the Combustion Institute*, Vol. 30, pp. 775–783.

69. R.P. Lindstedt (1994) in: H. Bockhorn (Ed.), *Soot Formation in Combustion: Mechanisms and Models*. Springer-Verlag, Berlin, pp. 417–441.

70. P. Donde, V. Raman, M.E. Mueller, and H. Pitsch (2013) "LES/PDF based modeling of soot–turbulence interactions in turbulent flames", *Proceedings of the Combustion Institute*, Vol. 34(1), pp. 1183–1192.

71. I.M. Khan and G. Greeves (1974) "A method for calculating the formation and combustion of soot in diesel engines", in: N.H. Afgan and J.M. Beer (Eds,), *Heat Transfer in Flames*, Chapter 25. Scripta Book Co, Washington, DC.

72. P.L. Roe (1986) "Characteristics-based schemes for the euler equations", *Annual Review of Fluid Mechanics*, Vol. 18, p. 337.

73. R. Mittal and P. Moin (1997) "Suitability of upwind-biased finite-difference schemes for large-eddy Simulation of Turbulent Flows", *AIAA Journal*, Vol. 35, p. 1415.

74. H.K. Versteeg and W. Malalasekera (2007) *An Introduction to Computational Fluid Dynamics, the Finite Volume Method*, 2nd Ed. Pearson Education Limited: London.

75. S.V. Patankar and D.B. Spalding (1972) "A calculation procedure for heat, mass and momentum transfer in three-dimensional parabolic flows", *International Journal of Heat and Mass Transfer*, Vol. 15, pp. 1787–1806.

76. S.V. Patankar (1980) *Numerical Heat Transfer and Fluid Flow*. Hemisphere Publishing Corporation, Taylor & Francis Group: New York.

77. R.I. Issa (1986) "Solution of the implicitly discretised fluid flow equations by operator-splitting", *Journal of Computational Physics*, Vol. 62, pp. 66–82.

78. S. Vilfayeau, N. Ren and Y. Wang (2015) "Trouve numerical simulation of under-ventilated liquid-fueled compartment fires with flame extinction and thermally-driven fuel evaporation", *Proceedings of the Combustion Institute*, Vol. 35, pp. 2563–2571.

79. H.R. Baum, K.B. McGrattan and RG Rehm (1997) "Three dimensional simulations of fire plume," *Fire Safety Science: Proceedings of the Fifth International Symposium*, Tsukuba, Japan, pp. 511–522.

80. T.G. Ma and J.G. Quintiere (2003) "Numerical simulation of axi-symmetric fire plumes: Accuracy and limitations", *Fire Safety Journal*, Vol. 38, pp. 467–492.

81. V. Babrauskas (1983) "Estimating large pool fire burning rates", *Fire Technology*, Vol. 19, pp. 251–261.

82. A. Tewarson and R.F. Pion (1976) "Flammability of plastics. I. Burning intensity", *Combustion and Flame*, Vol. 26, pp. 85–103.

83. N. Jarrin (2008) "Synthetic inflow boundary conditions for the numerical simulation of turbulence", PhD thesis, The University of Manchester, United Kingdom.

84. M. Klein, A. Sadiki and J. Janicka (2003) "A digital filter based generation of inflow data for spatially developing direct numerical or large-eddy simulations", *Journal of Computational Physics*, Vol. 186, pp. 652–665.

85. X. Zhou, K.H. Luo and J.J.R. Williams (2001) "Large-eddy simulation of a turbulent forced plume", *European Journal of Mechanics - B/Fluids*, Vol. 20, pp. 233–254.

86. ANSYS (2009) *ANSYS Fluent, User's Guide*. ANSYS: Canonsburg, PA.

87. Trettel (2013) "Outflow boundary conditions for low-Mach buoyant computational fluid dynamics", Master of Science, University of Maryland.

88. J.X. Wen, K. Kang, T. Donchev and J.M. Karwatzki (2007) "Validation of FDS for the prediction of medium-scale pool fires", *Fire Safety Journal*, Vol. 42, pp. 127–138.

89. Y. Wang, P. Chatterjee and J.L. De Ris (2011) "Large eddy simulation of fire plumes", *Proceedings of the Combustion Institute*, Vol. 33, pp. 2473–2480.

90. B.J. McCaffrey (1979) Purely buoyant diffusion flames: Some experimental results, NBSIR 79–1910, National Bureau of Standards.

91. L. Audouin, L. Rigollet, H. Prétrel, W. Le Saux and M. Röwekamp (2013) "OECD PRISME project: Fires in confined and ventilated nuclear-type multi-compartments - Overview and main experimental results", *Fire Safety Journal*, Vol. 62, pp. 80–101.

92. H. Prétrel, W. LeSaux, and L. Audouin (2012) "Pressure variations induced by a pool fire in a well-confined and force-ventilated compartment", *Fire Safety Journal*, Vol. 52, pp. 11–24.

93. T. Beji, F. Bonte and B. Merci (2014) Numerical simulations of a mechanically-ventilated multi- compartment fire, *Fire Safety Science (IAFSS Symposium)*, Vol. 11, pp. 499–509.

94. S. Suarda, C. Lapuertaa, F. Babika and L. Rigollet (2011) "Verification and validation of a CFD model for simulations of large-scale compartment fires", *Nuclear Engineering and Design*, Vol. 241, pp. 3645–3657.

95. A. Shabbir and W.K. George (1994) "Experiments on a round turbulent buoyant plume", *Journal of Fluid Mechanics*, Vol. 275, pp. 1–32.

96. X. Zhou (2013) "Characterization of Interactions between hot air plumes and water sprays for sprinkler protection", *Proceedings of the Combustion Institute*, Vol. 35, pp. 2723–2729.

97. X. Zhou (2015) "PIV measurements of velocity fields of three hot air jet plumes impinging on a horizontal ceiling", *10th Asia-Oceania Symposium on Fire Science and Technology*, October 5–7, Tsukuba, Japan.

98. S. Ebrahim Zadeh, G. Maragkos, T. Beji and B. Merci (2015) Large eddy simulations of a ceiling-jet induced by the impingement of a turbulent hot air plume on a horizontal ceiling, *Proceedings of the Second IAFSS European Symposium of Fire Safety Science*.

99. E.J. Weckman and A.B. Strong (1996) "Experimental investigation of the turbulence structure of medium-scale methanol pool fires", *Combustion and Flame*, Vol. 105, pp. 245–266.

100. E. Gengembre, P. Cambray, D. Karmed and J. C. Bellet (1984) "Turbulent diffusion flames with large buoyancy effects", *Combustion Science and Technology*, Vol. 41, pp. 55–67.

101. K.D. Steckler, J.G. Quintiere and W.J. Rinkinen (1982) "Flow induced by a fire in a compartment", National Bureau of Standards, Centre for Fire Research, Report Number NBSIR, 82–2520, September. Available at http://fire.nist.gov/bfrlpubs/fire82/PDF/f82001.pdf.

102. J. Li, H. Prétrel, S. Suard, T. Beji and B. Merci (2021) "Experimental study on the effect of mechanical ventilation conditions and fire dynamics on the pressure evolution in an air-tight compartment", *Fire Safety Journal*, Vol. 125, p. 103426.

103. K.V. Meredith, X. Zhou, S. Ebrahimzadeh and B. Merci (2015) "Numerical simulation of spray-plume interactions", *9th U.S. National Combustion Meeting Organized by the Central States Section of the Combustion Institute*, Cincinnati, Ohio.

104. R.L. Alpert (1985) "Numerical modeling of the interaction between automatic sprinkler sprays and fire plumes", *Fire Safety Journal*, Vol. 9, pp. 157–163.

105. H. Prétrel and J.M. Such (2007) "Study based on large-scale experiments on the periodic instabilities of pressure and burning rate in the event of pool fire in a confined and mechanically ventilated compartment", *Third ECM Meeting*.

106. C.A. Lautenberger (2007) "Generalized pyrolysis model for combustible solids, PhD Thesis, University of California, Berkeley.

Index